Earth
Odyssey

*Around the World
in Search of Our
Environmental Future*

Mark Hertsgaard

BROADWAY BOOKS
New York

This book is dedicated to the memory of

my editor and friend, William Shawn,

with whom it began.

BROADWAY

A hardcover edition of this book was published in 1998 by Broadway Books. EARTH ODYSSEY. Copyright © 1998, 1999 by Mark Hertsgaard. All rights reserved. Printed in the United States of America. No part of this book may be reproduced or transmitted in any form or by any means, electronic or mechanical, including photocopying, recording, or by any information storage and retrieval system, without written permission from the publisher. For information, address Broadway Books, a division of Random House, Inc., 1540 Broadway, New York, NY 10036.

Broadway Books titles may be purchased for business or promotional use or for special sales. For information, please write to: Special Markets Department, Random House, Inc., 1540 Broadway, New York, NY 10036.

BROADWAY BOOKS and its logo, a letter B bisected on the diagonal, are trademarks of Broadway Books, a division of Random House, Inc.

First trade paperback edition published 1999.

Designed by Amanda Dewey

The Library of Congress has cataloged the hardcover edition as:

Hertsgaard, Mark, 1956–
Earth odyssey / Mark Hertsgaard. — 1st ed.
p.   cm.
Includes bibliographical references and index.
1. Global environmental change.   2. Nature—Effect of human beings on.   3. Hertsgaard, Mark, 1956– —Journeys.   I. Title.
GA149.H47   1999
363.7—dc21   98-28202
CIP

Portions of the book have previously appeared in *The Atlantic Monthly, Outside,* and *The Nation* magazines.

ISBN 0-7679-0059-6

03   10   9   8   7

Contents

Introduction to the Paperback Edition

Of all the hundreds of people I met during six years of traveling the world to research *Earth Odyssey*, it is Zhenbing Zhou who remains most vivid in my mind. Zhenbing was my interpreter in China; we traveled and worked together for six weeks before parting at the Hong Kong border in January 1997. He was a delightful and valuable traveling companion, for he had an easy laugh and exuded a charismatic good cheer that enabled him to strike up conversations with everyone from high government officials to gap-toothed peasants. And while he knew little about environmental issues per se, Zhenbing ended up teaching me more about the human environmental predicament than many of the august policy experts I consulted during my global adventure.

One day, as the train he and I were riding across Shanxi province in central China was delayed at a remote rural station, a middle-aged woman in rags crawled into a garbage can on the platform to scavenge for food. Such a scene was common in China, but it caused Zhenbing to reminisce about the pinched circumstances of his own early life growing up amid the economic deprivation and political chaos of the Cultural Revolution. The second of three sons, he had been raised in a small village about 120 miles northwest of Beijing. Like most Chinese peasants, his family was too poor to afford coal to heat their mud hut in winter. Instead, in a climate where temperatures often dropped well below zero, they burned crop stalks or dried leaves.

Zhenbing insisted the arrangement had been "comfortable enough," but he also confided that the inside wall of his family's hut was often "white with icy waterdrops" from November to March. "In my village," he added, "when a girl was preparing to marry, the first thing the parents checked was: 'Will the back wall of the would-be son-in-law be white or not?' If not white, they approved the marriage, because that meant his family was wealthy enough to keep the house warm."

The life of China's peasants began improving in the late 1970s as the market reforms of Deng Xiaoping transformed the nation's economy. More money began trickling into the hands of Zhenbing's family, and one of the first things they, and many other families, bought with it was winter coal. Who could blame them? As a result, white walls grew less common, and for the first time in their history, the Chinese people felt somewhat warm in winter. Zhenbing even got his first pair of shoes.

But the additional coal burning carried a heavy environmental cost. Multiply the story of Zhenbing's family by the nearly one billion people who lived in China at that time and you understand why China soon suffered some of the worst air pollution on earth. When I arrived in 1996, after fifteen more years of coal-powered economic development, there was so much coal dust in the air over cities like Beijing and X'ian that the midafternoon sky looked as dark as dusk. At the time, China boasted five of the world's ten most air-polluted cities (the ratio would increase to nine of ten by 1998), two million Chinese were dying every year from air and water pollution, and acid rain was affecting 29 percent of the nation's land area. Outsiders had cause for concern as well, for China's coal burning had made it the world's second-largest producer of the greenhouse gases that were causing global climate change.

But as I say, who could blame Zhenbing's family for wanting to shield themselves from winter's cold? Anyone else would have done the same thing. Which is why I'm grateful to Zhenbing; he unwittingly helped me understand that the biggest environmental problem in the world today is not climate change or ozone depletion or disappearing species or any of the other threats I set out to investigate when I left San Francisco in 1991 to travel the world. No, the biggest environmental problem is poverty—or, more precisely, the urge of billions of people to escape a level of poverty inconceivable to Americans. No one can begrudge the poor a better life, but the environmental consequences of their ascent from misery will be profound, particularly since most of the six billion humans now living on this planet are indeed distressingly poor.

If poverty is the biggest environmental challenge of our time, however, wealth is the biggest environmental burden. The consumption patterns of the nearly one billion people who live in the affluent world of Europe, North America, and other industrialized countries cause much more environmental damage—more greenhouse gas emissions, more forest cutting, more soil, air and water pollution—than do the strivings of the impoverished human majority. China again illustrates the point. Measured by population, Chinese outnumber Americans nearly five to one. Yet the United States dwarfs China's total environmental impact because Americans consume fifty-three times as many goods and services per capita.

All this gives Americans a special responsibility regarding the question I set out to investigate in *Earth Odyssey*: Will the human species act quickly and decisively enough to save itself from the many environmental hazards crowding in on it at the dawn of the twenty-first century?

Opinion polls indicate that most Americans consider themselves to be environmentalists. Nevertheless, for most of us, the environment remains an abstraction. Problems like global warming and ozone depletion may worry us, but we assume they lurk well off in the future or in distant lands and pose no real danger to us. In the following pages, I have tried to pierce that complacency, and make the abstract concrete, by writing a different kind of environmental book. Rather than the dry bureaucratic prose found in most such volumes, I have grounded my narrative in stories about the people and places I encountered during my global adventure—people like Zhenbing and his family; Garang, my guide through famine-plagued southern Sudan; Valodya, my driver in western Siberia; and dozens more. To these people, the environmental crisis is no mere abstraction; it is a punishing reality they live with every day.

And that reality, alas, continues to deteriorate. Just in the months since the hardback edition of *Earth Odyssey* was published, scientists have discovered that the Amazonian rainforest is being destroyed twice as fast as previously thought—ominous news, considering that the current global rate of loss will, if unaltered, leave the planet with no rainforests at all by 2050. Also by 2050, global emissions of greenhouse gases are expected to triple—another bad sign, considering that two South Pacific islands, Abanuea and Tebua Tarawa, disappeared beneath rising sea levels in 1999. Meanwhile, the Red Cross reported that deforestation and climate change are responsible for a sharp increase in what used to be called "natural disasters," and that these tragedies are driving more and more people around the world into the poverty that spurs more environmental damage. The Red Cross estimated the number of so-called environmental refugees at 25 million in 1998. But that is a conservative figure. After all, the floods that roared through China's Yangtze and Songhua river valleys in 1998 displaced, at least temporarily, 56 million people.

All these examples come from overseas, but Americans make a big mistake if they think they are immune to the gathering environmental crisis. It is an axiom of ecology that everything is connected to everything else, and perhaps nowhere is this more apparent than regarding climate change. With its huge coal reserves and ambitious development plans, China alone could doom the rest of the planet to severe climate change. Global sea levels are already expected, under "business as usual" projections, to rise an average of one meter by 2100, submerging large parts of not only Shanghai and Guangzhou but also New York, Washington, and San Francisco. And the future is closer than we think. Babies born today could very well live the one hundred years necessary to experience this hot and stormy new world, thanks to humanity's continuing advances in medical and genetic technology.

Americans should therefore care a great deal about whether China triples or merely doubles its coal use over the next twenty years. And we should care even more about the behavior of our own nation, the world's single largest source of ecological stress.

The good news is, there are solutions to our environmental problems—solutions that are not just practical but profitable. I detail many of them in the closing chapters of this book, but I emphasize the point here because so many people assume that any environmental book must carry a message of gloom and doom. That may be because mainstream economists, politicians, business executives, and journalists have been proclaiming for years that cleaning up the environment must involve economic sacrifice. But the truth is that restoring our planet's ravaged ecosystems could become the biggest business enterprise of the new century, a bountiful source of jobs, profits, and opportunity.

In his new book, *Cool Companies*, former Assistant Secretary of Energy Joseph J. Romm documents how such firms as Dupont, 3M, Xerox, and Compaq are fattening their bottom lines while dramatically reducing the amount of greenhouse gases their facilities unleash into the atmosphere. Many "cool" companies enjoy "a return on investment that can exceed . . . 50 and even 100 percent," reports Romm. The key is efficiency: not doing without but doing more with less. Toyota Auto Body Works, for example, a facility in Long Beach, California, that manufactures and paints the rear deck of Toyota pickup trucks, was consuming 2.5 million kilowatt hours (kwh) of electricity in 1991. By 1996, the plant had doubled its production volume while cutting electricity consumption by a third, thanks to such efficiency improvements as better motors, lighting, and air compressors.

With the right kind of presidential leadership, the federal government could encourage companies and individuals throughout the nation to join the efficiency crusade. The government could also get products like "green" cars off the ground by using the government's purchasing power. Americans have personal computers on their desks today largely because Washington got the computer industry up and running with a steady stream of purchases by the Pentagon in the 1960s. Those purchases helped companies climb the learning curve and bring down costs to where average consumers could afford to buy their own computers. Washington could do the same today with cars, the ultimate symbol of the environmental crisis. Every year, the government buys fifty thousand vehicles for official use. If Washington told Detroit it would keep buying those cars only if they were powered by hybrid-electric motors or hydrogen fuel cells, Detroit would surely comply, if only to keep the federal money flowing. And given Detroit's ample engineering talent, who can doubt that the industry would soon be producing green cars at competitive prices?

Americans tell pollsters they want environmental protection even if it means less economic growth, but the happy truth is they need not choose between the two. The same holds for the rest of the world. We humans know what to do to save ourselves. The question is: Will we act in time? Ten years ago, experts began warning that humanity had to have environmental solutions in place by 2000 to avoid eventual catastrophe.

Today, most environmental trends are instead still heading in the wrong direction, and some are accelerating. This makes it hard to be optimistic. But it is important to distinguish between optimism and hope. Optimism is a rational calculation, while hope is an expression of faith—faith in the human capacity to surprise, to unify and sacrifice, to innovate and to overcome what appear to be insurmountable odds. "With hope, one activist told me,"You can have magic." Hope has triumphed more than once in recent years, as Václav Havel, whom I interview near the end of this book, so eloquently personifies. For my part, I see no reason hope can't triumph again, provided, that is, we all do our part.

Playing the Sorcerer's Apprentice

I am where all light is mute
With a bellowing like the ocean
Turbulent in a storm of warring winds
The hurricane of Hell in perpetual motion
—DANTE, *The Inferno*

The light is mute in Chongqing nearly all the time in winter. The city sits at the confluence of the Yangtze and Jialing rivers and is encircled by mountains that block any cleansing winds, so it is a naturally foggy place. As the industrial center of southwestern China, Chongqing also happens to suffer some of the worst air pollution in all China, which makes it a strong candidate for the most polluted city in the world. When the fog and pollution are both at their thickest, say the locals, "If you stretch your hand out in front of your face, you cannot see your fingers."

Visibility was somewhat better than that during my visit to Chongqing in December 1996. Perched high above the Jialing one morning, peering into the dank grayness before me, I could dimly make out a black-and-white tugboat hugging the far shore of the river and, beyond that, the outlines of what looked like office buildings. This was the view from the back of the Chongqing Paper Factory, a massive, state-owned facility that local environmental officials had singled out as evidence of how well they were cleaning up Chongqing. Built in the 1940s, the factory had long been a terrible polluter, discharging enough chlorine and other toxic chemicals into the Jialing "to cover the entire river with white foam," according to an official of the Chongqing Environmental Protection Bureau who must remain nameless. Now, however, the factory had been "basically shut down," the official had bragged in an interview.

At the factory itself, though, it didn't look that way. The official had discouraged me from trying to visit: "I myself would have to seek permission to enter," he scolded. But when my interpreter, Zhenbing Zhou, and I arrived at the factory the next morning, we found the front gate open. Since no one stopped us, we simply walked in, kept moving, and tried to look like we belonged, passing ourselves off as just another Western investor and his trusty interpreter, checking out business opportunities in modern China.

The factory's entrance road descended to the right, past neatly stacked bamboo poles and cardboard boxes waiting to be pulped. A five-minute walk brought us to the back of the factory, adjacent to the cliff high above the river. Though clearly startled by my white face, a young worker feeding the coal furnaces cheerfully informed me that the factory was indeed still operating, though only about one-quarter of its eight thousand employees was working these days. He wasn't sure what had caused the layoffs—maybe the market economy?

A long set of concrete steps led out the back of the plant down to the river some eighty yards below. Halfway down the steps, Zhenbing and I cut left across the exposed riverbank, our shoes leaving clear prints on the dark, sandy soil. Within seconds we saw ahead of us a broad stream of bubbling water cascading down the hillside. The astringent odor of chlorine soon attacked our nostrils, and once we reached the stream's edge the smell was so powerful we immediately had to back away. Downstream, where the factory's discharge was emptying into the Jialing, a frothy white plume was spreading across the slow-moving river.

After walking another fifty yards, we encountered a second stream, this one a mere foot wide but clogged with bizarre clusters of dried, orange foam the size of pineapples. Up ahead was a third small creek. Its stench identified it as household sewage (workers in China's state-owned factories generally live on-site or nearby), but its most extraordinary feature was its color—as black as used motor oil. Not ten yards from the creek, a grizzled peasant in a dark-blue Mao jacket and trousers (an outfit still worn in China by the poor) was bending over a tiny vegetable patch to pick some greens for his midday meal.

Yet all this was dwarfed by what lay ahead. It was the vapor we saw first—wispy white, it hung low in the air, like tear gas. Stepping closer, we heard the sound of gushing water. Not until we were mere footsteps away, however, could we see the source of the commotion: a vast, roaring torrent of white, easily thirty yards wide, splashing down the hillside from the rear of the factory like a waterfall of boiling milk. Again the scent of chlorine was unmistakable, but this waterfall was much whiter than the first. Decades of unhindered discharge had left the rocks coated with a creamlike residue, creating a perversely beautiful white-on-white effect. Above us, the waterfall had bent trees sideways; below, it split into five channels before pouring into the unfortunate Jialing. All this, and the factory was operating at only 25 percent capacity.

Hoping to leave the factory by a different gate than the one we had entered, Zhenbing and I headed uphill on a muddy trail that soon led us to a pathetic structure of brick and corrugated plastic. Vapor from the waterfall wafted through the building's doorways, past walls that had been stained a sickly white over the years. A legless red couch sat out front; a clothesline held a pair of blue nylon sweatpants. Curiosity compelled me to peek inside, where I saw not a stick of furniture nor electric light, only a couple filthy sleeping mats and thin blankets thrown in a corner beside what looked like cooking utensils. It seemed impossible that human beings lived in such a place, but no sooner had I snapped a photo than a young man with unkempt hair emerged from the far end of the building to urinate. "This is the poorer class," explained Zhenbing.

The muddy trail eventually led up to a service road that appeared to offer us a way out. But Zhenbing and I had proceeded only a few yards on the road when a man wearing the olive-green, ankle-length greatcoat of the Chinese military suddenly came running toward us. It seemed our unauthorized factory tour was about to end badly after all.

But military greatcoats turned out to be a bit like Mao jackets—lots of poor Chinese wore them because they were cheap and functional. In any case, this man had worries of his own. Liquid was spilling from two large, loosely connected hoses by the side of the road, one that led back up to the factory and another that stretched down to the river below. The man barked orders at two workers straddling the hoses, and they stepped back. Then, without a word of warning to Zhenbing and me—though we stood a mere five feet away—the man knelt down and tightened the connection between the two hoses.

Instantly, he was engulfed in an explosion of gas. But he had expected it, and in one fluid motion he straightened up and started sprinting back up the service road, vanishing into the billowing cloud of chlorine after two steps. Zhenbing and I were not ready for the blast, but forward was the only way out, so we quickly held our breath and plunged after him. Six running strides brought us past the worst of the gas, but when we slowed to a brisk walk to avoid inhaling more than necessary, we were still surrounded by huge puffs of it. We started coughing fiercely and were still sputtering thirty yards up the road when we passed three dump trucks parked against the factory wall. A dozen workers were lounging in the backs of the trucks while the man in the greatcoat, who had run all the way here, bent down to tie his shoe. Chlorine is deadly (it was used in the poison gas attacks of World War I), yet the men in the trucks showed no concern about the vapors now floating past their heads. They did, however, nudge one another and stare at the foreigner trudging through their factory, evidently a far more unusual sight.

Zhenbing and I walked in silence to the plant's side exit and left without further incident. We were in the middle of a six-week trip through China, investigating the environmental crisis in Zhenbing's homeland, and it was not a cheering task. In Beijing, Xi'an, and other cities of the north, we had walked amid air so thick with coal dust and car fumes that even sunny days looked overcast and foggy. In the bone-dry province of Shanxi, a day's journey west of Beijing, we rode by train all afternoon without seeing anything resembling woods—only a few scattered, spindly trees that looked ready to expire any minute. Everywhere, it seemed, the land had been scalped, the water poisoned, the air made toxic and dark.

Despite having lived with China's pollution for decades, Zhenbing was not exactly a militant environmentalist. Born into a very poor rural family thirty years before, he, like most Chinese I had met, was quite willing to put up with filthy air and dirty water if it meant better pay, more jobs, a chance to get ahead. But our visit to the paper factory had shaken my new friend. Later, outside the factory, we were waiting for the bus back downtown. I was scribbling in my notebook when, behind me, I heard Zhenbing murmuring, as if in a dream, "My poor country. My poor country."

When I left China in January 1997, several issues were making international headlines—human rights, China's possible entrance to the World Trade Organization, the government's alleged Washington influence buying. But before long, China's environmental crisis was bound to command equal attention. According to its government, China's population at the end of 1996 was 1.22 billion people. The true number was surely higher (as I will explain later, population figures are routinely underestimated in China), but even the official figure indicates that nearly one of every four humans on earth lives in China. What's more, the Chinese economy was the seventh largest in the world and some analysts expected it to be number one by 2010. Incomes have doubled since Deng Xiaoping initiated his marketplace reforms in 1979, and the environmental effects have been devastating.

Five of the world's ten most air polluted cities are in China. Sixty to 90 percent of the rainfalls in Guangdong, the southern province that is the epicenter of China's economic boom, are acid rain. Since nearly all the gasoline in China is leaded (Beijing switched to unleaded gas in 1997, and Shanghai and Guangzhou are expected to follow suit) and 80 percent of the coal is not "washed" before being burned, an extraordinary volume and variety of poisons seep into people's lungs and nervous systems. Water and air pollution are killing more than two million people a year, according to the World Bank. Lung disease, aggravated by air pollution and the increasingly fashionable habit of cigarette smoking, is responsible for one-fourth of all deaths in China. Suburban sprawl and erosion gobbled up thirty-five million hectares of farmland between 1950 and 1990, as much as all the farmland in Germany, France, and the United Kingdom. Farmland losses increased in the 1990s, as did water shortages, raising questions about China's ability to feed itself in years to come, especially as rising incomes lead to more meat-intensive diets.

Even the government's pronouncements, which invariably overaccentuated the positive, admitted that environmental degradation would get worse before it got better. After all, China's newfound wealth had only whetted people's appetites for more. The Chinese people wanted to join the global middle class, with all that entailed—cars, air conditioners, closets full of clothes, jet travel. Already, rising consumer demand was causing chronic electricity shortages. China planned to build more than one hundred new power stations during the next decade, adding 18,000 megawatts of capacity every year (roughly the equivalent of Louisiana's entire power grid). China's coal consumption was projected to double, if not triple, by 2020. All this will not only worsen the country's acid rain and air pollution but also endanger the entire planet by accelerating global warming and ozone depletion.

China's large population and grand economic ambitions make it the single most important environmental actor on the global stage in the late 1990s, with the exception of the United States. Like the United States, China can all but single-handedly guarantee that climate change, ozone depletion, acid rain, and other hazards become a reality for people all over the world. Virtually every key aspect of the environmental issue—from population growth and greenhouse gas emissions to air and water pollution; from the ecological impacts of communism and capitalism to the role of public opinion and the potential of technological fixes—is in play in China. What happens in China, like what happens in the United States, is therefore central to one of the great questions of our time: Will the human species survive the many environmental pressures crowding in on it at the end of the twentieth century?

By the time I left China in 1997, I had spent the better part of six years trying to answer that question. My quest had taken me on a trip around the world that included extended (and sometimes repeated) stops in nineteen countries and interviews with everyone from heads of state like Václav Havel in Prague to starving peasants in war-torn Sudan. I had left the United States in May 1991, eighteen months after the Berlin Wall fell and three months after a U.S.–led army drove Iraqi invaders from Kuwait to maintain the flow of oil that modern economies crave like lungs crave oxygen.

Leaving San Francisco and traveling west to east, I began my global tour in Europe. After two months in Holland, France, Italy, Germany, and Sweden, I went to what was still the Soviet Union for five weeks. I continued on to Czechoslovakia, Greece, Turkey, Kenya, Sudan, Uganda, Thailand, and Brazil, where I visited the Amazon and attended the UN Earth Summit in June 1992. I later returned to Europe and the United States before concluding my travels with six weeks in China. I financed my wanderings by traveling light, living low on the food chain, and writing occasional magazine articles from the road.

Scientists had long studied whether elephants in the wild and dolphins in the deep were heading for extinction. I wanted to shift the gaze and turn the binoculars on my fellow humans. Just as scientists compare a given animal's behavior with the dynamics of its habitat to determine whether it is endangered, I planned to analyze human behavior in relation to the earth's ecosystems to gauge the environmental prospects of *Homo sapiens.*

In *The Naked Ape,* his provocative study of the human animal, zoologist Desmond Morris observed that humans "suffer from a strange complacency that . . . we are somehow above biological control" and that our collapse as the earth's dominant species is therefore impossible. Such complacency, Morris pointed out, flies in the face of all we know about the natural world. Biologists have estimated that 99 percent of all species in the history of the planet have ended in extinction. These 99 percent have been unable to survive the ceaseless competition—against the elements, against other species—that is the biological essence of life. The best known example is the dinosaurs, which, if current scientific thinking is correct, were doomed by a dramatic shift in the earth's climate some sixty-five million years ago, perhaps brought on by an asteroid colliding with the planet. Dinosaurs flourished for one hundred million years before meeting their demise, but the average species lasts no more than one million years before expiring. That bodes well for humans if one dates the birth of our species at 200,000 years ago, as recent DNA studies suggest; it is less comforting if one begins the count with the earliest cases of stone tool creation, between 1.5 and 2.5 million years ago. In any case, *Homo sapiens* are part of the lucky 1 percent of species that have survived so far, as are the millions of other species currently in existence. But survival is a constant challenge. Ecosystems are forever in flux, and the scramble for life takes unexpected turns.

"The main piece of bad news at the end of the twentieth century is that we humans can now destroy ourselves, in either of two ways. We can destroy ourselves quickly, through nuclear weapons, or slowly, through environmental degradation," Hubert Reeves told me in Paris near the start of my global journey. Reeves was a cosmologist and bestselling author—a sort of French Carl Sagan. His appearance was dominated by a full gray beard that hung down to his chest and gave him the look of a wizard from the dim past who had miraculously been reincarnated and fitted out in modern garb. Yet in fact, he was the director of research at the Centre National de la Recherche Scientifique, the French government's main science institute. With the Cold War over, Reeves was optimistic that humanity could avoid nuclear self-destruction. He was less sanguine, however, about the threat posed by global warming, excessive population growth, and other more gradual forms of environmental overload. "This problem will be much more difficult to solve," Reeves said, "because it is so much more complex. You can't just have two men sit down at a table and agree to stop being stupid."

Indeed, many modern environmental hazards are rooted not in the collectively suicidal "logic" of nuclear weapons deployment but in economic activities and technological choices that bring pleasure, profits, paychecks, or simple survival to millions: the production and use of automobiles, the felling of rainforests by landless people, the relentless advertising and consumerism that boost sales figures the world over. Averting global warming, for example, could require phasing out fossil fuels altogether in favor of solar and other renewable energy sources, a shift that even solar advocates like Christopher Flavin of the Worldwatch Institute in Washington, D.C., acknowledge is "inconceivable" to most people (not to mention anathema to some powerful economic interests).

Of course, human activity has always imposed burdens on the earth's ecosystem. But the scale and technological power of twentieth-century civilization are many times greater than those of earlier generations, and so are the environmental side effects. Historically, sewage disposal has been the great challenge for human societies trying to maintain clean water supplies. That challenge remains today, especially in poor nations, but modern humans also live in a world awash in man-made chemicals. Global production increased 350 times between 1940 and

1982; the U.S. alone produced 435 billion pounds of such chemicals in 1992. Dioxin and other hormone-disrupting chemicals persist in the environment for decades and can travel thousands of miles; contamination and fertility declines have been detected even among Arctic polar bears. Likewise, the explosion at the Soviet Union's Chernobyl nuclear power plant in 1986 was arguably the most destructive accident in industrial history. The blast left the surrounding countryside uninhabitable for decades and brought death and disease to thousands of civilians. (The precise number of victims is still uncertain; see chapter 4.) The fact that different wind patterns could have sent most of the explosion's radiation across western Europe, where much of it blew in any case, attracted attention across the continent and generated new respect for environmental issues among masses and elites alike. "It was Chernobyl that caused the big change in public opinion about the environment," Antonio Cianciullo, the environmental reporter for the Italian daily La Repubblica, told me in Rome. "Our readers were more interested in these questions after the accident. This caused editors to take environmental issues more seriously and increase coverage of them."

Chernobyl made clear the irrelevance of national borders to modern environmental problems, a theme underlined by other key developments of the 1980s. Scientists had suspected since 1974 that the stratospheric ozone layer, which protects the earth from excessive ultraviolet radiation, was being damaged by man-made chemicals, especially chlorofluorocarbons (CFCs), the active agent in air conditioners and refrigerators. Epidemics of skin cancer, weakened immune systems, and damage of the marine food chain were but some of the potential consequences of ozone layer destruction. But definitive proof of the problem did not come until 1985, when scientists observed a large hole in the ozone layer over Antarctica. The hole was so large that at first it was dismissed as impossible and blamed on a faulty sensor. But after subsequent observations confirmed the initial finding, and large ozone losses were also reported over much of the northern hemisphere, an international agreement ordering a phaseout of CFC production was signed in 1987. The negotiators of this so-called Montreal Protocol breathed a sigh of relief— prematurely, it soon turned out.

Another atmospheric threat making headlines in the late 1980s was global warming. Once again, the scientific community had long known about this danger; the first scholarly analysis appeared in 1896. But not until 1988 did global warming become a household term, thanks to the combination of an extremely hot summer in the United States and some remarkably frank congressional testimony by a prominent government scientist, Dr. James Hansen of the Goddard Space Institute. "It's time to stop waffling," Hansen said. ". . . The greenhouse effect is here."

Svante August Arrhenius, the Swedish chemist who authored the 1896 paper, had theorized correctly that the carbon dioxide released when fossil fuels were burned could have a warming effect on the planet's atmosphere. Like glass in a greenhouse, carbon dioxide traps heat from the sun that otherwise would reflect off the earth and back into space. Perhaps because Arrhenius hailed from a cold weather country, he speculated that the greenhouse effect might produce "more equable and better climates." Modern scientists, however, saw trouble ahead. Higher global temperatures could melt glaciers and expand oceans, causing sea levels to rise and flooding such low-lying capitals as Amsterdam, Shanghai, and Washington, D.C. Since one-third of the world's people lived within thirty-five miles of a coastline, the potential loss of life and property was enormous.

Global warming would also likely cause more extreme weather events in general: more droughts, hurricanes, blizzards, and the like. The reasons are complex but boil down to the expectation that higher temperatures would increase water evaporation around the world. The extra evaporation would lead to more rainfall and storms for many regions even as it caused dry, inland areas like the Great Plains—the world's breadbasket—to experience more dryness. Although a warmer world might bring some benefits, most experts worried that it could disrupt global food production and price millions of the world's poor out of their daily bread. A hotter planet would likely be more disease-ridden as well, as insects and other agents of infection spread malaria and other tropical diseases to areas previously protected by their relatively cool temperatures.

To complicate matters, many of the emerging ecological threats seemed to reinforce one another. The same CFCs that widened the ozone hole also intensified the greenhouse effect. And the greenhouse effect, by raising temperatures too rapidly for plants and animals to adjust to, threatened to hasten species' extinction, another hazard attracting notice

In the 1980s. According to Harvard biologist Edward O. Wilson, civilization was wiping out species thousands of times faster than the usual "background" rate of extinctions that had pertained over the previous ten thousand years. This did not bode well for humans, Wilson wrote, because biodiversity "supports the natural ecosystems on which human life ultimately depends [by] enriching the soil, purifying the water and creating the very air we breathe." (Earthworms, for example, help make soil fertile by loosening it enough for oxygen and water to penetrate it.) Since tropical rainforests are home to more than half of the world's organisms, species loss has been driven most powerfully by tropical deforestation, which in the late 1980s was occurring so rapidly that, if the pace was maintained, the world would have no rainforests at all by 2050. Bringing the problem full circle, deforestation also boosted global warming because it released additional greenhouse gases into the atmosphere even as it deprived the planet of millions of trees that, through photosynthesis, could absorb excess carbon dioxide.

A further complication: although it is hard for humans to feel much urgency about problems far in the future, many of these problems have short fuses. The long lag time between cause and effect means that ozone depletion, climate change, and population growth could acquire so much momentum that they cannot be halted, much less reversed, quickly. The ozone layer, for example, is certain to remain depleted for decades, despite the CFC phaseout mandated by the Montreal Protocol, for the simple reason that CFCs remain in the atmosphere for decades. Not until 2050 are atmospheric concentrations of CFCs projected to return to the levels of the late 1970s, when the ozone hole first appeared. (And that assumes, naively, that everyone obeys the protocol.) Meanwhile, there will be costs. The U.S. Environmental Protection Agency estimated in 1991 that some twelve million Americans would develop skin cancer over the coming fifty years.

Likewise with global warming: the world's automobile tailpipes and industrial chimneys have been spewing greenhouse gases for decades, producing atmospheric concentrations of carbon dioxide 50 percent higher than what existed before the Industrial Revolution. According to the Intergovernmental Panel on Climate Change (IPCC), the 2,500 scientists and other experts commissioned in 1988 by the United Nations to study the problem, to stabilize concentrations at even this level—which

← other too

might or might not deter significant climate change—global emissions would have to be cut by 50–70 percent, taking them back to 1950s levels. Carbon emissions have instead been growing by almost 1 percent a year, a trend that will concentrate twice as much carbon dioxide in the atmosphere by 2100 as existed during the preindustrial era, thus increasing the chances of severe climate change.

Like the captain of an oceanliner who has to turn the helm miles ahead of where he actually intends the vessel to change course, humans will have to alter their environmental behavior years in advance of seeing much positive effect. But is such farsighted behavior consistent with human nature? Millions of years of evolution have left humans capable of responding to immediate threats—the rustle of leaves or the sudden shadow overhead signaling the approach of a predator—but less inclined to react to dangers in the distant future. "We have a saying in Russia," Moscow television reporter Sergei Skvortsov told me while explaining why Russians cared more about the economic collapse and political conflicts convulsing their nation than about their disasterously polluted environment. "We say, 'Even the old grandmother does not cross herself until the lightning strikes.' Which means that people don't worry about bad things until they start to happen."

Yet the more time that passes without taking action against hazards like global warming and population growth, the harder it will be to change course. Indeed, when I left on my global trip in 1991, some prominent environmental figures were warning that humanity was nearing a point of no return—that within ten years the momentum behind major environmental problems could become too powerful ever to reverse. Mostafa Tolba, executive director of the UN Environmental Program, declared that the 1990s would "determine the shape of the world for centuries." Thomas Lovejoy, assistant secretary for external affairs at the Smithsonian Institution in Washington, D.C., told me, "The key environmental problems are so big, and so synergistic, that at a certain point they get beyond bringing under control. You can't solve them in the 1990s, but you have to get a grip on them during that time and position yourself to bring them under control later on." Lester Brown, president of the Worldwatch Institute, also endorsed the ten-year action deadline, telling me, "We're faced with the need for an enormous amount of change in a short period of time if we're going to get the world onto a sustainable path."

Skeptics smirked that environmentalists were forever crying wolf and that the approach of a new century (not to mention a new millennium) often called forth apocalyptic prophecies. Yet the voices urging action were too sober and respectable, too much a part of the status quo, to dismiss so easily. In 1989, British prime minister Margaret Thatcher, perhaps the least sentimental politician of her era, had taken the lead in calling for a quicker phaseout of ozone destroying CFCs. In 1991, dozens of the world's largest corporations at least paid lip service to the need for more ecological business practices when they formed the Business Council for Sustainable Development. The U.S. National Academy of Sciences and the Royal Society of London issued a joint report in 1992 warning that "if current predictions of population growth prove accurate and patterns of human activity on the planet remain unchanged, science and technology may not be able to prevent either irreversible degradation of the environment or continued poverty for much of the world."

Defenders of the status quo tended to discount such warnings as overwrought and unproven. The OPEC nations and oil and coal companies, for example, launched an aggressive public relations campaign to discredit the IPCC's findings about global warming, claiming that the panel's computer models were little more than guesswork. The campaign met with success in part because there was a kernel of truth to this objection: the computer models in question, like many of the tools scientists used to analyze environmental problems, did contain elements of uncertainty. But uncertainty could cut both ways, a point often overlooked by environmental skeptics.

The Montreal Protocol, for example, has been celebrated as a momentous achievement from the moment it was signed in 1987, for it seems to demonstrate that countries could cooperate to reverse a deep-seated environmental problem. The truth is more modest. The protocol actually greatly underestimated the ozone problem, largely because of inadequate scientific knowledge. When governments realized this, in 1990, they toughened the treaty; CFCs and other ozone-destroying chemicals would instead be phased out entirely by the year 2000. But in 1992, this target too was recognized as insufficient; findings of much greater than expected ozone depletion led the United States and other countries to pledge a complete halt to CFC production by 1995. And even this dramatic step was overtaken by further events. The ozone hole detected over Antarctica in 1995 was again much larger than predicted—bigger than North America—and its size remained essentially the same in 1996 and 1997. Thus, the problem continued to fester, even as many people assumed it had been fixed.

Humans have accumulated an impressive body of knowledge about the environmental crisis, but there is no escaping the fact that our knowledge is incomplete. "In effect, we are playing the sorcerer's apprentice with the planet," Reeves told me in Paris, referring to Goethe's poem in which a wizard's assistant borrows the master's tricks, creates a deadly mess, and ends up fleeing for his life. "There are those who point to the uncertainties to argue against taking action, but this, I think, is a dangerous approach," Reeves added. "If you smell smoke, you don't wait until your house is on fire to look for the reason."

I had spent the hours prior to meeting Reeves strolling across Paris, visiting the Luxembourg Gardens, the Île de la Cité, and other favorite spots. By the time I ascended to his sixth-floor apartment near the Boulevard St. Michel, the setting sun was casting the city's pale stone facades and black window grilles into late afternoon shadow, and the light had attained that sparkling depth and clarity filmmakers revere as "magic hour." Amid such resplendent testimony to the complementary beauties of the natural and man-made worlds, Reeves's comments about the two types of self-destruction humanity was courting seemed almost blasphemous to me. But not at all, Reeves replied. For there was a third possibility: that humans would learn to live in balance with the natural systems that make their existence possible. Which path humans would take was an open question, he added, which made the late twentieth century a very exciting time to be alive. "The fact that the summit of complexity in the known universe is now threatening itself with extinction is a cosmic drama of enormous proportion!" Reeves exclaimed.

It was that drama I hoped to observe and record during my travels. To circumnavigate the globe therefore seemed essential. Library research and telephone reporting are invaluable, and there is much of each in this book, but they are no substitute for observing things firsthand. I wanted to see for myself the rainforests that were said to be disappearing at such an alarming rate from tropical regions. I wanted to talk with the people whose farmland was supposedly turning to desert or highways before their very eyes. I wanted to walk the streets of the cities whose pollutants were threatening atmospheric disaster. I wanted to interview the scientists, activists, businesspeople, and government officials who were researching these issues and fighting out their policy implications.

I especially wondered what average people around the world thought about environmental problems. How much did residents of Prague, for example, the capital of the most polluted country in Europe, know about ecological threats? In the wake of the Cold War, at a time when Czechs and Slovaks were sorting out their national identity and struggling with the transition to liberal democracy and a market economy, how much of a priority was the environment? In a world overflowing with immediate crises like the Middle East conflict and the war in Bosnia, not to mention everyday issues of governance like taxes, jobs, and crime, how much urgency did people anywhere feel for ecological threats whose worst effects might not be felt for decades, if ever?

And were things really as bad as environmentalists claimed? Even if the gloomiest scenarios of global warming, topsoil loss, or chemical poisoning were realized, would that necessarily spell the end of the human race? Or, like the Black Plague that struck fourteenth-century Europe, would it perhaps "merely" thin out the population by killing one of every three people? And what about solutions? Was the environmental story an endless litany of gloom and doom, or was there good news as well?

In short, how much of a danger did environmental hazards pose to the future well-being of the human species, and how was humanity faring in its struggle against these hazards? Would human civilization still exist one hundred years from now? Or would our species have been wiped out, partially or completely, by ecological disasters of its own making?

Of course, none of the ecological hazards in question threatened to end all life on earth—just human life. Newspaper headlines notwithstanding, it is not a question of "saving the planet." It might take thousands or even millions of years for the earth to recover from such man-made catastrophes as runaway global warming or full-scale nuclear war, but that is barely the blink of an eye in geological time. Modern humans have inhabited this planet for only the last 200,000 years of its estimated five-billion-year lifespan; the earth could obviously exist perfectly well without us. The real question is whether humans will act quickly and decisively enough to save themselves.

I trust it implies no disrespect for the rhinoceros or the aztec ant to confess that I am partial to my own species—I would like to see it survive and flourish. Yet in my travels, and in this book, I have tried not to let that bias color my views. I have sought to investigate our ecological future without being swayed by an emotional attachment to the outcome, approaching the question almost as if I were a visitor from another universe. This approach is more unusual than it might seem. Overtly or not, most environmental authors seek to persuade readers to think and act in certain ways: to recycle bottles, or to worry about population growth, or not to worry about population growth. That method is valid, but it can end up compromising truth to the vagaries of human psychology; a given situation is typically portrayed in worrisome enough tones to grab readers' attention, but not so darkly that it risks plunging them into a paralyzing despair. The aim of this book is different. It does not so much seek to promote solutions (much less guarantee happy endings) as to describe our collective behavior and ask where such behavior is likely to lead.

As the twentieth century draws to a close, many scientists, business leaders, government officials, and citizen activists see handwriting on the global wall warning of impending ecological collapse. I have done my best to examine the validity of that view, working not as a scientist but as an investigative reporter, armed with an open mind, a restless curiosity, and the freedom to explore. I freely concede that the ambitious scope of my inquiry has made omissions inevitable. I could not examine every environmental hazard in the detail it deserves, anymore than I could visit every country I would have liked. Travel is like knowledge: the more you see, the more you know you haven't seen. But I saw more of this planet's people and places than I ever dreamed possible, and I am deeply grateful for the privilege. I only hope that this account of my journey illuminates enough of our environmental predicament to help light the way forward for the people who live it.

One

"... We Are Still Here."

> The biggest problem of prejudice we face today is
> not black versus white ... but rich versus poor.
> —JIMMY CARTER

Was it a dream, a trick of the night? Or the coming of dawn? Inside the hut, the darkness was total. There was only sound: the insistent keening of a great multitude of birds—not cheerful songsters but agitated complainers whose hoarse moans rose and fell, rose and fell, like a chorus of the dead jealously trying to hold back the day. The hut sat on a riverbank, and as the lamentation continued it became evident that the birds were in the trees across the river, in Ethiopia. Outside, it was still too dark to see, but the sky above the treeline was showing its first streaks of gray. By midday the sun would strike the earth so forcefully that neither man nor beast would venture far from shade, but for now the air was surprisingly cool and utterly still; not the slightest breeze grazed the skin. As the gray light advanced, the birds' cries receded, as if the approaching sun was driving the flock inexorably back to the netherworld of night. But the birds had succeeded in rousing a rooster, and its cawing began to awaken the people.

There was murmuring, coughing, rustling of limbs—the sounds of a large mass of humans emerging from slumber. They were members of the Dinka tribe, and they numbered in the tens of thousands. Their square, low huts of grass and saplings extended well into the bush; to walk from one end of the village to the other would require the better part of a morning. In the distance, far back from the river, a trumpet bleated an unsteady string of notes, reveille for the garrison of soldiers who only last month had repelled armed attackers from the settlement's southern border.

Above the trees, the gray had now turned silver, pink, and ivory. Perched on the wall of the hut by the river, a foot-long lizard cocked his head from side to side, then dashed the length of the structure in a single burst of clattering motion. This hut and six others nearby were surrounded by a high, wire fence that kept the villagers away from the supplies of food, medicine, and fresh water inside; the enclosure housed four relief workers from the International Committee for the Red Cross, as well as the occasional visitor.

As morning fires were lit, the acrid but not unpleasant smell of burning cow dung wafted past. Like a fleshy fist tapping a tabletop, the dull *thump-thump-thump* of a *dura* log announced the preparation of the day's first meal. The log stood taller than the youth who hoisted it, but he grinned with grown-up pride as he lifted the tool skyward before thrusting it point first into the hollowed-out stump before him. The kernels inside jumped and splashed against the sides. Black fingers gripping tan wood, the boy set a slow, patient rhythm. Two girls had gone to fetch water in which to boil the shelled kernels. In six hours, the *dura,* a tasteless gray mush, would be ready to eat.

The rising sun had now cleared the treeline, but its yellow gleam was visible for only seconds before a pale-blue cloud cover obscured it. In groups of threes and fours, Dinka now began passing by on their way to the river, which marked the border between Sudan and Ethiopia. On maps, this river, the Akobo, flows north to feed the White Nile, which merges with the Blue Nile to become the mighty tributary that irrigates

Egypt. But during the dry season parts of the Akobo are little more than muddy ditches. Now, along its Sudanese bank, brilliant pink and yellow butterflies flashed among the low grasses. On the Ethiopian side, a monkey scampered down to the water's edge, then hurried back to the trees. Everything and everyone was awake now. Here in the village of Pochala, on the eastern edge of southern Sudan, deep in the continent of Africa, the daily cycle had begun anew.

It was ten minutes past seven when Garang arrived at the Red Cross compound for breakfast. Garang was serving as my translator in Pochala, but he had once been a librarian, which seemed a fitting occupation for a man with his patient manner and keen, understated intelligence. Perhaps it also explained the care with which he transcribed his name when we first met. Although he had introduced himself simply as Garang, he insisted, upon learning that I might write about him, on spelling out his full name: Daniel Garang Atiel. The written record is important.

The breakfast menu, as usual, was limited: a choice of tea or coffee colored with UHT milk, the tasteless, boxed variety that requires no refrigeration. There was no sugar, but—surprise!—there were finger bananas, squat green fruits the size of a man's thumb. Garang accepted two and munched them slowly, savoring each bite. Just under six feet tall, Garang was short for a Dinka, a tribe whose members are known for their long, stringy builds; heights of seven feet are not uncommon among adult males. I myself am six-feet-two, so back in the United States I rarely have to raise my gaze to look someone in the eye. In Pochala, I was slightly unnerved by having to look upward time and again to return smiles even from teenagers, boys and girls alike.

When I asked Garang how he had slept, he smiled and answered, "Not bad." But not good either, apparently—barking dogs kept waking him up. "The dogs sensed wild beasts in the bush," he explained, "and became agitated."

One of the things I had come to marvel at about Garang was the breadth of his vocabulary in English, which was but one of five languages he spoke. He learned English in primary school and used it in his librarian's job; he also spoke Swahili, the pantribal language used throughout eastern Africa, as well as the tribal languages of his mother and father and a workable amount of Arabic, the official language in northern Sudan.

Without affectation, Garang frequently used words like "agitated" rather than the more common but less precise "excited." At the moment, though, I was less interested in his word choice than in learning exactly what kind of wild beasts had been prowling nearby last night.

"Hyena and lion," Garang said. "Perhaps also leopard."

"Weren't you afraid they would attack?" I asked.

"They would not attack in such a situation. The dogs made them uncertain, so they dared not try."

His matter-of-fact tone suggested that this was a topic of no special concern to him, but prodded by my questions he was soon calmly outlining different tactics for defending oneself in the bush, tactics that varied according to the beast at hand. When he assured me that an unarmed man could kill a leopard, I asked if he had ever done so himself. He smiled and shook his head.

"Only very brave men do this," he replied. "The leopard is the fiercest animal of all. And he is completely fearless. If a lion comes upon a large group of people in the bush, he will be uncertain and run away, unless he is almost dying from hunger. But a leopard will attack, even if he is one against a hundred. He will come straight at you. To kill him, you must wait until the very last moment, when the leopard has made his final spring toward your face. Then you throw your blanket at his head. He thinks the blanket is his prey and he grabs it with both hands and pulls it toward him. At this moment, the man must hit him across the face with a heavy stick. If he can break the leopard's nose, the leopard can die quite quickly."

"And if the man misses?"

"Then the man will die quite quickly."

Lions, on the other hand, employ subterfuge, Garang continued. "The lion moves from one side to the other, stirring up the dust with his hands so you cannot see. Then he springs." The only animal a lion would not attack is the elephant, said Garang, but "a lion can be defeated by a pair of monkeys." This astonishing assertion made me wonder if Garang was pulling my leg, a suggestion that seemed to mystify him; any child who had grown up in the bush knew these things. "Monkeys are very fast," he explained, "with strong arms, and teeth as sharp as the lion. One monkey grabs the lion's tail while the other leaps on his back and bites him. The lion becomes confused, loses blood, and they overcome him. Monkeys are feared by all the other animals because they work as a team."

Garang was in the middle of telling me about a band of monkeys that had attacked a military garrison in Sudan and left three soldiers dead when our conversation was interrupted. A young village woman had appeared at the gate of the Red Cross compound. She entered the compound timidly, as if unsure she had the right to be there. She asked to see the nurse and sat down in the shade to wait. Sunrise had been less than two hours ago, and already the freshness of dawn had given way to the heavy, blanketing heat of the tropical day. On the relief compound tape player, John Lennon was singing "Nobody Told Me There'd Be Days Like These."

At last the nurse, a stocky German woman in her late twenties, appeared. Only then did the village woman unfold the tattered cloth she had slung from her neck. Inside was a nine-month-old baby girl, a tiny, doomed creature with a hideously large skull protruding from a wrinkled body with sagging skin and legs no thicker than a man's finger. The nurse laid the child on its back, using sacks of donated wheat from Canada as a makeshift examination table. The child opened its mouth to cry, but no sound came out. The nurse squeezed some milk into its mouth with a dropper, gave the mother a few additional packets of dried milk, and delivered a flurry of instructions in rapid-fire English. The mother nodded in polite incomprehension, smiled thank-you, and headed back to the village.

Like one out of every eleven African children, this unfortunate youngster would not live long enough to see her first birthday. The child had a slim chance of surviving, the nurse said after the woman had left, "but it will be better for everyone if it does not. It will never develop properly now. It will always be weak, always catching diseases. The mother cannot cope with that, and neither can the family." The nurse had seen severe malnutrition too many times to be sentimental. So had Garang. "In the village we have a number of such cases," he remarked in his quiet, even way. In fact, Garang looked far from healthy himself. His clean, angular features gave his face a certain handsomeness, but their gauntness made him look a decade older than his forty-two years. Especially across his forehead and around his eye sockets, the skin was stretched so tightly that his face seemed a quarter size too small for his skull. The veins in his forearms stood out so prominently that I could silently picture him as a breathing anatomy lesson.

Six months earlier, however, Garang had been even thinner, and many of the village children were as wretchedly malnourished as the baby girl we just observed. That was in June 1991, when the Dinka had just reached Pochala after a forced march of sixty miles, compelled by Sudan's long-standing civil war.

The original homeland of these Dinka was near Bor, a town approximately two hundred miles west of Pochala, in central southern Sudan, on the White Nile. There they had lived as farmers who raised cattle, caught fish, and grew a variety of crops. They fled Bor in the mid-1980s, when the civil war swept through the area, and eventually settled in a UN relief camp in western Ethiopia called Pinyudo, where they regained a kind of life. Then, in May 1991, they again found themselves victims of political violence after the overthrow of the government of Haile Mariam Mengistu in Ethiopia. Armed attackers intent on forcing the Dinka back to Sudan drove them out of Pinyudo. A neighboring relief camp was attacked first, which gave these Dinka enough warning to escape unharmed. Or so they thought.

"People ran in fear" upon hearing of the attack against their neighbors, Garang told me. "After a few days, we reached the Gila River. Some people thought they were safe then, so they stopped to rest and collect leaves in the thick forest there. Those were the ones who got disaster." Unbeknownst to the Dinka, their assailants had pursued them. Garang allowed his wife and four young children less than a day's rest before ferrying them across the Gila in a rough-hewn raft. Saved by their speedy departure, they could only look back in horror when the second attack began. The Dinka who had waited before crossing the Gila, said Garang, "tried to join those of us already on the far side, but many did not make it. Some drowned, some were shot by gunfire. I lost two cousins that day."

By the time the Dinka reached Pochala, they were frightened and exhausted. It was the rainy season in southern Sudan, and the humidity and insects were at their worst. Recalling that time, one Red Cross worker told me, "As soon as you woke up in the morning, you were covered with sweat, before even moving a muscle! The humidity must have been 100 percent. And the daytime temperature was always over 40 degrees, sometimes 50 [that is, between 104 and 122 degrees Fahrenheit], so it made people even weaker. The dampness encouraged mosquitoes and the spread of disease. I remember thinking, 'If someone wanted to believe in hell, the literal hell they talk about in church, with fire and steam and suffering people, this is the place.' "

global warming

Most of all, the Dinka who made it to Pochala were extremely hungry. What food they had brought with them from Pinyudo had been consumed during the journey. Most families were eating only once every two days and surviving mainly on leaves and wild, poisonous fruit (which they soaked in the flood-swollen river for a couple days to purge the toxins). During the first week of July, the Red Cross managed to begin airdrops of food. Quantity was limited by the lack of an airstrip, however, and supplies fell woefully short of demand. Conditions among the Dinka deteriorated rapidly.

"The weaker children were lying on the ground, not moving, waiting for something to happen to them," the Red Cross nurse told me. "When food bundles were air-dropped, our monitors had to race to reach them before the villagers did, or the bags would be torn apart and gone. After the drops, the healthier children would crawl around the drop zone, collecting the kernels left behind." At the end of our conversation I asked the nurse what would have happened if the emergency food shipments had not arrived last July. She looked at me sharply, as though I hadn't listened to a word she had said. "There would have been many people who died from starvation, of course," she replied.

I had come to Pochala as part of my environmental journey around the world, but what did starving Africans have to do with my investigation? Some analysts drew a link between famine and the greenhouse effect, arguing that the drought punishing Sudan and the rest of the Horn of Africa was one of the many consequences of global warming. Indeed, the Intergovernmental Panel on Climate Change had predicted that Africa would be the region hardest hit by global warming, both because of the continent's susceptibility to drought and because its farmers were too poor to rectify their dependence on rain-fed agriculture. Epidemics were also likely to increase, according to Paul Epstein, associate director of the Center for Health and the Global Environment at Harvard Medical School, because global warming would both raise temperatures and increase extreme weather events. "Since the late 1980s, the Horn of Africa has experienced very significant increases in the range in which mosquito-borne diseases like malaria are occurring, as well as a sharp increase in floods that have precipitated outbreaks of both mosquito-borne diseases and water-borne diseases like cholera," said Epstein, who added, "Both of these trends are consistent with the predicted effects of global warming. I'm afraid we're in for some dark days."

If these predictions turn out to be true, the plight of the Dinka offers a sobering preview of the punishing lives awaiting many Africans in the twenty-first century. Yet drought has plagued Africa for thousands of years, long before industrialization raised the specter of man-made climate change. The deeper relevance of the Dinka to the global ecological future, it seems to me, extends well beyond specific cause-and-effect scenarios. The Dinka are a living reminder of the enormous environmental challenges human beings have faced on this planet since our emergence as a species untold thousands of years ago. At the end of the twentieth century, the Dinka are still living the way that virtually all of us *used* to live—as hunter-gatherers and small-scale agriculturalists on the edge of survival. Extracting from the physical environment enough food to survive and reproduce has been a basic challenge for human beings since time immemorial. Indeed, life on the brink of starvation has been the fate of the vast majority of humans throughout history; only in the last two centuries have most people enjoyed adequate nutrition.

The Dinka do not have the luxury of worrying about the environmental dangers of the twenty-first century, even though they are likely to suffer disproportionately from them; they have enough problems simply surviving from one day to the next. And the environment is no abstraction to them, like it is to so many people in the United States, Europe, and the rest of the wealthy, industrialized world. The Dinka experience the natural world directly, unmediated by electricity, running water, refrigeration, antibiotics, motor vehicles, and other modern technological marvels. Wildlife is the leopard that attacks their cattle or children, not something seen in books or at the zoo. And weather is no mere irritant to be neutralized with raincoats or central heating; it is an omnipotent unpredictable force whose whims determine whether there is enough food to eat.

None of this imputes a morally superior "noble savage" status to the Dinka; they are people like anyone else. But their material circumstances do encourage a type of consciousness that has been largely forgotten in the world's prosperous societies, where people have gotten used to buying their way past inconvenient environmental facts by, say, simply turning up the air-conditioning. Many Americans and Europeans, especially those living in cities, have grown so distanced from the natural world that they seem to think they could live without it. But let thunderstorms knock out the electricity for their computers and televisions, or let political turbulence shut off the supply of gasoline to their cars, and those same individuals would soon be as helpless as bugs in a jar. The Dinka, by contrast, know better than to take nature for granted. Their relationship with the environment is a vital concern to them—literally a matter of life and death. In this, they differ from more affluent members of the human species only in degree—in the starkness of their situation and the immediacy of its consequences.

I had left the United States wondering whether the human species would survive the next hundred years, but in Africa I encountered huge numbers of people for whom surviving even the next hundred days was no sure thing. According to the UN Food and Agriculture Organization (FAO), the number of chronically malnourished people in sub-Saharan Africa increased from ninety-six million in 1970 to over two hundred million in the early 1990s. The UN's World Food Program estimated the number of Africans at risk of starvation in 1992 at forty-two million; the countries most severely affected included Niger, Somalia, Ethiopia, Mozambique, Angola, Liberia, and Sudan, which alone accounted for seven million of the imperiled. Africa's largest country geographically, Sudan covers an area as big as the United States east of the Mississippi River. Its population numbers twenty-five million, which means that, if the UN figures were correct, more than one in four Sudanese were at risk of starvation during my visit in 1992.

African hunger has various causes, not least of which is the terrible poverty and poor climate that characterizes so much of the continent, not to mention the larceny often practiced by its rulers and the inequitable trade and financial arrangements imposed upon it by the global economy. Every year, the countries of sub-Saharan Africa pay some $12 billion in interest to service their debt to Western financial institutions; this money would more than cover the immediate food, health, education, and family planning needs of the entire continent.

But what often propels these underlying causes of poverty into full-scale famine is war. War not only kills people directly, it reduces the freedom to plant and harvest and disrupts the transportation networks needed to connect food growers and buyers. In Sudan, war has been a fact of life on and off since 1956, when independence was declared from Great Britain. The most recent phase of conflict began in 1983, when the Arab-dominated government in Khartoum adopted Islamic law. Blacks in the mainly animist and Christian south of Sudan took up arms against this move, sparking civil war. It was this war that had driven the Dinka in Pochala to flight and starvation, and they were hardly its only victims.

In western Sudan, fighting in the breadbasket region of Darfur and Kordofan had made the 1991 harvest "a virtually total failure," despite normal rainfall, said Steven Green, a UN agricultural expert who toured the area and observed widespread and severe hunger. "In the central and northern areas of Darfur and Kordofan you don't see cows anymore," Green told me in 1992. "Any cows. People have been slaughtering their animals and gathering famine foods like *mukheit* [a bitter wild fruit] to survive. Food stocks at the village level have run out. Those stocks may be moldy, two-year-old millet—people have a habit of putting aside something from every harvest—but that is gone now, too. People have used up their reserves of body fat. Humans can live on 1,000, 1,300 calories a day, but they get thin, they get sick, and some die—the older ones from respiratory diseases like tuberculosis and flu, the younger ones from gastrointestinal ailments. There were large numbers of deaths in Darfur. I visited every major village in Darfur and Kordofan, and we heard the same thing everywhere: death rates of up to five children per week."

I wanted to see conditions in Khartoum and Darfur for myself, but the Sudanese government refused to give me a visa. An authoritarian regime, it had never welcomed reporters, and it had been embarrassed recently by journalists and international aid workers who documented a murderous famine that the government had claimed did not exist. This rebuff limited my travels in Sudan to rebel-controlled areas in the south, which I visited by flying in on a Red Cross relief flight from a base camp in Lokichokio, in northern Kenya.

The airstrip where I landed in Pochala was a vast open space from which all brush had been cleared, leaving only brown, pebbly soil. As the plane touched down, it kicked up huge clouds of dust, a spectacle wildly applauded by the hundreds of children who had gathered in the shade of the airstrip's supply depot. When the plane taxied to a halt in front of a makeshift depot, the children tried to surge forward, only to be forced back by young men holding long slender tree shoots, which they brandished like whips. The men strode up and down, yelling and swatting at youngsters who dared to step out of line. The kids shrank back en masse, like pigeons scattering before a pedestrian on a city sidewalk. The switches must have stung, but the children treated the whippings like a game, laughing and teasing those who got hit. Clearly their condition had improved since the days when they were scrambling on the airstrip on their hands and knees for leftover kernels of grain.

Nevertheless, what riveted my attention was the virtual nakedness of Pochala's inhabitants. The clothes the villagers wore were so torn and filthy that they scarcely qualified as clothes. Many of the youngest children went without clothing altogether; older ones were clad in T-shirts and shorts so dotted with holes they exposed more skin than they covered. Many adult women wore only threadbare, dirty skirts, while men wore shirts too grungy to pass as cleaning rags in Paris or San Francisco. It was hard to discern what color these clothes were originally; they appeared to have never been washed. They were all the color of old, grimy dirt—a smudged brownish-gray.

Virtually the only fully dressed people were the village chiefs, whom I met the morning after my arrival in Pochala. One chief was wearing a brown polyester African leisure suit; the other, who had a wildly crooked front tooth that curved like a half moon across his lower lip, sported a black athletic jacket with red trim. Their image of prosperity was undercut by their tennis shoes, however, which were in a pitiful state of decomposition.

Garang, too, wore laceless, rotted sneakers, as well as a scarlet knit sweater that would have been far too warm for this weather were it not ripped in so many places.

The cause of this sartorial deprivation became clear later that morning, when Garang and I toured the village. Choosing at random, I asked if I could speak with a large-boned man I saw playing with some young children. It turned out that the man spoke polished English. In fact, his educated manner and soiled dress shirt and trousers were nearly all that remained of his past as a government civil servant. He greeted me warmly, gave his name as Abot Awan Gaidit, and invited me into his hut, out of the sun.

Bending over to fit through the opening, I found the inside of the dwelling pleasantly cool and quite dark. The ceiling was too low to allow anyone to stand upright, but if one could, it would take four strides to cross the room. The floor was dirt, swept clean with a broom made of leaves and twigs. Occupying the left third of the room was a mat, covered by a thin blanket; Awan slept there. His wife and the three smallest of his six children slept on a larger mat on the right, while his other three children slept on a mat at the back of the hut, next to a table that had a bucket of grain beneath it.

Recalling his arrival in Pochala last June after the flight from the refugee camp in Pinyudo, Awan said, "When we got here, my children looked like skeletons. You could see their ribs through their skin. Their skulls stood out like dead people's. To get food, I sold our clothes to the Anuak, the local people here. First, our bedsheet. That fetched two tins of maize, enough to last my family for three days, if my wife and I ate only at night."

Given the Dinka's desperate condition, it was a seller's market. Awan recalled that after some Anuak offered him a mere three tins of maize for his wife's brightly patterned dress, "I asked them, 'Why do you cheat me? Dinka and Anuak come from neighboring regions, we are brothers. Today disaster strikes my house, but tomorrow it may be you who need help from me.'" The Anuak did not budge, said Awan: " 'It is for you to decide,' they said. 'Do you want the maize or not?' "

Now, in late January, many if not most of the Dinka seemed to have regained something approaching normal body weight. But if they were no longer at death's door, they were still far from well. One morning Garang and I headed to the south end of the village to visit the hospital—if "hospital" is not too grand a word for a facility without beds, anesthetics, or much more than limited supplies of vitamin and antimalarial pills to treat patients who huddled under ratty blankets, eight to a room, inside six canvas-roofed shelters whose sole benefit seemed to be the protection they offered from the broiling sun.

The hospital was run by the Sudanese Relief and Rehabilitation Agency (SRRA), the civilian wing of southern Sudan's main rebel force, the Sudanese People's Liberation Army (SPLA). In the first tent we entered, Garang and I encountered a man whose attire—a navy-blue T-shirt and red trousers, clean and in good repair—marked him as the person in charge. Ajan Mabior Chol, the senior medical assistant in Pochala, said he had come here six years ago, when the village contained only five hundred families of the local Anuak people. The invasion of the Dinka had overwhelmed his clinic, yet no hint of despair intruded upon his unruffled composure. He spoke English well and needed no translator to tell me about the ailments he was treating. Tuberculosis was especially prevalent, he said, escorting me to a smaller hut where a patient was being kept in isolation. The man lay on the ground, shivering and sweating beneath his blanket, his only voluntary movements monosyllabic grunts that he gave in answer to Mabior's softly phrased questions. The man stared at the wall, looking weak and very afraid. Mabior said the man could live another ten days and perhaps even recover if he were transported to a proper treatment center outside the war zone. "But there is no real possibility of that," he conceded.

As Garang and I accompanied Mabior back to the main patient area, my eye was caught—seized, really—by a boy whose head and neck were swollen to twice their normal size. I tried not to stare, but Mabior saw my reaction and without a word strode over to the young man. Sitting on the ground with his eyes puffed shut and his torso seemingly immobilized by the enormous protrusion of flesh between his ears and collarbones, the boy looked like a miniature Buddha with an unspeakably pained expression on his face. He was not eager to be introduced; through Garang, he muttered that he realized how ugly he looked and was embarrassed by it. I replied that bad fortune was no cause for shame; indeed,

I was awed by his endurance of such adversity. As we conversed, he opened up a bit, responding to Garang's translations of my words with the tongue-clicks that Dinka use to signal comprehension or assent. He gave his age as thirteen and said he had had this sickness before, in Pinyudo, but the doctors there gave him medicine that cured it. Then, after the forced march to Pochala, the disease had reappeared. It got worse and worse, and now it hurt to walk, it hurt to eat, it hurt all the time. When I urged that he not give up his fight because the illness could still be defeated, his reply was categorical: "No, this thing will kill me."

Mabior entered the next tent and beckoned me over to a woman in the corner, saying she was one of the patients whose test results he was awaiting. Stepping gingerly around the sprawled bodies of nearby patients, I knelt beside the woman, who looked up at me with wide, anxious eyes. Nodding to reassure her, Mabior drew back the blanket from her body, revealing a belly so grotesquely swollen it looked like a watermelon had somehow gotten lodged between her sternum and pelvis. Adding to her freakish appearance were two apple-sized growths bulging out of the swelling, near her navel. "What you must realize," said Mabior, "is that this woman is not pregnant. She has symptoms of tuberculosis and suffers other diseases as well. I would like to send her to [the SRRA hospital in] Kapoeta for examination, but I am told by the Red Cross that there is no transportation for her."

Mabior was called away, so Garang and I wandered around by ourselves. Turning the corner of the supply shack, we encountered half a dozen teenage boys, crouching in the dust, stoically awaiting treatment. During the past days I had noticed white bandages wrapped around the shins and ankles of many village youngsters; now I was face-to-face with what lay beneath those bandages—tropical ulcers. A tropical ulcer is a ghastly thing, a vicious, open wound the size of a man's palm that oozes blood, pus, and body matter. The skin literally explodes from the inside out, giving the impression that the top layers of flesh have been scraped away; left behind is a shock of red moistness that attracts a constant swarm of flies. A tropical ulcer can eat all the way down to the bone and eventually cause cancer. Garang and I watched a nurse remove the gauze from one boy's wound using only scissors and squirts of disinfectant. Not once did the boy so much as wince. He and his friends were used to the procedure. Two of them had borne their ulcers for a month, two others for three years; a youth of nineteen had had his since he was ten.

In my final chat with him, Mabior identified malnutrition as the underlying cause of most diseases he encountered. Tropical ulcers were caused by a combination of poor hygiene (no soap and clean water amid the endless dust of the bush) and chronic vitamin deficiency. Severe cases were treated with a daily iron tablet, but the boys I saw were not in that category. Mabior explained that there simply was not enough medicine for all who needed it. The Red Cross had been reducing its shipments to Pochala in recent weeks, he added; last July, five tins of antimalarial pills were supplied each week, but now it was five tins every two weeks. As a result, Mabior had begun turning away all but the most serious cases. This he found difficult, he said, "but what choice do we have? If we give treatment to everyone who needs it, our supplies will run out in a matter of days, and then what do we do?"

As Garang escorted me along the dusty paths that crisscrossed Pochala, he seemed to glide more than walk. This impression stemmed partly from his erect posture and graceful stride but also from the incredible slowness of his pace. Everyone here moved slowly, except the kids in moments of rambunctiousness; the physical environment allowed for nothing else. Newcomers in Africa sometimes sneered at what they saw as the laziness of the locals, but they soon learned that only a fool hurried in this climate.

Social patterns in rural Africa also discouraged rushing: as a rule, Garang was unable to walk more than fifty yards without pausing to greet one friend or another, and our walk back from the hospital was no exception. The Sudanese in general had a reputation as the most hospitable people in all of Africa; if the Dinka were typical, this reputation was richly deserved. Cheering cries of "Welcome, welcome!" were always being addressed to the visitor, and attempts to mimic the greeting in the Dinka tongue produced gales of laughter, as did efforts to work the dura log (a task rather more difficult than it looked). Despite the poverty, smiles were warm and frequent. Often these began as shy grins, only to blossom into a display of shining teeth that would melt the world's hardest heart. (Teeth happened to be very important to the Dinka. They were forever cleaning and picking at them with small twigs, and they commonly removed one or more of the front ones as a beauty symbol.)

As Garang and I headed toward the Red Cross compound, we passed a gaggle of small children walking the other way, each of whom, even the two- and three-year-olds, smiled and reached up a little hand to shake hello with the visitor. I marveled at the contrast between this poised behavior and what I had witnessed while traveling among the Turkana, Kikuyu, and other tribes of Kenya. Despite enduring far worse deprivation, the Dinka in Pochala remained calm and controlled, and not one of their children engaged in anything like begging. "No, no," replied Garang, wrinkling his nose as if the mere mention of begging emitted a bad smell. "They are taught this way. To beg is a bad habit. You lose your dignity. The Dinka believe that in a time of suffering you must be patient. You must bear the suffering until others see it and come to your aid. Not because you ask them to, but because they choose to do so themselves."

Although outsiders had indeed come to the Dinka's aid in the past, they were now withdrawing. The reduction in medical supplies lamented by Mabior was matched by a cut in food shipments; a psychological message was being sent to the Dinka. Ole Sorensen, a Red Cross organizer back in Lokichokio, had told me, "We're gradually reducing our stocks so as to tell them, 'Look, we're not going to be bringing in so much anymore. It's time for you to move.' " In effect, the Red Cross was pushing the Dinka out of the nest, urging them to take advantage of the dry season to move to a more habitable, permanent location.

The official explanation for these cutbacks was that the Red Cross did not want to encourage dependency; its role was to provide emergency aid, not long-term relief. But some relief workers also seemed to harbor a suspicion that the Dinka were taking advantage of the outside world's generosity. "For half a year, the grain literally falls from heaven for them," Sorensen told me. "So now [their attitude is], 'Where's my bag?' " The nurse in Pochala went further. To her, it was obvious the Dinka had more than enough food now; after all, they weren't rummaging on the airstrip for extra kernels of grain anymore, were they?

The relief workers' suspicions reminded me of the comments of some fellow journalists I had had dinner with a few weeks earlier in

Nairobi. It was the day after Christmas, and a dozen reporters represent-
ing high-profile news organizations in Europe and the United States were
seated around a table laden with a mouth-watering selection of holiday
food and drink. Shop talk dominated the conversation: travel schedules,
homeside editors, who was writing what. When asked, I said I was doing
a story about hunger in the Horn of Africa, and I cited the UN's warnings
about forty-two million Africans being at risk of starvation over the coming twelve months. My colleagues, with few exceptions, were non-
plussed. Old hands at reporting on Africa, they argued that the UN
routinely exaggerated the hunger problem to justify its bureaucratic existence. "You can't take those UN numbers too seriously," one reporter
asserted. "You don't want to criticize the UN too much, because every-
body remembers being caught unprepared last time [in 1984, when
American news organizations ignored the Ethiopian famine for nearly a
year]. But there's no famine in Ethiopia this year. People aren't walking
miles and miles to find food. Basically, the UN is covering its ass and
padding its budgets."

If by "famine" one meant masses of starving people dropping dead at
the same time—the definition embraced by the news organizations represented at that table—this reporter was correct: there was no famine,
and thus no breaking news story, in the Horn of Africa. But the realities
of Pochala illustrated how flawed that standard of newsworthiness was.
Mass starvation was actually the culminating tragedy in a complex chain
of events. By failing to cover hunger until large numbers of people were
dying, the news media ended up missing most of the story; in effect, the
media waited until the battle was over before sounding the alarm and it
rarely probed deeply enough to identify the real roots of the problem. "If
you aren't talking about ten thousand people falling over dead from
hunger, it's not a story in the newspaper," Brigette Menge, a Red Cross
press officer, told me in Nairobi. "If instead you need a vaccination program to keep farmers' cattle alive so they and their families can avoid
falling into the destitution that leads to starvation, it's not sexy enough
to interest the media."

Conventional media coverage also left the impression that Western
food aid was the most important solution to famine, partly because by
the time TV crews arrived on the scene, it was indeed too late to do much

except dispatch emergency food. But aid that left people perpetually two steps from death was hardly worthy of the name. In truth, charity could never be more than a stopgap solution to hunger. Beyond an end to war and the disruption of economic activity it caused, what was needed were practical, revolutionary things that enabled people to better their own lots: clean water, education, healthcare, and jobs; honest, responsive government; more equitable distribution of land. In short, justice and development, not last-minute handouts.

"The only hope is somehow to empower the rural poor," Ben Foot, director of the Nairobi office of Save the Children, told me. "The people who suffer in famines are the poor. There's usually food available during a famine, but the shortage of supply pushes prices up beyond what the poor can afford. The problem in Sudan, as it is in many parts of Africa, is that it is basically still in feudal times. A tiny elite owns most of the land, while the masses sharecrop. There has been no redistribution of wealth, because governments have been loyal to the elites. The only hope is to get governments that are more accountable and concerned about what their people say—democracy, in other words."

In his classic study of the Nuer tribe of central Sudan, Edward Evans-Pritchard recounted the tribe's myth about Stomach. At the beginning of time, Stomach lived on its own in the bush, subsisting on insects roasted by the spontaneous firing of grasses during the dry season. One day Man came along, spied Stomach, and decided to put it in its present place, but with unhappy results: When Stomach lived alone, it was satisfied with tiny morsels of food, but now it is always hungry. No matter how much it eats, it craves more.

Hunger has always been part of African life. Nature gave Africa ferocious pests, rainfall patterns that fluctuate unpredictably between flood and drought, and soils that tend toward low fertility and erosion. Colonialism brought cash-cropping and exploitation; seizure of the best land by white minorities further undermined Africans' ability to feed themselves. Self-preservation therefore compelled Africans to develop their own strategies against chronic food shortage long before Western rock stars and relief agencies arrived.

Simply eating less was the most important coping mechanism. "Hunger: Tie your stomach tightly and wait—it will leave you," said a peasant quoted in Alexander de Waal's study of the 1984 famine in Darfur that killed one hundred thousand people. De Waal reported that Western food aid played but a minor role in relieving most famines, accounting for only 10–12 percent of all food consumed. In Darfur, de Waal wrote, people often went without food for days rather than sell their livestock, tools, and other means of subsistence. To go very, very hungry was preferable to sinking into a state of temporarily sated destitution that left a family unable to survive the next disaster.

"You pull your belt very close, so your intestines are tight and your stomach stays small," Awan, the former civil servant, told me. Garang added, "A man can walk all day this way without eating. You only keep a little tobacco in your mouth to chew for juice." Above all, starving people sought to preserve their livestock and other assets. The Dinka tried not to kill their cows not only because cows produced milk for children and helped with chores like plowing but also for cultural reasons. "For the Dinka, how many cattle a man has decides his status in the community," Garang explained. "They can only be killed at times of communal need, such as weddings or special celebrations."

Or imminent starvation. Garang did not volunteer the point, but when I asked, he confirmed that he used to own a cow in Pinyudo. "It survived the journey here," he said, "but then I was forced to kill it. This was last July, when we were starving." Kinship rules compelled Garang to share this meat with his entire extended family: his wife and four children, plus an uncertain number of additional brothers, cousins, and so forth. Such sharing was yet another coping mechanism; in subsistence-level communities, sharing maximizes the chances of collective survival. In this case, it kept Garang's clan fed for seven days, he said.

How that beef could remain edible for seven days in the heat of Pochala became clear the next day, when Garang and I visited the open area that passed for the marketplace in Pochala. The hundred-odd people conversing in the midmorning sun were drawn together more by their habit of socializing than any real opportunities for buying and selling; commerce was all but nonexistent. The single item for sale was displayed on a waist-high wooden table surrounded by men who seemed to pay no attention to it. From a distance of ten feet the item's shape and color reminded me, ludicrously, of a ladies' large black handbag tipped on its side. As I stepped closer, one of the men gestured and the surface of the item seemed to shiver. Then the man grabbed the item and flipped it onto its back, causing parts of the surface to fly off. Only then did I realize that the surface was a covering of flies so black and dense that it completely obscured the redness of the meat beneath. At that instant, my nostrils were assaulted by a stench as powerful as that emanating from the village long-drops, the narrow holes in the earth that served as toilets here.

"What is that?" I gasped to Garang.

"It is meat," he replied, unflappable as ever.

"Is someone going to eat it?"

"You can be sure of it."

Garang guessed that the cow had been slaughtered yesterday and that the owner hoped to sell a hunk of it before it spoiled beyond recovery. Whoever bought the beef would have to carry out a laborious cleaning process to try to rid it of its smell and germs.

"You boil it in water, pour out the water, boil it again in new water, pour out the water, boil it again, pour out the water, then you cook it," recited Garang.

"How long does that take?" I asked.

"It depends how far away water is. If water is nearby, then one day. If you start in the morning, you can eat at night."

"How does it taste?"

"It does not taste. It is too dry."

"But is it safe? Don't people worry that the flies will make them sick?"

 "Yes, they worry. But what choice do they have?"

Accusations of freeloading to the contrary, the Dinka in Pochala did not think life on the dole was so wonderful; they would have preferred to be back in Bor fending for themselves. "The [relief] grain keeps us alive, and for this we are grateful," said Awan. "But to eat the same thing twice a day, every day, is very tedious." There were problems with returning to Bor, however. For one thing, it was two hundred miles away—two hundred miles of desert, filled with armies and bandits through which the Dinka would have to walk. And according to the Red Cross, Bor had just been the site of another civilian massacre. Nevertheless, the Red Cross was preparing to kick the Dinka out of the nest and on to Bor. My conversations with the Dinka revealed that they knew nothing about these plans, however, nor about the recent violence in Bor.

One man was helping his wife milk a cow when I asked to speak with him. Powerfully built, with a solemn, intelligent face, the man wore only a single article of clothing: an olive-green, military-style trenchcoat that was open down the front and extended to the middle of his shins. He stroked the cow's back as his wife squeezed the last drops of milk from each teat, employing the same smoothing motion one might use to push the last bit of toothpaste out of the tube. The man walked with a cane and a bad limp; looking down, I saw that his left foot was twisted down and outward at a grotesque angle. He told Garang he had stepped in a hole while running from gunfire during the attack on Bor five years ago. He never saw a doctor, and the ankle froze in its fractured position.

A crowd gathered as the man answered my questions. No, he had no idea that the food aid was ending. No, he didn't know where he would go then; he would rely on what the SRRA told him to do. Since there had been accusations that the SRRA was exploiting the refugees' plight to cadge extra food for the rebel army, I asked the cowtender whether he trusted what the SRRA told him. Garang translated my question, the man replied somberly, and the crowd exploded in laughter. Even Garang, generally a pretty composed fellow, chuckled before translating the answer: "He says, 'When a man gives me food when I have none, I trust what he says.' "

A few days later, back in Lokichokio, I spoke to the man in charge of giving the cowtender his food. Michel Minnig, the acting director of the Red Cross base camp, was a slim, quietly intense man whose neat silver hair and efficient formality seemed, amid the heat and lassitude of Africa, the epitome of Swiss order and reserve. (Like most senior Red Cross officials, Minnig was a Swiss national. In 1996, he would attract international media attention by entering the Japanese embassy in Lima, Peru, to try to defuse a hostage crisis.) Minnig was too intelligent and compassionate to utter anything like the comments some of his staffers had made about the Dinka ripping off the Red Cross. Instead, he told me how the Red Cross would help the Dinka return to Bor. Tracing the routes on the map on his office wall, he pointed out that the walk south from Pochala to Kapoeta was 150 miles, plus another 250 to Bor; or people could head directly west and reach Bor in 200 miles, though that route involved a greater risk of encountering combat. The Red Cross would stock some food and water along the way, and trucks would carry the very youngest and the disabled.

"Of course such a walk will be very difficult, but that is a situation faced throughout the Third World," Minnig said. "Once we can bring these people up to the general Third World level, we have to stop our work, because after that it becomes development, which is beyond our mandate. But it's a very bad mistake if the outside world thinks that because these people received some relief aid, they somehow have a good life now. They do not."

I had been warned before leaving Nairobi (by an official of the SRRA, no less) that "southern Sudan is like the Stone Age." But I did not truly appreciate the comment until I visited southern Sudan a second time, some days after my Pochala trip. On this second visit I was again a passenger of the Red Cross, this time traveling in a truck that would evacuate wounded soldiers from the war zone.

We departed from the Lokichokio camp in northern Kenya at 10:30 in the morning, heading for Torit, a rebel army stronghold that lay 160 miles northwest. The road was an abominable dirt track, deeply rutted and pocked with large rocks and moguls of hard dirt. It led past landscape that would remain unchanged until we reached the town of Kapoeta, seventy miles ahead: dry, dusty bush, with brittle grass bleached white by the sun, head-high shrubs, and thorn trees dried into shades of silver and gray.

Because we were leaving so late in the day, the heat was already scorching. Nevertheless, Peter, a local Turkana who was the driver, informed me that the windows had to stay rolled up; otherwise, the dust from the Red Cross jeep preceding us would choke us. This directive was all the harder to accept given our cramped quarters. There were four of us in the truck's cab: Peter, his assistant Samuel, another journalist, and me. Peter had his own seat; the rest of us were crammed into a space meant for two, which could accommodate us only if I hunched toward the windshield while the other journalist leaned back against the seat. Even so, we were all pressed tightly against one another, skin to skin, and since I am tall and lanky, I dreaded the hours of discomfort ahead.

Within half an hour we crossed the unmarked border between Kenya and Sudan, and half an hour after that we encountered the first sign of human activity: a rebel army checkpoint, where our way was blocked by a slender tree trunk slung across the road. An SPLA youth wearing nothing but filthy torn underpants emerged from the bush, swung the tree trunk aside, and waved us on to a second "gate" at the far end of the outpost. There, equally young men crowded around the truck while Phillipe, the Red Cross doctor in the jeep ahead, made the necessary explanations. Were it not for the rifles that two of these young men carried, one would never have guessed that they were soldiers; like their comrade down the road, they wore essentially nothing. They thrust their heads inside the cab and, in pidgin English, roughly demanded soap, pens, cigarettes.

Ten miles farther on was a second checkpoint, where we went through the same routine. At a third checkpoint farther on, the travel permit issued me by the SRRA office in Nairobi was taken inside to be approved. As we pulled away from the outpost, Peter said the soldiers had advised him to keep an eye out for bandits between here and Kapoeta. Nevertheless, except for growing irritation between me and the other journalist as we struggled against leg cramps and inadvertent elbows in each other's ribs, the rest of the journey to Kapoeta passed without incident. Not once did Peter or Samuel betray the slightest annoyance at the tight quarters, which is all the more impressive considering that, had we two interlopers not barged aboard at the last minute, they would have had this cab to themselves.

Downtown Kapoeta consisted of little more than a single block of redbrick buildings with white fronting columns—ghostly colonial relics crumbling into the constant swirl of wind and dust. There was no one on the street, not a vehicle in sight, nothing but searing heat and gusting sand. We stopped for two minutes outside the SRRA hospital, just long enough for Philippe to rush in and out. As we pulled away, I saw a lone figure lying out front beneath the shaded portico and was amazed to recognize the woman from Pochala with the swollen belly. So Mabior had gotten her to Kapoeta after all! No one seemed to be helping her, though; she looked as abandoned as the town itself.

At the town's roundabout, a wooden sign indicated the road to Pochala: a path of white sand opening into an expanse of sunblasted nothingness. It was from that expanse and on that "road" that the Dinka in Pochala were invited to make their journey back to Bor. As we pressed on toward Torit, I wondered yet again how the Dinka would manage. Here I was, riding in a truck, with a proper breakfast and luncheon sandwich inside me and favored by good health, yet I found myself drained after a mere five hours of travel. The sun's glare and the blazing heat had glazed my mind into blankness. Reaching for the bottled water I had been saving since we left Lokichokio, I discovered that despite the bottle's sealed plastic top and the truck's rolled-up windows, the water had turned brown from windblown dust. I sipped it anyway and made sure to save some for later.

After Kapoeta, the road, unbelievably, got worse. Ruts and gulleys two feet deep became routine. Peter had to slow the truck to an almost complete halt and ease the front wheels down into the gullies, then tap the accelerator just enough to lift the wheels out the other side. We crossed numerous hundred-yard stretches in which we inched forward, bucking up and down like a launch fighting its way through heavy waves. Inside the cab the jostling was at once bone jarring and monotonous; I lost count of how many times my head banged against the roof.

Outside the cab, though, things looked much worse. The road, poor as it was, was almost the only trace of human activity in sight—that, and the occasional traveler. Bumping down a long straightaway shortly before sunset, we passed a man walking back toward Kapoeta who was wearing absolutely nothing but a scowl on his face. Tall, erect, moving slowly, and carrying a rifle over one shoulder, the man did not return the waves of greeting that were customary between travelers here. A few miles farther on, we passed a young man, also heading the other way, whose pouring sweat and stumbling gait suggested to Peter that he had probably been walking all day from Torit, which was still thirty miles away.

Travel here was exclusively by foot. Except for the jeep ahead of us, we observed not a single functioning vehicle during our entire ten-hour trip from Lokichokio. Indeed, other than a handful of trucks and Land

Rovers belonging to Western relief groups that we would find in Torit, there were almost no signs of twentieth-century life in southern Sudan: no paved roads, no running water, no electricity, virtually no schools or public buildings. That much was true in other parts of the rural Third World, but the deprivation in south Sudan went deeper. There were no shops here of any kind, not even the usual run-down roadside stands selling drinks and fruit. At the few open-air markets, the only commodity available in any quantity was local tobacco, sold in small, mudlike clumps. Paper money was useless; exchange was all barter. There were no telephones or books, no radio or newspapers, no postal service. Contact with the outside world was sporadic at best, consisting of whatever went in and out along this primitive highway.

Two hours past nightfall, we finally arrived in Torit. We made camp in the grassy courtyard of an old Catholic mission and retired without eating. As I unfolded my mosquito net, I recalled the SRRA man's comment about the Stone Age; he hadn't exaggerated much. I crawled into my tent and lay down, but despite my exhaustion I did not sleep. There was a breeze, and soon it was bringing to my ears the unmistakable sound of human voices joined in song. The singing seemed to come from far away; when the breeze shifted, the sound faded like a radio station losing power. I remembered a night in Pochala when the Dinka were singing like this. Deep and rhythmic, almost a chant, the voices here in Torit conjured up a sound from the deep human past, a sound conveying not so much sadness as simple human steadfastness. I listened as long as I could before the voices lulled me to sleep.

By 8:00 the next morning we were at Torit hospital, where the truck was being loaded with SPLA soldiers whose wounds were too serious for treatment here—everything from broken bones to bullet holes to smashed heads and limbs. Most of the men were conscious, swathed in white bandages marred by large red patches of seeping blood. Some cried out as they were lifted into the truck and hung along the walls like bunkmates in a submarine. Of course, this jostling was nothing compared with what they would suffer during the ride back to Lokichokio, but Phillipe said not to worry: "They are used to bad conditions."

The soldiers were casualties of the very fighting near Bor that made it so dangerous for the Dinka in Pochala to return to their old homes. An English health worker at the hospital said that the fighting had displaced an estimated 150,000 people and killed an indeterminate number of them. In addition, virtually all the Dinka's cattle had been looted. The refugees had fled south in panic, and some were now gathered in a town called Ngangala two hours north of here by car. A Norwegian relief worker volunteered that she was going to Ngangala tomorrow at dawn; I could join her if I wanted.

I returned to the Catholic mission, where I spent the afternoon beside a water borehole that boasted a concrete platform and sturdy metal pump. By local standards, this was true luxury, and it drew a procession of visitors through the day, virtually all of them female. Ranging in age from toddlers to seniors, they invariably offered a warm smile and greeting before stepping up to the handle, pumping vigorously, filling their crude metal pots to the brim and then balancing them atop their heads for the walk home. One little girl who looked barely two years old was with her mother and three sisters. As she watched them leaving for home, each with a correspondingly smaller pot on her head, she grabbed her own tiny, flat can, knelt down, and with ferocious concentration set it atop her head without spilling more than a couple drops. She then stood, removed her hands haltingly, and found to her delight that the can remained in place. She had joined the grown-ups.

Near sunset, I joined the mission's priest, Father Matthew, on a dilapidated, screened porch to watch the sky fill with a murky collage of red, gold, and violet. Father Matthew had lived among the African poor for over twenty-five years, first in Zaire and now in Sudan, and he spoke with the patience and passion of a man who saw salvation as a matter not only of heavenly reward but of justice on earth. "It is very easy to succumb to despair," he agreed. "But the spirit of the people saves me from that. Their strength and perseverance, the joy they express amidst terrible adversity, is very inspiring. Our relationship is very much a two-way street. In fact, I get far more from them than they get from me." He rose, went to the kitchen, and came back with two plastic cups and a small cardboard box, which I assumed contained UHT milk. It turned out instead to hold bad sherry—sour and flat. As the day's last light disappeared, he lit another cigarette. "This is something many Western aid groups seem not to understand, I'm afraid. But there is a big difference between seeing our work as charity, which is how most groups see it, and seeing it as solidarity. Charity is about hierarchy, imposed solutions, and a short-term perspective. Solidarity is a partnership of equals, committed to struggling together."

By now it was so dark I could not make out Father Matthew's face three feet away from me. But he was used to lampless evenings, so our conversation continued in darkness. As he sipped the warm sherry in the sultry night, he reminded me of a character in a Graham Greene novel. Sporadically he inhaled his cigarette, and for a moment his eyes and nose glowed eerily in its reddish flash before vanishing back into the night. It was rehabilitation and development that southern Sudan needed, he said, though he conceded in the next breath that it was difficult to move beyond emergency aid as long as the war continued.

Though generally described as a religious conflict between Islamic fundamentalists in the north and rebel animists and Christians in the south, the Sudanese civil war was, according to Father Matthew, also rooted in racial and economic oppression that went back centuries. The north was the province of Arabs, whose traditional relationship with blacks in the south had been as master to slave. "The Arabs have practiced a form of apartheid here in Sudan," said Father Matthew. "The West has paid more attention to the apartheid in South Africa because that one was imposed by our own brothers, but the system here is quite similar. Before the war, the south took orders from the north. Southerners were not given equal access to jobs, to education, housing. All the development and infrastructure were kept in the north. Many times I have heard southern Sudanese say, 'This war will end when we are no longer slaves.'"

In the meantime, the price of war was horrific. During the drive the next morning to Ngangala, we passed the overturned skeletons of six trucks that had been destroyed by land mines—mute testimony to the fighting that had ruined the economy of this region and shattered the lives of its people. When we pulled into the Ngangala refugee camp, we found approximately one hundred people crowded under the shade of the largest tree in sight. It was immediately clear that this situation was far more grim than in Pochala. All the children here were naked. With their bodies and faces covered with gray cow-dung ash to deter flies, they looked like spirits from another world. Seeing my white face, some parents approached and pointed repeatedly to their open mouths and tugged gently at my sweaty shirt, asking to share my food and clothing.

The crowd had assembled to hear a lecture on how to prevent diarrhea, a leading killer of children under these conditions. The lecturer, an African nutritionist from Torit, spoke in English while a local medical assistant translated into Dinka, but it was hard to imagine much of the message getting through. A cacophony of phlegmy, hacking coughs and sobbing tears from the children and desultory chattering by adults made it very hard to hear. The nutritionist had brought a classroom picture chart that incongruously showed a loaf of bread, a bottle of milk, a piece of corn on the cob, and other store-bought items. At the end of the talk, a woman raised her hand and said, "We know how to boil water like you say, but you tell us to mix salt and sugar with it before we give it to our children. You know we don't have these things. So why do you tell us this?"

"I realize that," he replied. "Still, you must do the best you can with what you have."

The presentation over, the people drifted back to their huts and another day of aimless waiting for the sun to set. On my way to see the water supply—a stagnant pool of green, fetid liquid a mile away—I passed a shrunken, knobbly kneed toddler whose malnourished frame reminded me of the doomed infant Garang and I had seen back in Pochala. A few steps later I met a girl of ten whose left arm hugged a baby to her hip and whose gleaming smile caused me not to notice at first that her right arm had been amputated at the shoulder. And so it went, one sad sight after another, for longer than I care to remember.

My host in Ngangala was Abraham Jok Aring of the SRRA, and after I toured the camp he invited me to his hut. The gesture surprised me, for his manner had been brusque so far. The day was extremely bright, so my eyes were useless once I crouched through the hut's opening. I felt my way through the darkness to a rough bench and sat down. It was very hot and stuffy inside; I wished I could leave. Jok was delayed for a couple minutes but soon joined me. Only then did I realize that I had not been alone. Out of the shadows in the corner, not five feet away, a figure materialized, a servant of uncertain age who unfolded himself from a crouch, shuffled

over to me, and with an indistinct murmur removed the lid from a straw
basket and offered me its contents.

"What is it?" I asked, looking at Jok.

"It is goat," he replied. "Take some."

This struck me as a bad idea. I begged off, thanking him but explaining that I could not possibly take what little food they had.

"Do not worry, it is for us," Jok said. "Take some."

My eyesight had adjusted sufficiently by now to see that the meat
was nearly raw. This reinforced my reluctance and I again declined, but
Jok repeated his invitation, more forcefully this time. After we went back
and forth some more, I tried the only other excuse I could think of: "I'm
sorry, but I grew up on a farm where we raised goats, so I could never eat
a goat now."

This happened to be true, but it only convinced Jok that I took him
for a fool. In his world, only an idiot would raise goats and not eat them,
so I had to be a liar.

"Do you *fear* it?" he thundered.

Now I had no choice. I answered no and, as calmly as possible, removed a piece of goat from the basket. It was slippery and dark. I swallowed it quickly, after one chew, and hoped for the best.

The servant bowed and turned away to offer the basket to Jok. Jok
had made his point, but his frustration at the outside world was too fierce
to be so easily sated. Helping himself to a handful of meat, Jok declared,
"Sometimes we think outsiders come here to laugh at us. They take pictures, ask questions, make notes. Then they go back to their comfortable
lives and nothing changes. To them, it is just a joke. But we are still here."

The civil war has dragged on mercilessly since I left Sudan, reinforcing the
country's reputation as one of the most godforsaken places on earth. By
1996, an estimated 1.5 million Sudanese had died and 85 percent of the
southern population had become refugees. By late 1997, the rebels had regained control of most of southern Sudan, compelling the government to
join in peace talks that, in 1998, yielded a mutual pledge to accept an internationally supervised vote on self-determination for the south. Under
the circumstances, it would be a miracle if the refugees in Ngangala— or

Pochala, or anywhere else in the war zone—had managed to stay in one place. Indeed, many of the people I met have probably died. But others will have taken their places, so Jok is right: he and his people are still there, locked in a world of bottomless bloodshed and misery.

During my time in Sudan, I often thought about how lucky my loved ones and I were to have been born into the circumstances we were. As far as I could see, the suffering that plagued the Dinka was no particular fault of their own; had I been born into their world, the same fate would have befallen me. Since leaving Sudan, my thoughts have returned most often to my esteemed companion, Garang, who so ably served as my Virgil there. I have wondered how the endless patience of Garang and the other Dinka had been rewarded, and I feared I knew the answer. Terrible food shortages have continued to haunt Sudan since my visit, condemning hundreds of thousands to lives of destitution and suffering. The situation grew particularly desperate in February 1998, when the Khartoum government refused to allow UN planes carrying emergency food supplies to land in hunger-stricken southern towns. By May, aid workers were warning that as many as 350,000 people could die if relief flights were not reinstated.

Of course, famine and crushing poverty are by no means confined to the Horn of Africa. Every day, about eleven thousand children around the world die of starvation. In India, fewer than half the children under age five have enough to eat. Throughout the Third World, nearly 40 percent of rural people live in absolute poverty. Conditions of Third World city dwellers differ in details but not results, as anyone familiar with the slums of Rio de Janeiro or Bangkok could attest. Six hundred million city residents in poor countries occupy such shabby housing that their health is in constant danger. According to the World Bank, 20 percent of humanity exists on the equivalent of less than one U.S. dollar a day. The FAO estimated in 1996 that 841 million people were chronically malnourished—that is, more or less constantly hungry.

Such numbers may seem too huge and abstract to mean much, but what they amount to is this: nearly one out of every six human beings on the planet lives a life like the Dinka's—often famished, perhaps ill, a whisper of bad luck away from death. This ratio is central to the ecological predicament of the human species, yet it is often glossed over in discussions about "saving the planet." As Mabub ul-Haq, the former finance minister of Pakistan, observed at the Earth Summit in 1992, "Although global warming has yet to kill a single human being and may not do so for centuries, it has received enormous attention and resources. At the same time, silent emergencies that are killing people every day—the fact that 1.3 billion people around the world lack access to clean water, the fact that 150 million people live in desertifying areas—do not attract the same kind of screaming headlines and well-funded action plans."

For the well to do to ignore the poor is as commonplace as it is callous. But, morality aside, it seems to me a grave intellectual error to assume that the fate of the world's poor can be kept separate from the human species's larger ecological prospects. Some of the most illuminating conversations I had during my trip around the world were with people on the lower rungs of the human family: not only the Dinka but working men and women on the streets of Istanbul, London, and Bangkok; peasants scratching out a living on Brazil's central highlands and Russia's vast steppes; young students in Prague, Beijing, and Thessaloníki yearning for a better future. In nearly all my conversations with such people, one common theme was sounded. Even in remote areas, there was a clear recognition of environmental threats, and it was taken for granted that such threats had to be vigorously countered. But in every case, a caveat was added: first, bread must be put on the table; one cannot starve today to preserve the environment for tomorrow.

This perspective, writ large, will shape much of the human species's interaction with the global environment in the decades to come. "In the [wealthy countries of the] North, ecological degradation is linked to the industrialization of their societies and thus to economic benefits, but in the South it is linked to economic survival," Maria da Graca de Amorim, director of the regional office for Africa of the UN Environmental Program, told me in Nairobi. "People in the South know that cutting down the trees is ecologically bad—they are very aware of the problem—but they will tell you straightaway, 'What can I do? My family and I must have that wood to survive.' That is why it is impossible to deal with the environmental problems of Africa—soil erosion, deforestation, desertification, and, of course, [lack of clean] water—without confronting issues of poverty and development."

And issues of poverty and development inevitably raise questions about distribution and fairness. If the planet's atmosphere has space for only a certain level of greenhouse gases, then humans must work out a means of sharing that space. If the earth's fisheries, forests, and farmland can supply only a certain amount of resources without collapsing, then humans must agree on how those resources should be divided. The same point applies to most environmental threats; experts refer to it as the problem of sharing "environmental space."

Right now, the bulk of the earth's environmental space is being claimed by the United States and the other wealthy nations of the North. But this cannot last. The nations of the South will develop economically, and even if solar power and other environmentally enlightened technologies are soon introduced to aid this progress, the economic development of the South will still generate substantial amounts of pollution. For example, China and India, with their gigantic populations and ambitious development plans, could by themselves all but doom the rest of the world to severe global warming if they chose.

Technology, both existing and that still to come, can help tremendously in using the planet's environmental space more efficiently. But that space is what it is, and technology by its very nature is never a 100 percent cure-all. With population and per capita consumption levels both rising, the pressure on the earth's environmental space will inevitably increase in the years to come. The human species has been divided into rich and poor since at least the Agricultural Revolution of ten thousand years ago, but the issue is taking on new resonance at the dawn of the twenty-first century. As humanity seeks to create an environmentally sound future, no challenge will be more fundamental, or more difficult, than bridging the ancient gap between rich and poor.

Two

Bicycles, Churchill,
and Evolution

> It is paradoxical: people in the age of science and
> technology live in the conviction that they can
> improve their lives because they are able to grasp
> and exploit the complexity of nature and the
> general laws of its functioning. Yet it is precisely
> these laws which, in the end, tragically catch up
> on them and get the better of them.
>
> —Václav Havel

Where the Nile River enters Lake Albert, in the northwestern corner of Uganda, lies a tiny fishing village named Wanseko. It is literally the end of the line—the last stop on the public bus route from Kampala, the run-down capital nestled among the hills above Lake Victoria, 160 miles to the south. The journey took nine hours the day I traveled it, crammed inside a 1960s vintage American school bus that for some reason had been painted chocolate brown. The bus was scheduled to leave at 8:00 in the morning, and by a quarter before the hour its seats were already overflowing with passengers. The bench seats that were originally meant to accommodate two American schoolkids each were now packed with four and five Ugandans of all ages, the small sitting on elders' laps amid much high-pitched chatter and good-natured jockeying for space.

Paper bags holding food, clothing, and other belongings blocked the aisle, obliging the bus driver to crawl over and around the passengers while making his way up and down the aisle collecting fares, a spectacle that seemed to amuse everyone involved, not least the driver himself.

Confounding African stereotypes, the bus departed from Kampala five minutes ahead of schedule. Apart from the close quarters, the drive was pleasant and cool for the first two hours. We passed through gently sloping valleys boasting lush foliage and splendidly bizarre trees, including one I silently nicknamed "the sausage tree," on account of the dozens of long beige pods hanging from its branches. The contrast with southern Sudan and northern Kenya, where I had spent most of the past three months, was breathtaking. What a difference water makes! Even in this, the dry season, Uganda was more green than brown, thanks to its abundant system of lakes and rivers. The farther north we traveled, the drier the land became, yet it remained beautiful and apparently fertile. On either side of the road stretched plains of golden grass, dotted by cone-roofed huts and oblong structures whose simple white crosses identified them as the schools and churches bequeathed by European missionaries a century ago.

All was bliss until our bus was abruptly halted by half a dozen men with rifles a couple hours shy of Wanseko. The passengers were ordered to disembark and line up along the side of the road. All the men on the bus except me (the only white) were asked for their identity papers and aggressively questioned. After half an hour of tense waiting and profuse supplications on the part of our driver, we were allowed back onto the bus and on our way. The men with rifles, it turned out, were government soldiers searching for outlaws who had ambushed some comrades a few hours before.

By the time we reached Wanseko, it was late afternoon and I was one of only three passengers still on the bus. Wanseko was little more than a few low-slung shacks grouped around a dusty clearing the size of a football field. To the west, across Lake Albert, I could dimly make out the mountains of Zaire through a bluish haze. I entered one of the shacks to eat a dinner of fish, tea, and chapati inside a smoky room lit by a tiny oil lamp the size of a child's fist. There was nothing like a hotel in town, so I paid the equivalent of a single U.S. dollar to spend the night inside a dusty concrete room behind the general store. I slept poorly, awakened repeatedly by the chickens (or was it rats?) that rustled and scratched against the wall outside, inches from my head.

I had come to Wanseko while retracing a trip that Winston Churchill made across Africa in 1907. At the time, the future British prime minister had just begun his first significant government job, as the parliamentary undersecretary of state for the colonies, a post that naturally included Africa among its concerns. Churchill's expedition took him by ship across the Mediterranean, through the Suez Canal, and around the Horn of Africa to the old Arab port city of Mombasa, located on the Indian Ocean coast of what is now Kenya. The newly constructed Ugandan railway carried Churchill west to Nairobi and on to Lake Victoria, the presumed source of the Nile. He then crossed the great lake and followed the Nile through Uganda, Sudan, and Egypt to Cairo, from whence he returned to London. The expedition was a combination of business and pleasure for the thirty-three-year-old Churchill, undertaken during Parliament's autumn recess and paid for in part by sales of a book he wrote about the experience, *My African Journey*.

Part travelogue, part policy paper, *My African Journey* is a short, impassioned book of keen observation and dazzling prose; reading it, one understands why Churchill was later (in 1953) awarded the Nobel Prize for Literature. The book's account of Churchill's stalking and slaying of a rhinoceros, for instance, is as taut and intelligent a story of man versus beast as one could ever wish to read. Coming across "a scene unaltered since the dawn of the world," Churchill wrote of spotting a creature that was "not a twentieth century animal at all, but an odd, grim straggler from the Stone Age" grazing placidly beneath Mount Kilimanjaro. Firing his rifle from 120 yards away, Churchill heard the "thud of a bullet which strikes with an impact of a ton and a quarter, tearing through hide and muscle and bone with the hideous energy of cordite." As the wounded rhino charged him, Churchill had a last moment to reflect that "after all, we were the aggressors; we it is who have forced the conflict by an unprovoked assault with murderous intent upon a peaceful herbivore; that if there is such a thing as right and wrong between man and beast—and who shall say there is not?—right is plainly on his side." Philosophizing complete, he then finished off the animal with a final blast from his weapon.

My African Journey had a deeper significance as well. In vivid, compact form, it articulated virtually all facets of the ideology that would shape industrial man's impact on Africa in the twentieth century—the values, fears, goals, and justifications that animated European efforts to recast the human and physical environment of this most foreboding and seductive of continents. Churchill saw Africa through the eyes of an inveterate colonizer, an unashamed imperialist who believed that colonialism benefited colonizer and colonized alike. Even more than his white skin, what set Churchill apart from the Africans he encountered was the technology at his disposal—guns, steamships, railways, telegraphs, and other emblems of the industrial era. Technology had brought wealth and progress to the people of Britain, argued Churchill, and it would do the same for the naked savages still mired in the "primary squalor" of Africa. Thus, when passing through the "dense and beautiful forests" of western Kenya, Churchill was aghast to discover that the forests were being cut by hand to fuel the railway. He wanted to install a steam-run sawmill, asserting that:

> It is no good trying to lay hold of Tropical Africa with naked fingers. Civilization must be armed with machinery if she is to subdue these wild regions to her authority. Iron roads, not jogging porters; tireless engines, not weary men; cheap power, not cheap labor; steam and skill, not sweat and fumbling; there lies the only way to tame the jungle.

It was Churchill's unqualified enthusiasm for technology that convinced me to retrace his African journey, for technology lay at the heart of humanity's relationship with the environment. But, oh, how views had changed since Churchill's day! To many modern environmentalists, "technology" was almost a dirty word. In their view, the twentieth century—the century of Hiroshima and acid rain, satellites and superhighways—had revealed technology to be a mixed blessing at best. With prosperity and convenience had come life-threatening pollution and a disorienting acceleration of the pace of daily life, not to mention the threat of nuclear war. The root of the problem was the arrogant belief that modern man could, by virtue of his technology, live separate from, and be superior to, nature, "to tame the jungle," as Churchill put it.

I sympathized with this critique. But my travels in Africa had made me wonder whether the issue was so black and white. Of course humans had to respect the laws and limits of nature. But visiting southern Sudan made me realize that a certain amount of separation from nature was a good and necessary thing. The Dinka had such hard lives precisely because they lived in a foreboding environment with few means of shielding themselves from drought, predators, pestilence, and other forces of the natural world. They seemed to need more technology, not less.

Churchill made the same argument in *My African Journey,* and not just because he coveted the rubber and other raw materials that technology would enable Europeans to extract from Africa. Churchill's generation had good reason to regard technology as a liberating force, for the evidence of how technology had improved human health, productivity, and comfort was all around them. For millennia, the vast majority of humans had lived stunted lives of endless labor and want, struggling against natural forces beyond their control—lack of water, the vagaries of weather, the limits of human muscle power, the uncheckable spread of disease. The industrial ascent of the nineteenth century—notwithstanding the often abominable working and living conditions it imposed on the laboring classes—produced a rise in overall living standards that suggested humans might at last transcend their brutish past. By imposing their will on the natural world, by turning its power and resources to their own purposes through the application of knowledge and technology, humans could become masters of their fate and increase wealth and comfort for all. The know-how of technological society would inaugurate a new era, liberating humankind from a history of scarcity and endless toil. Indeed, so pervasive was the faith in science and machines at the turn of the twentieth century that it was embraced both by capitalists like Churchill and communists like Lenin (who reveled in the glories of Taylorist efficiency).

Like other champions of the industrial order then and now, Winston Churchill had big ideas about what technology, properly applied, could achieve. I was following in his footsteps because I wanted to see how those ideas compared with African reality eighty-five years later and what that implied about the modern challenges of overcoming both poverty and environmental degradation.

Determined to remain as faithful to Churchill's itinerary as I could, I rented a bicycle in Wanseko for the trip to Murchison Falls, the waterfall lauded by Churchill as the most spectacular splash the Nile makes in its

3,400-mile journey from Lake Victoria to the Mediterranean Sea. In *My African Journey,* Churchill wrote that, astonishing as it might seem, a bicycle was "the best of all methods of progression in Central Africa," for it offered both speed and mobility. In my case, I had little choice but to rent a bike if I hoped to reach Kabalega Falls, as they were known locally. I had been told back in Kampala that I could catch a bus to the falls from Wanseko, but that turned out to be false. Walking was not advisable; the distance was twenty-seven miles, and the area was frequented by wild animals, not to mention the bandits whom the soldiers had been hunting the day before. Begging a ride from a passing vehicle was a possibility, but it could be anywhere from five minutes to five days before a vehicle passed; motorized transport was rare in Uganda in general and very rare here in the northwest, the nation's poorest region. On the other hand, there were plenty of bicycles around; the Ugandans seemed as fond of them as Churchill had been. Many of the local men and boys rode them, and they never traveled without passengers or large quantities of goods perched over their back wheels.

How I managed, amid such plenty, to select the singularly pitiful specimen of bicycle I ended up with is something I cannot easily explain. Some people just have a sixth sense about these things. I do recall telling the owner after my test ride that something was wrong with the left pedal; it was cocked at a funny angle, and my foot kept slipping off. Besides that, the back tire was treadless, the front had no brakes, and the rusted metal seat offered a standard of comfort unknown since the Middle Ages. But the owner, a teenage boy with a round, eager face, assured me that the pedal was no problem, and I was in such a hurry (always a mistake in Africa) that I didn't doubt him. In Africa the traveling day begins at six o'clock. It was already half past nine, and I wanted to keep my exposure to the equatorial sun's midday brutality to a minimum. For my haste, I was rewarded with a penetrating lesson in the value of reliable technology.

The first six miles of hard dirt road passed quickly enough, and in half an hour I reached the turnoff for the falls. I pedaled east, twenty-one miles to go. The road became a dusty track through clusters of thatched huts where children played in the shade beneath mothers' watchful eyes. A teenager in a torn white T-shirt who introduced himself as Robert began riding his bike alongside me and appointed himself my new best friend for life. The track began to climb and climb through sparse, dry bush. I was in good physical shape, but after three or four miles on my one-speeded stallion with a thirty-pound rucksack on my back, I was feeling the strain. Robert was, too, I think, but the smile never left his face as he casually but directly asked whether I had an extra T-shirt or notebook I could spare.

Suddenly, as if to mock my exertions, a white jeep barreled past us in a blizzard of dust. It was no doubt heading for the same campsite I was; there was nothing else up ahead. It was a chance in a thousand, but if I had waited at the turnoff with my thumb out, I could have been in that jeep. Instead, I faced another sixteen miles of hard labor beneath a sun that, in Churchill's words, "even in the early morning . . . sits hard and heavy on the shoulders. At ten o'clock its power is tremendous." It was now after eleven; the sun was a hazy white mass making my insides ache.

The jeep's passing out of sight seemed to be the signal my bicycle was waiting for. I had gone no farther than two hundred yards when its left pedal abruptly collapsed beneath my foot like a cliffside after too much rain. The bike keeled over sideways, and my pack and I went sprawling. As I lay in the dust trying to collect my wits, Robert looked down and helpfully observed, "Your bike is faulty, I think."

I reassembled the pedal and banged it back into place, but the lug nut had vanished in the crash and I had no tools, so there was no means of securing it. I climbed back on anyway and got about five feet before the pedal gave way again and I toppled over a second time. I banged it back into place, climbed on again, and toppled over again. After a couple more rounds of this sport, I devised a crabbed method of pedaling that enabled me to travel as much as five hundred yards before the pedal collapsed and had to be reset. Progress was slow. Still, I began to entertain thoughts of piloting my stricken vessel to safety. This conceit was quickly extinguished when the path turned from navigable clay to wheel-swallowing sand and I was flung to the ground once more.

It was at this point that I began to suspect Churchill of overstating the attraction of Murchison Falls, not to mention the virtues of bicycle travel in Africa. The next five miles I covered on foot, pushing my bike before me through the sand like a Bedouin trudging behind a reluctant camel. Finally I saw the gate to Kabalega Falls National Park. The outpost seemed deserted, but it turned out that the ranger was merely napping inside the hut by the gate. Wearing ragged cutoffs and no shirt, he examined my bike, ducked inside his hut, and returned with one of the most beautiful pieces of technology I have ever seen: a pair of battered pliers. He took my U.S. $10 park entrance fee and for no extra charge restored my bike to semiworking order by binding the pedal together with a spare piece of wire. I shall never forget him.

When I finally arrived at the campsite an hour later, weak and light-headed, the first sight to greet me was the white jeep that had left me in the dust, now resting comfortably beneath a leafy tree. I stumbled off the bike into the shade and collapsed on the ground, whereupon one of the jeep riders, an Englishman with uncommon powers of deduction and tact, gasped, "Was that you on the bike? We almost stopped to pick you up!" Reading that line over now, I find it almost incredible that I soon became that man's friend. But such are the intoxicating powers of Murchison Falls: they were every bit as magnificent as the bike ride to reach them was miserable. Churchill was lucky enough to observe them in the first light of dawn:

> The river was a broad sheet of steel grey veined with paler streaks of foam. The rock portals of the Falls were jetty black, and between them, illumined by a single shaft of sunlight, gleamed the tremendous cataract—a thing of wonder and glory, well worth traveling all the way to see.

It happened that the jeep riders had arranged for a park ranger to ferry them upriver later in the afternoon so they could see the "tremendous cataract" up close. As if to make amends, they invited me to join them. What followed were three of the most exhilarating hours of my entire global journey.

I felt as though I had entered another world, a world more tranquil, beautiful, and vibrant than the one I had left behind. We did not see another human being the entire trip. Indeed, we saw no signs that humans had ever been here: no buildings, no trailheads, no docks, no litter. In a sense, this stretch of river was the mirror opposite of the harsh landscape I had traversed in Sudan while riding in the Red Cross truck to Torit. Here, too, evidence of human hands was nil, but instead of the parched desperation of southern Sudan, there was the pristine fecundity of a healthy ecology humming with activity.

The river Nile, as the locals called it, was often hundreds of yards wide and straddled on both sides by steep hillsides covered with thick greenery. The morning haze had cleared, and the river looked amazingly blue and clean, its rippling surface sparkling in the afternoon sun. The park's wildlife population was said to have been all but eliminated by rampaging soldiers during the Obote and Amin dictatorships of the 1970s, but if so, the subsequent recovery had been remarkable. I saw more wild animals along this thirteen-mile stretch of the Nile than I had seen in many weeks of wide-ranging travel in neighboring Kenya. There were literally hundreds of hippopotamuses—some plodding up the riverbanks, others squatting in the shallows with only their bulging eyes visible, still others disappearing underwater (where they are quite agile swimmers) only to reappear half a minute later on the other side of the boat. Sharing sandbars with the hippos were dozens of plump, brown crocodiles. Nearly all of them were stretched out on their bellies with their jaws open wide, revealing long rows of nasty-looking yellowish teeth. This open-mouthed posture was actually a cooling reflex, like a dog's panting, but it lent the reptiles a peculiar aspect, at once menacing and lazy.

Perhaps because humans had had so little presence here in recent years, the animals rarely shied away from us. Often the boat came close enough to the hippos and crocodiles that I could have reached over the railing and touched them. On the shore, back among the trees, I spied numerous graceful giraffes and self-possessed elephants, as well as a few shaggy, skittish waterbucks. And we were surrounded by an extraordinary array of waterfowl: goliath herons, fish eagles, saddle-backed storks with yellow, orange, and black striped beaks that resembled miniature Ugandan flags, and, most entertaining of all, pied kingfishers, which hovered forty feet above the water like hummingbirds for minutes at a time before diving straight down to snag their unsuspecting prey.

After two hours of steady chugging, our boat passed a long, calm stretch of water and rounded a bend. Suddenly the waterfall burst into view. Even from a kilometer downriver, the raw energy of Murchison Falls was fearsome to behold—a glistening cascade of white fury that carried such force our boat could not advance against the current. This extraordinary power stemmed from the fact that, as Churchill explained, above the falls the banks of the Nile "contract suddenly till they are not six yards apart, and through this strangling portal, as from the nozzle of a hose, the whole tremendous river is shot in one single jet down an abyss of a hundred and sixty feet." Transfixed, the jeep riders and I admired this sight for I don't know how many minutes before the captain finally turned the boat around and, with the surging current at our back, returned us to camp in half the time it had taken to get there.

The river's effect was equally powerful the next morning, when the jeep riders invited me to ride with them overland to the top of the falls. Soundings we took along the way indicated that Churchill may have been lubricating his tale somewhat when he claimed the falls could be heard from ten miles away, but they were certainly audible from five, and when we finally clambered down to the actual waterline, the roar was fantastic—like the fiercest windstorm imaginable. In the last kilometer of its approach to the falls, the Nile seemed to sprint so impatiently forward that the foamy green water got ahead of itself and leaped exuberantly upward as if ascending an invisible escalator. Just before the fall line, the river separated into three flows. The one feeding the cataract was over the edge in an instant, crashing against the sides of the chasm and down to the bubbling pool below. The other two flows looped around a massive stone outcropping and supplied a completely separate waterfall, shorter but far wider than its famous brother. The spray, the noise, the water's irresistible force and volume were as overwhelming to the senses as the knowledge of its distant destination was to the mind. One felt in the presence of something eternal, blessedly beyond human enterprise, categories, and pretensions.

Murchison Falls was without doubt one of the most glorious natural wonders a modern-day African traveler could still behold in its original state—but only because Winston Churchill did not get his way. Astonishingly enough, Churchill wanted to build a dam at Murchison Falls. Its "terrible waters" itched at his restless nature. They had to be put to some productive purpose:

> I cannot believe that modern science will be content to leave these mighty forces untamed, unused or that regions of inexhaustible and unequalled fertility, capable of supplying all sorts of things that civilized industry needs in greater quantity every

year, will not be brought—in spite of their insects and their climate—into cultivated subjection.

Of course, the dam whose construction Churchill was advocating would have covered up forever the very falls he had praised as one of the great wonders in all Africa. Prudently, he ignored this contradiction. He did seem to sense there was something unholy about his proposal, however; he wrote that he was roused from his reflections on damming the Nile by "an ugly and perhaps indignant swish of water" that nearly drenched him.

It is easy for a late twentieth-century observer to condemn Churchill's boorish insistence on conquering nature to serve the needs of "civilized industry." Yet there is no denying that technology in its most basic sense—the application of human knowledge to the natural world—has been inseparable from human progress since time immemorial. From the moment our first human ancestor picked up the first stone tool some two million years ago, the fate of our species has been inextricably linked to the creation of technologies that give us more efficient means of extracting food, water, shelter, and other essentials from the physical environment. Eastern Africa was a particularly fitting place to explore this history, for it was here that that first stone tool had been used. Moreover, in eastern Africa one could still sense the deep past of life on this planet—in the bloody maw and languid gaze of a lion fresh from a kill, in the perfect stillness of the endless rolling plains of Masai Mara at sunrise, in the stoic endurance of tribespeople like the Dinka and Masai, whose daily lives are more similar to those of their Stone Age ancestors than not.

Only by exploring the deep human past could one appreciate how utterly different our current living standards and environmental status are from those that have prevailed for 99 percent of human history. Africa taught me that gazing only a few thousand years into the past—to the civilizations of Rome and Athens, whose remains I had visited earlier in my travels—was not enough. To grasp the full trajectory of human life on earth, one needed to go all the way back, to the very birth of the species.

According to current scientific consensus, it was in eastern Africa that the first *Homo sapiens* walked the earth some 200,000 years ago, after five to seven million years of evolution that separated them from their original ancestors, the apes. Of the many discoveries that established this chronology, few have been more influential than those of the Leakey family. In 1964, Louis and Mary Leakey uncovered in northern Tanzania a two-million-year-old fossil of *Homo habilis,* a human ancestor who lived two evolutionary stages prior to *Homo sapiens. Homo habilis*'s chief distinction was as the first toolmaker. Mary Leakey later found evidence that our evolutionary forebears were already walking on two legs some three and a half million years ago.

Mary and Louis's son Richard filled in a later piece of the puzzle with his discovery of "Turkana Boy" in northern Kenya in 1984. Turkana Boy was the most complete specimen ever uncovered of *Homo erectus,* the species that preceded *Homo sapiens.* The *Homo erectus* who lived near Lake Turkana one and a half million years ago were already "becoming human in a way [modern humans] would instantly recognize," wrote Richard Leakey, which demonstrated that "the emergence of the essence of humanity [was] a gradual process—one that links us to the deep past."

Personally, I found it hard to imagine that the Turkana area could ever have supported much life—it was so barren and dry. I had traveled to Lake Turkana by truck from Nairobi, rumbling north along the edge of the massive Rift Valley, a depression in the earth so long and wide it is plainly visible from outer space. Along the way, I descended into the valley to visit Lake Nakuru, where I found the shore so densely packed with flamingos that it looked like pink heat waves were rising off the lake's surface. As I approached the waterline on foot through baking afternoon heat, thousands of flamingos fluttered about in the shallows on their long, pencil-thin legs, honking constantly. Scattered among them were smaller contingents of even less aeronautically impressive pelicans, their deep-pocketed mouths bobbing beneath the surface in search of refreshment.

Little did I know that this raucous explosion of life was already under severe environmental stress. Over the next three years, Lake Nakuru would shrink to a quarter of its former size, reducing the pink flamingo population from 1.5 million to a mere 100,000. The persistent drought plaguing the Horn of Africa was partly to blame, but it was exacerbated by the very civilizing forces set in motion by Churchill. Extensive land clearing had damaged the watersheds whose runoff fed the lake, and much of the reduced flow was being diverted to urban residents, whose numbers were rapidly expanding, due to Kenya's high (3.4 percent) annual rate of population growth.

The farther north and closer to Lake Turkana I traveled, the more forbidding and parched the surroundings became. The vegetation gave way to flat-topped acacia trees spaced well apart and knee-high, silver-gray brush that provided groundcover for dik-diks, creatures that looked like miniature antelopes and kicked up tiny clouds of brown dust as they darted from the bellowing sound of the truck. A cheetah also sought refuge in the bush, albeit at a much more leisurely pace; it seemed less frightened than annoyed. Stopping for lunch near the village of South Hore, I was told by a local man that no rain had fallen there for the past fourteen months, a comment amply borne out by the rest of the afternoon's drive to Lake Turkana. The road here was in some ways even worse than the roads in southern Sudan, for not only was it pocked with holes and gullies, it was strewn with boulders the size of footballs and even larger. Despite traveling at little more than walking pace, the chassis of our truck bounced up and down like a cowboy atop a bucking bronco.

Two hours shy of the lake, we crossed a vast plain that seemed to have been broiled black by the ceaseless assault of the equatorial sun. The soil was sand—or perhaps I should say ash, for we had entered a lava field. On both sides of the road, acres and acres of land were covered with rounded volcanic rocks that had been ejected from the bowels of the earth untold eons ago. Gazing across this bleak panorama, I wondered if the heat was causing me to hallucinate. The rocks, crammed tightly together, took the form in my mind of hundreds of thousands of deceased African tribesmen, buried up to their necks in the sand, their shining black skulls exposed to the merciless sun and wind for all eternity.

My vision was interrupted when the truck abruptly halted in the middle of the lava field. Far off to our right, the driver had somehow spotted a water jug being raised by a lone shepherd stumbling behind a flock of desiccated goats. It took the shepherd nearly ten minutes to reach the truck, and when he did, the young man had no energy to waste on the waves of greeting that two previous water recipients had extended that day. His earth-colored jug looked vaguely Egyptian, with red markings around the tubelike spout. As soon as it was filled, he began gulping. After thirty seconds, he paused for breath and tugged his wraparound cloth more tightly beneath his arm. He was still a teenager. The cloth, the jug, and a herding stick were all he carried, and this was the only existence he would ever know. The driver refilled the drained jug and got back in the cab. As we pulled away, the shepherd half-lifted his hand in response to our farewells, but he was still in a daze, staring dully into space. Back in Nairobi, while reading up on the region's nomadic tribes, I had been shocked to learn that they had a tradition of drinking cow and goat blood (mixed with milk on special occasions). Now I was beginning to understand how such practices might have originated; out here, there weren't many other options for quenching thirst.

We reached the lakeside village of Loiyangalani half an hour before sunset, time enough to pitch our tents before the heavens above Lake Turkana filled with a layered sunset of deep purples and oranges, a spectacle matched hours later by the brilliant expanse of stars that filled the night sky. The dawn, however, revealed an earthly landscape of harsh implacability. I headed first for the lake, where simple wooden skiffs resting on the beach hinted that fish was an important part of the local diet (despite the ubiquity of alligators in the lake). Conversations with villagers confirmed my guess but also indicated how bereft the villagers would be without the supplemental maize, greens, and other relief food regularly flown in via the village airstrip. Hiking above Loiyangalani, I found barren hillsides of pebbles and scrub, hardly the terrain for growing crops. Once again I marveled that this was where humanity had gotten its start; it looked as different from the Garden of Eden as one could imagine.

What I wasn't taking into account, of course, was that the climate of the Lake Turkana region had changed dramatically over the 1.5 million years since Turkana Boy and his fellow *Homo erectus* had lived here. After all, the nearby Sahara Desert had been covered by lakes and forests a "mere" six thousand years ago, and Lake Victoria had been a dry, grassy plain only twelve thousand years ago.

Envisioning such immense lengths of time was, for me, one of the most difficult parts of coming to terms with the history of the human species. To most Americans, history is anything that happened more than fifty years ago, and even more historically rooted peoples tend not to look much further into the past than the birth of Christ two thousand years ago or the beginning of written history four thousand years before that. The findings of paleontologists like the Leakeys reveal such dates to be mere yesterdays. To cast one's mind back the hundreds of thousands, indeed millions, of years necessary to reach the beginning of our species requires a special kind of mental gymnastics.

No such ruminations intruded on Churchill's itinerary, of course. Not because these issues would not have engaged his always voracious curiosity; the information simply was not yet available at the time of his journey. An Englishman of Churchill's class and education would certainly have known about Darwin's theory of evolution (*The Descent of Man* was published in 1871, over thirty-five years before Churchill's trip), but the theory was still far from universally accepted. Nor had Darwin's theory yet been supported with much archeological investigation. Only after such twentieth-century scientific breakthroughs as Einstein's theory of relativity, Edwin Hubble's astronomical observations of an expanding universe, and the development of carbon, uranium, and DNA dating mechanisms did it become possible to accurately measure the geological age of the earth and its component materials and life forms. Among the key dates derived from such measurements are these:

- The formation of the earth (4.5 billion years ago)
- The emergence of the first primitive cells, perhaps following a bombardment by asteroids that fertilized our planet with vital organic compounds (3.5 billion years ago)
- The so-called Cambrian explosion, when in a period of five to ten million years the ancestors of virtually all the creatures that would subsequently inhabit the planet arose as multicellular organisms (540 million years ago)
- The breakup of the planet's land mass from its original two supercontinents into what became today's configuration (400 million years ago)
- The appearance of the first land creatures (350 million years ago)
- The beginning of the dinosaur era (225 million years ago)
- The extinction of the dinosaurs, which enabled the emergence of mammals (65 million years ago)
- The evolutionary split within the genus of African apes that eventually gave rise to humans (5–7 million years ago)

Darwin worked out his theory of evolution in large part by studying finches on the Galápagos Islands off the western coast of South America. In the 1970s, a team of scientists returned to those islands to test his ideas. Led by Princeton University professors Peter and Rosemary Grant, the team spent twenty years monitoring generation after generation of finches, observing the birds' eating and mating patterns in minute detail, and then analyzing the data with supercomputers back in Princeton. As recounted in Jonathan Weiner's fascinating book *The Beak of the Finch,* the Princeton team's labors yielded an astonishingly exact picture of evolution in action that resoundingly confirmed Darwin's vision. Their central finding was that evolution is neither an ancient nor a slow-moving process but one in constant motion, occurring at lightning speed all around us.

The pressures of natural selection were observed to act *within a single generation* when a severe drought killed large numbers of *fortis* finches in 1977. Weiner wrote:

> During the drought, when big tough seeds were all a bird could find, these big-bodied, big-beaked birds had come through the best. . . . The average *fortis* beak before the drought was 10.68 mm long and 9.42 mm deep. The average beak of the *fortis* that survived the drought was 11.07 mm long and 9.96 deep. . . . What made the difference between life and death was often 'the slightest variation,' an imperceptible difference in the size of the beak, just as Darwin's theory predicts.

This preference for large beaks was passed on to the next generation, also as Darwin's theory predicts: "The average *fortis* beak of the new generation was four or five percent deeper than the beak of their ancestors before the drought." Extended over multiple generations, such inheritances could dramatically transform the entire species's appearance, even lead to the development of new varieties, with larger beaks, darker coloring, bulkier builds, or an infinite number of variations on these and other characteristics.

These same dynamics—the moment-to-moment interplay between an organism's characteristics and the pressures imposed by its surrounding environment—were also at work five to seven million years ago, when some of the great apes of eastern Africa descended from the trees and began to walk on the ground on two legs rather than four. Why the apes did this remains a subject of great controversy among scientists. The explanation most widely favored is that the tectonic convulsions that created the Rift Valley twelve million years ago separated the ape population into two subsets, with those east of the valley (in modern-day Kenya and Tanzania) eventually adopting bipedality to cope with a climate and terrain substantially drier and less forested than their traditional homeland to the west. In any case, while these newly bipedal apes were still apes, they had literally taken the first step—walking on two legs—that, in tandem with the different natural selection pressures found east of the rift, led to a cascade of further evolutionary adaptations that over many thousands of generations produced a fundamentally different animal: the modern human.

Walking on two legs was crucial to human evolution because it freed the hands for other purposes, particularly the next big development: the use of tools, which is thought to have begun two to two and a half million years ago. It takes a single sentence to relate that fact, but (again, the mental gymnastics) it took three to five million years before the shift took place. The use of tools coincided with a third major transformation, the enlargement of the brain, and within another half million years prehumans were not only using tools but creating them.

This marked the decisive separation of humans from the ape line—apes can use tools, but only humans can create them—and it was at this point that "man" first began to transform nature for his own ends, as evidence from the Lake Turkana region shows. When stone tools used at a campsite east of the lake 1.5 million years ago were analyzed under a microscope, scientists discovered that they had been used to cut wood, plants, and meat, indicating that the tool owners were already constructing rudimentary shelters from trees and grasses and converting animals into food and clothing.

Fast-forwarding to half a million years ago, prehumans had domesticated fire and perhaps developed language as well, two breakthroughs that enabled them to leave Africa and spread as far as northern Europe and even to present-day China, despite the cold encountered along the way. *Homo erectus* had now evolved into *Homo sapiens,* a creature with a larger brain, more erect posture, and generally more human appearance. But *Homo sapiens* should not be confused with fully modern humans, who are technically known as *Homo sapiens sapiens.* In fact, most *Homo sapiens,* including the Neanderthals, proved unequal to the evolutionary challenge and died out over subsequent millennia. Today's humans can be traced to a tribe of *Homo sapiens sapiens* who were still living in Africa approximately 200,000 years ago. A subset of this tribe left Africa about seventy thousand years ago, and they and their descendents eventually spread throughout the planet, colonizing all continents (except the Americas, which were settled later) over the next thirty thousand years and evolving into the various races and ethnic groups that constitute modern humanity.

Throughout millions of years of evolution, the ancestors of modern humans lived as hunter-gatherers. They sustained themselves by hunting animals and fish, gathering wild fruits and vegetables, and nomadically wandering from place to place as weather and food availability dictated. Hunter-gatherers have been called the original leisure society because only relatively limited amounts of labor were necessary to meet their basic needs of food and shelter (especially because human population density was still so low); modern environmentalists have also praised their intimate knowledge of the natural world. Such assessments should be tempered, however, with a recognition of the severe physical hardships and cultural limitations inherent in such an existence.

One of the last remaining hunter-gatherer societies on earth are the Huaorani, a tribe of Amazonian Indians who live so deep inside the Ecuadoran rainforest that not until the 1970s did they come in contact with the modern world of roads, radios, and wristwatches. Asked what time it is, a watch-toting Huaorani will cheerfully announce, "In Japan it is nine o'clock. In Europe it is ten. In Ecuador it is eleven. In America it is twelve." As Joe Kane chronicled in his fine book, *Savages,* the Huaorani make decisions by consensus, value group harmony above all, and live so completely in the present moment that they have no word for "later." They live in such exquisite symbiosis with their natural surroundings that adult males will set off on a two-week journey through the forest carrying nothing more than T-shirts, shorts, and rubber boots, confident they will find all they need to survive along the way. Yet the hunter-gatherer life is also brutal and harsh. The Huaorani's moral code allows for stealing and killing, and they are fierce, eager warriors; the preferred method of killing is to run victims through with heavy, nine-foot-long spears. They often go days between meals and suffer from tuberculosis and other diseases. Their lifespans rarely exceed fifty years.

The hunter-gatherer way of life pertained for everyone everywhere until about ten thousand years ago, when humans all over the planet gradually began cultivating crops, husbanding animals, and otherwise raising the food that they used to simply collect. That this so-called Agricultural Revolution occurred more or less simultaneously in Persia, China, and Mesoamerica would seem to suggest a planetwide cause for the shift, which began just after the last Ice Age, when temperatures were increasing and new types of vegetation were springing up the world over. "Revolution" is actually a somewhat misleading word here, for cultivation did not immediately replace hunting and gathering; it slowly displaced it, mainly because of agriculture's higher productivity. Agriculture required more work than hunting and gathering, but the reward was ten to one hundred times as much food output per acre, which meant that fewer members of the cultivating family or tribe died from malnutrition and related diseases—no idle concern among people living on the edge of survival.

The Agricultural Revolution was arguably the single most fundamental change in humanity's relationship to the environment in history. It required the clearing of large amounts of land and sparked a gradual rise in human population beyond the four million total that had prevailed for millennia. Above all, it wrought massive changes in social relations that made all future technological advances possible. The shift to agriculture encouraged humans to forsake their nomadic past in favor of a settled life, which led to the emergence of the first villages and towns some nine thousand years ago. It also undermined the collective ethos of hunter-gatherer societies, replacing it with hierarchy and class stratification. Agriculture's ability to yield a food surplus was central to this transformation, for no longer was the whole society compelled to labor directly and cooperatively to feed itself; some members could pursue other activities, so long as part of the food surplus was diverted to them. The first elites to benefit from this arrangement were religious; most early societies were dominated by priestly castes whose authority derived from their presumed knowledge of the natural cycles affecting agriculture. But eventually such societies were able to support craftsmen, builders, artists, scientists, and administrators as well. This division of labor laid the basis for a blossoming of human thought, creativity, and problem solving that dramatically increased humans' ability to manipulate the natural world for their own benefit.

Technological innovations began arising much faster now—in matters of centuries rather than multi millennia. By six thousand years ago, writing had developed, an exceptionally important step since it allowed knowledge to be stored and communicated over space and time. Within two hundred more years, amazingly precise astronomical schedules were being used to govern planting and harvesting in Babylonia. Three hundred years after that, the wheel was invented, revolutionizing transportation, fostering increased production and trade, stimulating the development of pottery, spinning, and other wheel-based crafts, and foreshadowing the exploitation of water power. With the invention of metalworking came the realization that matter could be taken apart and reconfigured to suit human purposes; humanity left the Stone Age and entered the Bronze and Iron ages (five and four thousand years ago, respectively). Metal tools and weapons multiplied economic and military power, strengthening central authorities' ability to appropriate the labor and resources of others. This gave rise to the world's first empires, in Persia and Egypt, five thousand years ago.

The Sumerian empire of Persia was the first literate society on earth; it was also the first to succumb to self-inflicted ecological collapse. Agriculture had been practiced in the so-called Fertile Crescent of the Tigris and Euphrates river valley for nearly five thousand years before the Sumerian empire arose. Irrigation had been invented as well (about 5500 B.C.), a technological innovation that eventually brought Sumeria grief. Irrigation increased crop yields substantially, but it also increased the salt content and water retention of the soil. These problems could be reversed by allowing the land to lie fallow for long periods, but the population growth that irrigation had made possible and the increasing competition Sumeria faced from neighboring city-states outweighed such long-term considerations; the Sumerians kept their production targets high. Crop yields remained high for a time, but they collapsed abruptly in 2400 B.C. The food shortfall made it difficult to support the empire's vast administrative structure, especially its army, and Sumeria was conquered by the neighboring Akkadian empire within a matter of decades.

The rise and fall of Sumeria illustrates a tendency that would assert itself time and again over the coming five thousand years: a given technological development increases humanity's ability to extract a higher level of comfort from the natural world, but it does so at the cost of greater environmental damage. True, not all early civilizations made such errors. Egypt, for example, which relied on the annual flooding of the Nile for its irrigation, maintained a relatively healthy ecological balance. On the other hand, the Chinese, Greek, and Roman empires all ignored vital ecological constraints, and each paid a price. The Greeks cut down the trees from the mountains above Athens to construct the ships that made them the world's leading sea power, but once the trees were gone, the mountains turned bare and dry from erosion and the Greek navy was fatally weakened by its enforced reliance on imported timber. The Romans likewise denuded the Italian peninsula (wood was the key resource of preindustrial societies, necessary for heat, light, cooking, and building), and though the consequences were evaded for a time by making North Africa the breadbasket of the empire, North Africa too was soon overexploited and reduced to desert, with predictable effect on imperial fortunes.

In his excellent book, *A Green History of the World*, Clive Ponting related dozens of similar examples where human societies failed to strike a sustainable balance between their own material demands and the environment's long-term health. One of the many virtues of Ponting's book is that it corrects the impression conveyed by some modern ecologists that environmental degradation was a rare and trifling thing in the days before plastics and multinational corporations. Ponting was careful to say that the environment was usually but one factor among many in causing a society to unravel, and that even in such cases "the decline and eventual collapse were usually prolonged . . . and generations living through this process would probably not have been aware that their society was facing long-term decline." (Sumeria, for example, enjoyed six hundred years of dominance before falling.)

If humans have despoiled their local environments throughout history, Ponting argued, it is not necessarily because they have been ignorant, selfish, or apathetic; most of them simply have not had the luxury to place long-term considerations ahead of their daily struggle for survival. "Since the rise of settled societies some eight to ten thousand years ago," he wrote, "the majority of the world's population have lived in conditions of grinding poverty. They have had few possessions, suffered from miserable living conditions and have been forced to spend most of their limited resources on obtaining enough food to stay alive."

Throughout the fifteen centuries following the fall of the Roman Empire, the human population grew very slowly—only one child in three survived infancy, and life expectancies hovered at around thirty years. Not just adequate food but clean water was a constant problem. As town life brought humans into closer contact with one another, lack of proper sanitation facilities resulted in widespread contamination of water supplies; by 1610, fresh water had to be piped into London because the city's inhabitants had irreparably fouled the Thames River. (When demand continued to outstrip supply, unscrupulous merchants made up the difference by drawing extra water from the Thames anyway.)

Not until the Industrial Revolution of the late eighteenth century did this gloomy reality begin to improve substantially. Learning to substitute machines for human and animal muscle power marked a gigantic leap forward for humans and produced unprecedented increases in production and wealth. A decidedly European phenomenon, the Industrial Revolution was made possible by two preceding, also Eurocentric, developments: the Scientific Revolution, which generated the knowledge needed to build modern machines, and the Capitalist Revolution, which amassed the capital needed to finance the machinery, in part through the age-old method of forcibly appropriating the labor and resources of others, especially Africans, Asians, and South Americans colonized by European powers from the fifteenth century on. (Why Europeans were the colonizers and not the colonized and the leaders rather than the followers of the Scientific Revolution had nothing to do with any supposed inherent ethnic superiority. As Jared Diamond explained in his seminal book, *Guns, Germs and Steel,* the relative fates of human societies since the rise of agriculture have been shaped above all by geographic and other environmental differences. Agriculture took hold in the Fertile Crescent rather than in sub-Saharan Africa because the former was blessed with wild plants and animals that could be domesticated, while the latter was not. Agriculture spread to Europe and became the basis for Europe's subsequent technological prowess because the innovations of crop and animal husbandry were more transferable along the east–west axis of Eurasia, which shared similar climates and growing seasons, than along the north–south axis of Africa, which did not.)

The melding of science and capital into industry began the most rapid acceleration yet of humans' control over the natural environment; humanity's place in the world would never be the same again. Modern science—as foreshadowed by the breakthroughs of the Renaissance and pioneered by such giants as Copernicus, Galileo, Kepler, and Newton—asserted that the world operated according to consistent physical laws, not the inscrutable will of supernatural beings. And these laws could be discovered through reason and experiment and applied to practical effect. To a rising class of capitalists whose fortunes had been made in part through overseas trade, inventions like the sextant and telescope vividly demonstrated the commercial value of the scientific method; their investment support duly followed. The consequent technological takeoff was neither smooth nor inevitable (neither China nor India experienced its equivalent, though they possessed their own scientific expertise), but in Europe the process began to feed on itself. Better knowledge begat better machines, which lowered production costs, pulling more and more people into the web of exchange and accumulating the capital needed to develop still more sophisticated production methods and machinery.

By the late 1780s this pursuit of technological advantage had culminated in the industrial steam engine, a machine whose ability to transform burned wood into mechanical energy powered the Industrial Revolution for the next hundred years and spawned countless additional innovations, including the other two leading symbols of industrialism, the factory and the railway. The steam engine was an epoch-making breakthrough, but the old trade-off still applied: better technology raised human welfare but inflicted greater environmental damage. Churchill, it will be recalled, wanted steam engines brought to Africa to fuel the British-built railway. His proposal was plausible only because Africa's forests had not yet been reduced to stumps and brush the way forests in countries reliant on steam engines had been. The reason wood was eventually displaced by coal as the primary fuel of the Industrial Revolution was not that coal was a manifestly superior product but that wood was becoming too scarce. Humans had burned coal for centuries but in marginal amounts, for it was a dirty, inefficient fuel source that was difficult to extract from the earth. With necessity the mother of invention, the dwindling of wood stocks led to technological breakthroughs in the 1840s that enabled coal to be converted into heat much more efficiently, and the industry took off.

But coal was no more immune to the traditional environmental trade-off than the steam engine was—quite the contrary. Coal sparked massive increases in production and efficiency, but it also brought the blackened skies, putrid rivers, and other undesirable side effects lamented in poet William Blake's passionate assault on the "dark Satanic mills" of industrial England. Nevertheless, the transition to coal gave industrial man access to vastly more energy than before, leaving him poised for the quantum leaps in command of the natural world that would make the twentieth century his most revolutionary yet.

Human living standards began to improve significantly about 1850, and with each passing decade the future looked brighter. "Progress" became the watchword of nineteenth-century philosophy and politics. More and more food was now available, thanks to the introduction of machinery and scientific methods in the fields and the discovery of refrigeration for food storage and shipping. The development of anesthesia in the 1840s and disinfectants in the 1860s made hospitals less painful, infection-laden places, and by the 1890s scientists were discovering the microorganisms responsible for major diseases. The maturation of a market economy based on mass production drew millions of peasants off the land, creating real cities—not the overgrown villages of the past but teeming centers of economic possibility and social fermentation. For so many humans to live so closely together would have ensured epidemics in the past, but the construction of sewers and other sanitation improvements now began to reduce such threats. The taming of electricity liberated the night for human activity and enabled construction of the skyscraping office buildings that came to symbolize modern industrial achievement. By late-twentieth-century standards, life was not easy for the average man, woman, or child in Churchill's England, but they had infinitely more comforts than their African contemporaries did.

Africans could improve their standard of living, too, promised Churchill, if only they would accept the guidance of Europeans and embrace industrial development. In truth, Africans were not given much choice in the matter. This was the era of untrammeled European imperialism in Africa, and Britain was determined not to lose out in the carving up of the continent. As author Charles Miller explained in his book, *The Lunatic Express,* London's fear of losing the Upper Nile Valley to the French or Germans (a scenario that would endanger British control of Egypt and, because of the Suez Canal, ultimately India) led Britain to undertake the costly and difficult construction of the railway to Lake Victoria. If the railway was to support itself, however, it had to transport a relatively high volume of trade, which required "high volume, high quality agricultural production, managed on modern lines and grown on European-owned farms." This in turn required the removal of such tribes as the Kikuyu and Masai from lands they had occupied for centuries, a task local British authorities pursued with relish. Author Peter Matthiessen has written that by 1939, "four-fifths of the best land in Kenya was the province of perhaps four thousand whites; a million Kikuyu were to make do with the one-fifth set aside as the Kikuyu Reserves." Africans eventually rebelled against such inequality, overthrowing British rule in 1962. Nevertheless, land ownership remained highly uneven in Kenya even in 1992 (in contrast to neighboring Uganda, where whites ruled indirectly and never became major landowners), and this inequality fueled the country's rampant poverty and hunger.

Churchill may well have sincerely believed that Britain's intervention in eastern Africa would benefit all parties, but in retracing his journey eighty-five years after the fact, I found the economic disparity between Africa and the industrial world as vast as ever. The forces of progress Churchill championed seemed to have changed everything and nothing about eastern Africa. The physical environment had certainly been altered by the arrival of industrial man, but the prosperity derived was limited and narrowly distributed. Of the feast of materialism Churchill had prophesied, the majority of Africans had tasted barely a bite.

Leaving Mombasa on a train to Nairobi, I passed numerous signs of a functioning industrial society: smokestacks, power lines, petroleum refinery tanks, and row after row of low, concrete warehouses awaiting replenishment from the half a dozen container ships moored in Kilindini harbor. Next to a chemical processing plant, clusters of silver piping thrust themselves skyward like industrial dandelions, while overhead a red and white jetliner screamed its approach to the international airport. But the lives of the people were another matter; I often felt as if an African version of *A Tale of Two Cities* was playing out before me. As the descending airliner disappeared above me, the train chugged slowly past a squalid shantytown whose tin-roofed shacks of rotted wood housed the shops and hovels of the urban masses. Sprawled on the ground not ten feet from our click-clacking wheels, a man in trousers and a short-sleeved shirt slept open-mouthed, as if poisoned or drunk, in the ubiquitous eastern African pose of unreserved somnambulism. Past the city limits, small children scampered from their mud and grass huts to gather along the track, wave and cheer, and plead with outstretched palms, "Give me pen! Give me sweet!" or, merely, "Something!"

Churchill would have been relieved to learn that these rural Africans had abandoned nakedness for trousers and dresses, and that some formal education and knowledge of English were now the rule rather than the exception among them. In other respects, however, the gulf among the races of eastern Africa remained as wide as when Churchill was writing that it was impossible to "travel even for a little while among the Kikuyu tribes without acquiring a liking for these light-hearted, tractable, if brutish children, or without feeling that they are capable of being instructed and raised from their present degradation." Black-on-black tribal violence remained common, there was no love lost between black Africans and the Asian merchant class, and the dominant emotions between Africans and Europeans were distrust and fear.

Whites I met knew effectively nothing about the lives of the black majority, whom they regarded as the unknowable "they." The closest interaction most whites had with blacks occurred within master–servant relationships. Spend an evening in the company of whites and one certain topic of conversation would be the relative honesty and competence of their maids, cooks, and gardeners. "You just never know what they'll fancy," one Nairobi matron, recalling alleged stealing, mused while being served Christmas dinner by a squad of middle-aged servants. The atmosphere of paranoia and isolation was such that almost all whites maintained round-the-clock security guard protection, even though it was commonly assumed that 90 percent of the robberies in Nairobi were inside jobs involving the security guards themselves.

There was no more revealing symbol of the chasm between blacks and whites than the *matatu*, a vehicle in which most whites never set foot but which was the primary means of transport for blacks. To be sure, there were good reasons not to set foot in a *matatu*—unless terrible overcrowding and a high risk of death or dismemberment were your ideas of excitement. *Matatu* is a Swahili word for privately operated minibuses that were much faster than public buses, far more numerous, and only slightly more expensive. They also had lots more personality. Every *matatu* in Kenya had a nickname painted in bright colors across the front and back of the vehicle, with speed the usual theme. The *matatu* I rode during the two-and-a-half-hour trip from Lake Victoria to the Ugandan border was called "The Singaha Quick." Other names I saw included "The Road Shark," "The Gusii Express," and, inexplicably, "The '90s Explainer."

The *matatu* stand in Kisumu, a bustling town on the eastern shore of Lake Victoria, was a beehive of cheerful chaos when I arrived at eleven o'clock one morning. While hawkers whistled, clapped, and shouted out their destinations, drivers fiddled beneath their hoods, chatted, and jockeyed for position within the squirming mass of vehicles angling to leave the yard. Passengers milled about, occasionally hoisting their belongings up onto the roof before boarding their *matatu* of choice. I was assigned to an older *matatu* that already looked more than full. Twelve adults sat facing each other on metal benches that extended in a horseshoe shape down both sides of the van. Each person's hips and shoulders were wedged firmly against his or her neighbors', making it impossible to move one's legs without kicking the person across the row. The day was hot, the air close, and it was evident that many of my fellow passengers had not encountered soap for some days.

We were getting accustomed to our sardinelike surroundings when the hawker decided that one last traveler deserved a seat among us. A broadly smiling young man wearing a chocolate brown wool suit and carrying a large cardboard box was directed to sit in a nonexistent space across the aisle from me. I watched him with my own eyes and still don't know how he did it. No one else in the *matatu* even raised an eyebrow.

While we waited to depart, the cheerful, skipping guitar riffs of eastern African pop music filled the air, and hopeful vendors approached the van. A hand would suddenly thrust its way inside the open back door, six inches from my face, and flash items—bottles of soda, boiled eggs, pineapple slices, sweets, cheap wristwatches, plastic bowls and cups, cassette tapes, wrench and screwdriver sets, handkerchiefs, earrings, or, most bizarre of all, packet after packet of unlabeled pharmaceutical pills. One man silently proffered a palm-sized manila envelope containing god knows what kind of red and black capsules. He found no takers.

When we finally departed, the crowding inside the *matatu* made it impossible for us to see much outside, which was just as well. Daredevil speeds and passing maneuvers were matters of honor among many *matatu* drivers. Grisly accounts of highway deaths were a staple of the region's newspapers; one story featured unforgettable photographs of passengers still in their seated positions who had been charred into blackened lumps when their *matatu* hit a petroleum tanker head-on while struggling to overtake another *matatu*. Africans I talked to were well aware of the dangers of riding *matatus*, but they accepted them with placid nonchalance. On a continent where one infant in seven did not survive to age five and women of fifty were considered old, death was regarded not as a distant stranger but a familiar companion. That did not mean Africans didn't take death seriously, anymore than their tolerance of crowded *matatus* meant they didn't mind the discomfort involved. It was simply that death and discomfort were ever-present facts of life for Africans. They accepted them because they had no choice, just as they rode *matatus* because the only alternative was to cover the same distance on foot.

Humans are resilient, adaptable, extremely clever animals. The technological innovations they have produced over thousands of years of evolution are the central reason *Homo sapiens sapiens* became the planet's dominant species. But seen within the historical context outlined in this chapter, what is most striking about automobiles, televisions, and other technological marvels taken for granted in modern industrialized societies is how new and unusual they are for our species. No other generation has enjoyed the sheer ease and freedom from want that late-twentieth-century industrialized humans do. The paradox is that these blessings have been achieved through an unprecedented degradation of the natural systems upon which future generations must depend. Today's generations have reason to desire the conveniences that industrialization offers, however, and any ecological rehabilitation program that fails to take that into account is doomed to failure. Today, only one in six humans live like the Dinka, whereas for most of human history it has been closer to nine out of ten.

Our dilemma is that, on one hand, more and more scientific evidence suggests that the consumption levels now prevailing in wealthy nations are ecologically unsustainable and that extending such a standard of living to all six billion humans on the planet could threaten the very survival of the species. On the other hand, social forces have been set in motion that will not be contained. Electricity, telephones, running water—once humans experience these things, they want more of them, no matter the ecological cost. As well they might. It is easy for outsiders to warn against the long-term costs of damming Africa's rivers, ruining its scenery, or destroying its woodlands, but it is akin to a glutton admonishing a beggar on the evil of carbohydrates—he lacks a certain moral authority.

My last *matatu* ride in Uganda took me to Jinja, a town on Lake Victoria that in Churchill's time was believed to be the source of the Nile. Churchill had ridiculed Jinja as an "outlandish name" for a town geography plainly destined for greatness; he wanted to rename it Ripon Falls, "after the beautiful cascades which lie beneath it, and from whose force its future prosperity will be derived." What was needed, he added, was to build a dam and "let the Nile begin its long and beneficent journey to the sea by leaping through a turbine." That may have seemed a "perfectly easy" task in 1907, but it was 1954 before it was actually accomplished. On the ride to Jinja, my *matatu* passed the electric power station that now hummed beside the dam. The material blessings forecast by Churchill—

"the gorge of the Nile crowded with factories and warehouses" and "crowned with long rows of comfortable tropical villas and imposing offices"—had yet to materialize, however. Local people remained so poor that, later that afternoon, when a few greedy *matatu* drivers suddenly raised the price of the return trip to Kampala by the equivalent of ten cents, more than half the passengers angrily disembarked and prepared to wait two hours for a later *matatu* rather than pay the higher fare.

The source of the Nile, where Africa's longest river emerges from its largest lake, should rank as one of the great scenic spots on earth. But the site was actually a bit disappointing; there seemed to be no *there* there. Because of the dam two miles downriver, Ripon Falls had disappeared beneath the waterline, so no one spot stood out as the beginning of the Nile. With a bit of logic and imagination, I could still make out where the falls used to be, and the illusion was maintained audibly by steel bars laid across the river to keep the current from overpowering the dam's turbines, but it wasn't the same.

Gazing down from the tidy park that overlooks the Nile, I watched a flock of long-necked, brilliantly white birds wheel lazily across the river before settling back among the branches of a half-submerged tree. On the far bank, swaying in the light breeze, were row upon row of rubbery-leafed *matoke* trees, the banana-like staple of the local diet. Off to my left, Victoria Bay, calm and spacious, curled out of sight to meld seamlessly into the great lake. Without question, this remained a place of uncommon beauty and peacefulness. Yet a feeling of loss and incompleteness was inescapable, especially when I thought back to the untrammeled glories of Murchison Falls. What this cosmic spot on the earth's surface looked like before the coming of industrial man could now only be imagined.

Leaving the park, I stopped to chat with the young man who had sold me my entrance ticket. Neatly dressed, wearing flimsy eyeglasses with black plastic frames, he lounged beneath a tree with a friend, taking refuge from the midday sun. Yes, he nodded, this was a very beautiful place to work, but day after day, week after week, it sometimes got boring. Spying his newspaper on the ground, I asked why he did not bring a book to read. It was a foolish question, but his answer was polite.

"It is very difficult to obtain books in Uganda," he explained. "Our shops are usually empty. And any book for sale costs a great deal of money."

When I marveled at how lovely this place must have been in its original state, he was again a step ahead of me, seeming to read my mind and discern the unspoken assumption it was making.

"Yes," he smiled, with the enchanting gentleness so common among eastern Africans. "But the dam has done much good for us, giving us electricity."

"You trade one for the other," I said.

He beamed with the pleasure of having communicated perfectly across our cultural divide. "Yes! You trade one for the other."

Compressed in that brief exchange, it seems to me, is the essential dilemma facing the human species as it approaches the twenty-first century. Can the material strivings of the entire human family be reconciled with the carrying capacity of the planet's ecosystems? Does my young Ugandan friend deserve the occasional book, and electric light to read it by? Will business as usual suffice to provide them? If not, will technical fixes be found and implemented in time? And what if technical fixes are not enough? What happens then?

Three

The Irresistible Automobile

> We find that everywhere in the world, people
> *want* cars. And they're going to work hard to
> get them.
>
> —JOHN SMITH,
> chief executive officer, General Motors

It was hard to leave Africa. No other place touched me as deeply. Is it because this is where we came from, how we began? Is it the awesome, sometimes terrible beauty of the place, its untamed vitality, its pulsating aliveness? One falls in love with Africa for many reasons, I think, but what made the strongest impression on me were the people I met—their warmth and ease, the joy they took in social interchange, their good humor in the face of appalling hardship. There was misery aplenty in Africa, but the people I met were not defined by it. Earlier in my journey I had spent a month in the Soviet Union. Africans had a far lower material standard of living than Russians did, yet it was the Russians who were prone to sour-faced complaints and self-pity. The Africans I knew tended to shrug and make the best of things, smiling, even laughing, amid adversity, if only because it was better than crying.

The pace of life in Africa was especially hard to leave behind. Despite some modernizing on the surface, most of Africa remained a preindustrial society whose pace and social patterns still reflected its rural foundations; in the nicest possible way, time seems to pass more slowly there. To hurry is foolish not just because of the heat but because life in Africa is less crowded with events, possibilities, and commitments. People take the time to speak with one another like human beings instead of machines; work is a part of life, but only a part.

And for me, there was still much to see in Africa. In four months of travel, I had covered thousands of miles in Kenya, Uganda, and Sudan, crisscrossing the region by plane, train, car, truck, boat, bicycle, and foot. Yet Africa's immense size made my efforts look puny; the map showed me never getting outside the northeastern corner of the continent. How I longed to travel south, to Zimbabwe, Mozambique, and South Africa! But the Earth Summit was scheduled to begin in Brazil in ten weeks, and if I was going to spend much time in Asia on the way, I had to get moving.

So it was with ambivalence that I finally made my way to the rundown international airport outside Nairobi one evening in March 1992. My plane ticket said I was taking the midnight flight to Bangkok by way of Bombay and Delhi. But to anyone living with one foot still planted in the nineteenth century (that is, to most of the people I had been traveling among the previous four months) this journey would qualify as something very close to magic. I would be seated inside a long metal tube that, despite its enormous weight, would lift off the ground, climb above the clouds, and travel thousands of miles, traversing in hours the same ocean that ancient Arab traders used to take weeks to cross in their windblown dhows. The Air India jet that would perform this feat epitomized what historian Eric Hobsbawn has called the "revolutionary and constantly advancing technology . . . which virtually annihilated time and distance" during the twentieth century. Indeed, the main reason time seemed to pass more slowly in Africa was that technologies like the airplane and telephone had not yet touched the daily lives of most Africans.

The same could not be said for Thailand. It was a country in rapid transition, dangling somewhere between the unhurried rhythms of impoverished Africa and the hyperspeed materialism of my American homeland. I had visited Thailand two years before, in 1990, and as the Air India jet headed eastward through the night, I found myself recalling a young man I had met during that visit who personified the contradictions of Thailand's passage to modernity.

At age twenty-five, Leno stood poised on the cusp of two utterly different cultures. I met him in Chiang Mai, a city in northern Thailand with a hundred thousand inhabitants and an airport that serviced a lucrative tourist trade. Leno rented his own apartment, drove his own jeep, and owned a stereo system and a TV set. He had visited Bangkok numerous times and was fluent in seven languages. Modest, capable, and genuinely friendly, he was a born leader who seemed destined for a bright future, perhaps as a diplomat or an entrepreneur within Thailand's booming economy.

In his heart, though, Leno remained a child of the forest where he grew up. Leno worked as a guide and interpreter, leading backpackers on tours of the hill country near the Burmese border, which is how we met. Leno was the nearest thing I have ever seen to a perfect physical specimen—sparkling smile, sleek torso, thighs that looked like they could run forever—and the joy he took in outdoor activity was boundless. A member of the Karen tribe, he had spent his boyhood in a village 120 miles northwest of Bangkok. Despite the village's relative proximity to the capital, no one there had ever seen an airplane or automobile; the villagers were subsistence farmers whose only contact with the outside world came during monthly visits to a nearby trading post. Life was simple, possessions few. Like his eight brothers and sisters, Leno went barefoot as a child. His entire wardrobe consisted of two homemade, hand-me-down cotton smocks, one red, one blue.

At night, everyone in Leno's village used to gather around the fire while the elders told stories. Leno's favorite, he told me, was the story of the eagle and the snake. "The elders said that one day a giant snake would appear in the jungle, flash its tail, and cut our village in two," he recalled. "This snake would have ten thousand legs. Then a big eagle would appear in the sky. The eagle would land on the ground, swallow people into its belly, and fly away."

No one knew what the story meant, not even the elders who told it. They had heard the tale from their elders, who heard it from their elders, and so on into the past for more generations than anyone could remember. Not until the late 1970s, when Leno was a teenager and the first paved road was built through his village, did the meaning finally become clear. This must be the giant snake, the villagers decided. For had not the ribbon of asphalt come flashing out of the jungle in a burst of noise and dust and cut the village in two? And the travelers and vehicles that began appearing on this road, were they not the ten thousand legs of the snake? Soon after, one villager traveled on the new road to Bangkok, where he saw an airplane land and take off at the airport. When the man described this sight back in the village, everyone agreed it must be the prophesied giant eagle, swooping out of the sky to gobble people up and fly off again.

Recounting the story to me years later, Leno seemed certain his ancestors had foretold the coming of the airplane and the automobile. "I don't know how they did it, but they did it," he said earnestly. "They saw the future." Leno had no trouble maintaining this belief, even as he spoke in the next breath of boarding one of the giant eagles soon to visit Europe, a plan that terrified his peasant mother.

Such incongruities were perhaps to be expected in a culture that was fast-forwarding from traditional isolation to high-tech overdrive. By 1990, the airplane, the automobile, and other modern marvels had revolutionized not only Leno's village but all of Thailand, opening it to the tourists, technologies, investments, and ideas of the relentlessly expanding industrial world. In less than two decades, Bangkok had been catapulted from an easygoing, Buddhist-flavored tropical capital to a bustling, global business center. Talk about magic!

From the moment I stepped off the plane in Bangkok in 1992, the contrast with Africa was bracing. The airport terminal was brightly lit, spacious, and air-conditioned, with all the amenities one would expect from its European counterpart: clean toilets, public telephones, plenty of newsstands and restaurants. The customs and baggage claim operations were models of efficiency, and within half an hour I was heading toward the taxi stand, where fixed-price rides downtown were offered. Since the airport was only fifteen miles north of Bangkok, I figured I might be checked into my hotel and asleep within the hour. It was, after all, past midnight, and I was exhausted after twenty-four hours of travel.

Outside, automobiles seemed to be everywhere—another culture shock. My taxi was a late-model BMW whose plush backseat was more comfortable than most beds I had had in Africa. The expressway was crowded with similar vehicles; Thailand was one of the world's leading markets for Mercedeses, and second only to the United States in purchases of pickup trucks. The highway was in good repair, too, more reminiscent of the autobahns of Germany than the crumbling pavement of Nairobi; three and sometimes four lanes traveled in each direction, with guardrails in between. High above loomed huge, lighted billboards with names like Sony, Siemens, and Samsung in sparkling colors, as if making clear to the newly arrived who (or rather what) was in charge here. The accompanying advertising slogans were presented not in the languorous script of Thai but the snappy authority of English, indisputable world language of the Technological Age.

And yet. Lowering my gaze from these celebrations of global consumerism, I peered into the shadows at ground level, where I saw that the highway was also straddled by slouching shantytowns that recalled the poverty of Kampala and Nairobi. Here in Bangkok, though, the shacks were crammed up to the very edge of the highway, which meant that the inhabitants, in addition to the other burdens of their existence, were treated to a ceaseless assault on their lungs, eyes, ears, and nervous systems by the apparently nonstop flow of traffic.

Did I say nonstop? Actually, within three minutes of leaving the airport, the taxi was engulfed in a traffic jam that reduced our progress to stop, crawl, and stop again. I remembered Bangkok's terrible traffic snarls from my visit in 1990, but I thought that arriving in the middle of the night this time would let me off the hook. Wrong. Here it was, nearly one in the morning in the middle of the week, and the highway looked like Los Angeles during Friday afternoon rush hour.

Undeterred by the tiny Buddha shrine glued to his dashboard, my taxi driver took the opportunity of one lull in the traffic flow to urge upon me brochures featuring photos of extremely young, naked Thai women. "I take you," he suggested, clearly hoping to drive me to the nightclub pictured. When I waved the brochures away with a murmured "no thankyou," he simply slipped them under his seat and turned his attention to our common predicament. Middle-aged and plump, he spoke very little English, but he knew the word for his primary occupational hazard.

"Traffic, bad," he announced.

"Traffic bad," I agreed. Aiming my words carefully, I asked, slowly, "Why traffic bad at night? Day, yes. But night?"

"Traffic same-same," he replied with a resigned shake of the head. "Day, night, same-same."

Though I smiled inwardly ("same-same" was a fetching piece of Thai English I recalled from my earlier visit), this was discouraging news. After poking forward in fits and starts, we finally reached the city proper some seventy-five minutes after leaving the airport. While still bumper to bumper, the traffic moved somewhat more easily downtown, perhaps because it included a higher proportion of motorbikes and *tuk-tuks.* Three-wheeled vehicles whose name comes from the sputtering sound produced by their horribly polluting two-stroke engines, *tuk-tuks* look like beat up golf carts with roofs and backseats and function as inner-city taxis. Unfortunately, the advantage that *tuk-tuks* and motorbikes offer in terms of mobility is undercut by their prodigious tailpipe exhaust. Both vehicles burn a fuel that is part gasoline, part benzene—benzene, of course, causes cancer—and each flick of a driver's wrist sends thick puffs of bluish-white smoke into the already souplike air.

I remembered from my previous visit that the pollution of Bangkok's air had been so extreme it seemed to have a tactile quality; you felt you could scoop up a handful of the stuff and splatter it against the wall like a dirty snowball. Curious, I lowered my window to check its current condition. Sure enough, it was the same viscous gunk as before, complete with that foul chemical odor that caught in the throat and used to give me a headache within two minutes of stepping onto the sidewalk. My rash gesture alarmed the taxi driver. Blurting "No, no," he quickly zapped my window closed by remote control and turned around to study me, as though trying to decide whether I was impossibly stupid or just uninformed. With a nervous, placating smile, he pointed at his dashboard and said, "Air condition."

It wasn't just the noxious fumes he wanted to block out but the awful noise. I got an earful of it when we finally reached my hotel, an ugly high-rise near the Chao Phraya River. Stepping onto the sidewalk while the driver hoisted my bags out of the trunk, I confounded the poor man again by climbing onto a nearby pile of bricks (like much of Bangkok, the property next door was under construction) to get a better view of the traffic congestion of which we had been a part. Off to my right, for as far as I could see, stretched three lanes of headlights, their beams semiobscured by the copious exhaust fumes hanging in the air like dense morning fog. To my left, perhaps ten car lengths away, was a rare bit of open space, an intersection, where the front of the traffic waited impatiently for the light to change. Quite a few drivers passed the time by intermittently revving their engines. The discordant whines and buzzing this generated was bad enough, but it paled next to the cacophony that erupted when the pack took off again. With everyone hitting the throttle at once (senselessly, for where was there to go?), it sounded like a cross between a chainsaw massacre and the Indianapolis 500.

Perhaps the most extraordinary aspect of the entire tableau was how little it seemed to bother most Thais. My hotel room overlooked the street —from eight floors up, thankfully—and when I stepped onto the balcony before I went to bed I saw on the sidewalk below an outdoor noodle shop that was doing an amazingly brisk business for two o'clock in the morning. Much of the clientele seemed to be young. They were casually dressed, giggling, eating, drinking, smoking—clearly enjoying themselves, while seated at tables no farther away from the traffic-clogged avenue than the shantytowns near the expressway had been. When the revelers were done, they would hop onto motorbikes, two and sometimes three to a machine, and, as if they hadn't inhaled enough poison already, expertly weave their way through the stalled traffic.

In the days to come, I encountered many more traffic jams in Bangkok, both as a passenger and as a pedestrian. I saw that it was by no means unusual for families of four to crowd onto a single motorbike, the father driving and balancing one child on his lap, the mother holding another child behind him. Conversations with such families revealed that, predictably enough, they were eager to trade up to a *tuk-tuk,* just as *tuk-tuk* owners yearned to trade up to cars and teenagers dreamed of their first motorbikes. All this helped explain why traffic speed averaged a mere seven kilometers an hour. I discovered I could usually cover the same distance faster on foot. I stayed off the main drags and took the river bus whenever possible, but most locals seemed indifferent to such self-protective measures. A few wore cotton face masks, but most went about their business impassively, seemingly unconcerned that they were being constantly gassed.

And traffic jams, it turned out, were only the most visible symptom of Thailand's environmental problems. The Chao Phraya River was virtually dead south of Bangkok. One afternoon, during a river bus ride, I saw three young boys, shirtless and happy, leaping off a rotting pier to swim in the murky waters. Twenty yards upriver, the bus passed a second pier, where an older man calmly lowered his pants, squatted over the side of the pier, and emptied his bowels. Normita Thongtham, a journalist who covered the environment for the *Bangkok Post*, told me that each of the dozens of luxury hotels crammed along the waterfront also flushed their waste into the river; they had no choice—the city had no sewage treatment plants. According to the Asian Development Bank, wastewater from factories, commercial buildings, and pig farms also received no treatment, pushing Bangkok's water pollution to "extreme levels." The groundwater was so badly overexploited thanks to soaring demand spurred by tourism, economic growth, and rapid population growth that the city was slowly sinking into the mud. Meanwhile, in the countryside, forests had been felled mercilessly. Thailand's tree cover had fallen from 60 percent of total land area in the 1950s to a mere 18 percent by 1991.

These assaults on the Thai ecosystem were side effects of the extraordinary economic expansion the country had experienced over the past two decades, especially in the late 1980s, when growth rates topped 10 percent. As recently as the 1970s, Thailand had been quite a poor country, with a standard of living not much different from that of Kenya or Uganda. Now, after massive foreign investment spearheaded by such latter-day Winston Churchills as executives of the World Bank, Thailand was an apparent economic success story. Per capita income in 1991 was U.S. $1,570—a stunning sixfold increase over the 1971 figure of U.S. $271.

Of course, it was that very surge in incomes that had put so many vehicles on the streets—enough to cause Thailand's workforce to lose forty-four work days stuck in traffic jams a year, at a cost of several percentage points of growth in the gross national product. Likewise, the factories that boosted the nation's GNP with their output were the same ones that were fouling its air and water with their wastes. One key growth area was the manufacture of air conditioners and refrigerators; a side effect was that Thailand's use of ozone-destroying CFCs had doubled between 1986 and 1989. Like many other environmentally destructive enterprises in Thailand, most of the air conditioner and refrigerator factories were foreign owned.

At the urging of the World Bank, Thailand had pursued a corporate-friendly, export-led development strategy since the 1960s. Thailand offered cheap labor and natural resources; the bank financed the roads, power plants, and other infrastructure needed to exploit those resources; foreign corporations supplied the capital—the factories and industrial-style farms—to turn the resources into salable goods. Thailand duly became a leading exporter of rice, timber, and electronics, but the costs were high. Forests and wetlands were cleared with abandon. National income rose dramatically, but so did social inequality; the World Bank itself admitted that, with half the nation's wealth in the hands of the richest tenth of the population by the mid-1990s, Thailand's income disparity was among the five most skewed in the world. Peasants unable to compete with global capital were driven off their lands and into the cities, where especially the young and the female worked in industrial sweatshops or the burgeoning sex industry. An estimated one-third of Bangkok's 6.5 million people were rural migrants who squatted in crowded shantytowns like those I had passed riding in from the airport.

Thailand received an environmental wake-up call in 1988, when rampant deforestation led to landslides that killed hundreds of people south of Bangkok. "The logs slid down the mountainside for two kilometers and crushed people," Thongtham told me. "That was the first awakening of environmental consciousness for both the press and the public in Thailand."

Popular anger led to passage of a ban on logging in 1989, plus a government pledge to increase the country's tree cover to 40 percent. But environmentalists complained that the government's reforestation program only repeated past mistakes, promoting not genuine reforestation but huge plantations where eucalyptus trees would be planted and harvested like so many rows of corn. These plantations required large-scale evictions of poor peasants; the Khor Chor Kor forest development program of 1991, for example, called for displacing 1.5 million people. Not surprisingly, the peasants resisted, which is more than could be said for most urban Thais.

"Villagers fight to protect the environment because they have no choice," Wittoon Permpongsacharoen, director of the Project for Ecological Recovery in Bangkok, told me. "The forest is the source of their food, their houses, the water for their rice fields. Middle-class people in Bangkok are concerned about air pollution, but they don't fight. It's such a money culture, people think money can solve any problem. So their response is to buy a better car and retreat inside their air-conditioning."

It was true. Everyone I talked to in Bangkok—from sidewalk food vendors and civil servants to businessmen, students, and retired folk—was happy to complain about the traffic problem, but no one was willing to reduce his own driving. Meanwhile, mass transit was virtually nonexistent. Bangkok's tram system, rejected as backwards, had been demolished twenty years ago. There was no subway, and the few buses I saw were fiendishly overcrowded and subject to the same delays as cars (but without the balm of air-conditioning). Years ago, Bangkok had been known as the Venice of the East because of the city's intricate web of canals. Most of the canals had been paved over into roads during the modernization drive, however, and still there was far from enough road surface to accommodate all the vehicles. There was talk of building overpasses to relieve the congestion. But how much chance did the construction crews have against the mounting tide of cars and motorbikes?

Bangkok's traffic conditions seemed certain to deteriorate after I left Thailand in 1992, and they did. In 1995, after three years of worsening congestion and pollution, the nation's beloved King Bhumibol Adulyadej complained publicly that the experts charged with solving Bangkok's traffic nightmare "only talk, talk, talk and argue, argue, argue." In 1992, I was told that the average Bangkok commuter spent three hours traveling to and from work. By August 1996, the standard commute was five hours, according to a report broadcast on CNN that showed the children of one family breakfasting in the backseat in predawn darkness. Their father explained that this was the only way to get them to school on time. With a tired sigh, he added that the car not only sapped his time and energy but 30 percent of his income; the only reason he did not leave Bangkok was that he wanted his children to get a good education.

CNN did not mention Bangkok's new auto accessory of choice: the portable toilet. Leaving home without one could be risky, especially at peak travel times. In 1995, one family tried to beat the rush out of town before a national holiday by leaving at ten o'clock the night before. But when they reached the expressway leading to the airport, they found themselves in a traffic jam sixty miles long. According to Thomas Friedman of the *New York Times,* it took the family twelve maddening hours to get as far as the airport—only a few miles from their house—where they gave up and turned around. The governor of Bangkok told Friedman he realized his city had to do something about its traffic. Foreign investors, he conceded, "won't come to live in a town where it takes three hours to travel somewhere by car . . . [and] kids grow up breathing bad air." Not all his colleagues felt the same way, though. One government minister interviewed by CNN actually called Bangkok's traffic congestion "a blessing in disguise." Without it, he explained, "we wouldn't have an automobile industry in Thailand."

The automobile may well be the ultimate symbol of the modern environmental crisis. With the exception of atomic fission, no other twentieth-century technology has done more to transform the human species's relationship with the natural world or raised as many troubling questions about reconciling human behavior with the planetary ecosystem. (I call the car a twentieth-century technology because, though invented in 1886 by Carl Benz in Germany, it came into its own as an item of mass production and consumption only in the 1910s and 1920s.) Like the steam engine in the eighteenth century and the railroad in the nineteenth, the automobile has been the principal engine of economic growth throughout the twentieth century, especially in the United States. It revolutionized not only transportation options but patterns of human settlement and stimulated such key industries as oil, steel, and construction. Without the car, the suburbs and the vast amounts of economic activity they represent would never have expanded so inexorably. Exxon and OPEC would not be household names. The Persian Gulf War, among other conflicts, would not have been fought. There would be no such thing as fast food or shopping malls.

The car greatly expanded humans' control over the natural world, providing those able to afford one with unprecedented mobility, speed, and a special kind of freedom. What is a car but a modern version of a magic carpet, ready to take its occupants wherever they command at the turn of a key? Make that carpet a convertible, cruising sun-drenched back roads with music pouring from the stereo, and the car is close to heaven; as the song says, you "get your kicks on Route 66." Besides being great fun, the car is also extremely convenient, genuinely useful, and a status symbol par excellence, not to mention almost impossible to live without in the American-style suburbs that are becoming increasingly common the world over.

But as ever, the environmental costs of this intoxicating new technology—pollution of the air, destruction of agricultural and wilderness land, intensification of the greenhouse effect—have turned out to be steep as well. Since the 1970s, when the environmental hazards of automobiles first began to be seriously addressed, human ingenuity has struggled to keep pace with human appetites. Impressive technical fixes have been devised—most notably, better fuel efficiency and reduced tailpipe emissions. But whether technical fixes can outweigh the effects of more and more people driving more and more miles (the world's auto fleet is expected to double, to one billion, by the year 2020) remains very much an open question.

A full reckoning of the environmental costs of automobiles may be beyond calculation, if only because human knowledge tends to lag behind the effects of human action. Not until the 1960s, for example, nearly fifty years after automobiles became a common commodity in the United States, did the health dangers posed by their exhaust begin to be recognized. And not until the late 1980s did the car's contribution to global warming—arguably the single greatest danger cars present—attract attention, because only then was global warming recognized as a serious threat. For all we know, additional surprises await us down the road. In the meantime, the current inventory is sobering enough.

The internal combustion engine is, after all, a prime culprit behind global climate change. According to the U.S. Department of Energy, 40 percent of the projected growth in greenhouse gas emissions through 2015 will come from cars and the rest of the transportation sector. It is already too late to prevent climate change, according to John Houghton, cochairman of the Intergovernmental Panel on Climate Change. The issue, Houghton told the *Guardian* newspaper in 1995, "is whether we can slow it down enough to avert the worst effects." What happens with cars will go a long way toward answering that question.

Houghton's remark came during a landmark year in the science of climate change. In 1995, the concentration of carbon dioxide in the earth's atmosphere reached 360 parts per million—30 percent higher than the 275 parts per million that existed as a "background" level before extensive fossil fuel use began during the Industrial Revolution. Nineteen ninety-five was also the hottest year in recorded history; all ten of the hottest years had fallen between 1980 and 1995. But what made 1995 especially significant was the release of the IPCC's *Climate Change 1995*, a dense, lengthy study that concluded for the first time that rising greenhouse gas emissions and higher global temperatures were connected. In the study's words, there was a "discernable human influence" on the earth's climate. Scientists had suspected as much for years, of course. But in 1992 the IPCC had predicted it would be at least a decade before a causal link could be observed. The fact that this "signal" was confirmed only three years later suggested that, as with the deterioration of the ozone layer, the alteration of earth's climate was proceeding faster than human understanding of it was.

Then again, global warming did not even exist, according to the fossil fuel lobby; self-interest can be a blinding thing. Representatives of auto, oil, and coal companies had complained for years that climate change was only a theory, a contention they supported by citing the work of a handful of university professors who, it was later revealed, had received funding from these companies, as well as from OPEC governments. Nevertheless, in 1989, one year after global warming made newspaper headlines around the world, these so-called greenhouse skeptics started receiving substantial media coverage of their own. Professor Richard Lindzen of the Massachusetts Institute of Technology ridiculed the IPCC's computer models as inaccurate. Other skeptics claimed that satellite records showed the earth was not warming after all, and that in any case climate was influenced by too many variables for science to distinguish between statistical "noise" and genuine cause and effect.

The skeptics had a real effect on public debate, at least in the United States. While passing through San Francisco in 1992 on my way to the Earth Summit, I listened as a friend blithely dismissed any worries about global warming. "Even the scientists can't agree if it's real or not," he snorted. My friend was no naïf; he was simply reacting to what he was reading and hearing in the media. By exploiting the media's scientific ignorance and its yearning for balanced presentation of issues, the Information Council on the Environment, the fossil fuel lobby's front group, succeeded in creating the impression that scientific opinion about climate change was hopelessly divided. The divide was, in fact, about 2,500 to 6—the 2,500 IPCC scientists who had been studying the problem since 1988 on one side, versus half a dozen mavericks on the other, most of whom had been recipients of fossil fuel industry funding.

"It's only in America that there is a debate over what's happening to the global climate. In [Europe] . . . the only debates are how fast and with what impacts the changes will happen," a journalist from the German newsweekly *Der Spiegel* told Ross Gelbspan, author of *The Heat Is On*, a book documenting how public discussion of global warming was distorted in the United States during the 1990s by the fossil fuel lobby's propaganda campaign. In the United States, industry front groups managed to foster the appearance of scientific uncertainty about climate change, thus "draining the issue of all sense of crisis," wrote Gelbspan. Even after Gelbspan exposed the propaganda campaign in 1995, news stories continued to quote the skeptics respectfully, thereby implying that the jury remained out on climate change. This was all industry needed; to undercut pressure for reform, it was enough to present the argument as unsettled, still open to dispute.

The fossil fuel lobby had reason to fear the IPCC's report. *Climate Change 1995* forecast that the average global surface temperature would rise 3.6 degrees Fahrenheit by 2100 if carbon dioxide and other greenhouse gas emissions were not reduced. That estimate was about one-third lower than the IPCC's previous prediction (largely because the 1995 assessment took into account the cooling effect of sulfates in the atmosphere). Nonetheless, such a rise would represent the most rapid temperature increase in the ten thousand years of human civilization, and its disruptions of social, economic, and environmental systems promised to be severe. "While there will be some beneficial effects of climate change," the report said, "there will be many adverse effects, with some being potentially irreversible." The warning that most beaches on the East Coast of the United States could disappear within twenty-five years caught the attention of the insurance industry, which was already reeling from a sharp increase in the frequency and severity of storms like Hurricane Andrew, which cost the industry some $17 billion.

One of the most troubling—and, inexplicably, overlooked—aspects of the problem was its inexorable historical momentum. Climate change was like a moving train. Even if the brakes were applied immediately, the train would not stop moving until it was well down the track, and the longer the delay before braking, the greater the distance it would travel. Thus, even if atmospheric concentrations were stabilized immediately (an impossibility, of course) global temperatures would still rise 1 to 3.5 degrees Fahrenheit by 2100, reported the IPCC. To reduce damages to a manageable level, the scientists had recommended in their first impact assessment in 1990 that greenhouse gas emissions be cut by 50 to 70 percent over the coming decades. It was a daunting challenge. As Angela Merkel, then the German environment minister, declared, such reductions would require a fundamental shift in patterns of production and consumption.

The car is the epitome of twentieth-century production and consumption. Cars consume one-third of the world's oil output. The average American car, assuming (generously) that it gets the federally required 27.5 miles per gallon of gasoline and travels one hundred thousand miles in its lifetime, ends up emitting nearly thirty-five tons of carbon dioxide. The world's five hundred million cars are responsible for between 20 and 25 percent of current greenhouse gas emissions; only electric power plants, with 25 percent, and deforestation, with 25 percent, are as damaging. But the car's share of global emissions is growing rapidly, as more and more people around the world join the auto economy. With motor traffic expected to increase by 60 percent over the next twenty years, the UN Population Fund has projected that, by 2025, "developing countries could be emitting four times as much carbon dioxide as the industrialized countries do today." That is a very long way from the 50 to 70 percent reductions proposed by the IPCC, and it demonstrates why taming the car is essential to defusing the greenhouse crisis.

In addition to hastening climate change, the car also fouls the air locally. Motor vehicles are the world's single largest source of air pollution. Despite recent design improvements, the car remains a deadly polluter, in part because so many people drive one and in part because human lungs are vulnerable even to small amounts of the toxins—chiefly, carbon monoxide, nitrogen oxide, and volatile organic compounds—produced by internal combustion engines. Automobile emissions cause an estimated $93 billion worth of damage to health and the environment every year in the United States.

Proponents of the car like to point out, as the Mobil Corporation did in a 1995 advertisement in the *New York Times*, that the cars and skies in the United States are much cleaner now than they were a quarter of a century ago. Mobil crowed that smog was much reduced since the mid-1970s, "even though the economy and the number of new cars on the road have grown substantially." Pollution fell, of course, because the federal government forced industry to meet tougher emissions and fuel efficiency standards, a fact left unmentioned by Mobil, which, like most oil and auto companies, bitterly resisted the standards in question. But even these lower emissions did not result in healthy air. A study released by Harvard University researchers in 1995 found that 30,000 Americans die every year from respiratory illnesses related to car exhaust, while another 120,000 people die prematurely because of such exhaust. And yet America has the strictest air quality standards in the world. (The United States banned leaded gasoline in the late 1970s, for example, something the European Community will not do until 2000.)

These statistics only begin to tabulate the stress placed on the planetary ecosystem by the automobile. Manufacturing cars requires huge amounts of steel (1,767 pounds for the typical American car of 1995), iron (398 pounds), aluminum (188 pounds), and plastic (246 pounds), four of the most resource-intensive, polluting industries in existence; steelmaking, for example, is the leading emitter of carbon monoxide in the United States, except for auto driving itself. Researchers at the Environment and Forecasting Institute in Heidelberg have calculated that the manufacture of a typical German sedan produced nearly as much air pollution as did the driving of the car. In addition, cars must eventually be disposed of, which involves further contamination of air, water, and soil.

The countless miles of streets, parking lots, and highways that carry cars gobble up still more natural materials in their construction while wiping out vast areas of forest and farmland. Additional habitats are damaged in the search for and delivery of the oil that fuels cars. Indigenous peoples like the Ogongi of Nigeria and the Huaorani of Amazonia are near extinction because their cultures and ecosystems have been brutalized by the modern world's relentless thirst for the oil beneath their lands. The Exxon *Valdez* crash, which spilled 10.7 million gallons of oil off the Alaskan coast in 1989, drew worldwide condemnation, yet routine industry practices— ruptured pipelines, deliberate dumping at sea—are far more destructive, releasing approximately ten times as much oil as marine accidents do.

"needs") Nevertheless, all these environmental concerns seem unlikely to turn humans against the automobile. After all, humans already accept quite placidly the more immediate and bloody risk of death and injury behind the wheel. In 1993 alone, an estimated 885,000 people around the world died from auto accidents, the equivalent of 2,400 deaths a day. Death rates are rising especially rapidly in developing nations where cars are still a relatively new commodity and many motorists have received little or no training. The way the *matatu* of eastern Africa is driven is but one example of the kind of speed and daredevil maneuvering that naturally results in frequent accidents. India, despite having a relative scarcity of cars, suffers sixty-five thousand road deaths a year. Brazil's death toll is twenty-seven thousand; South Africa's, ten thousand.

It may be hard for some Americans to believe it, but in my experience, Americans rank as some of the safest drivers in the world, perhaps because, as a European friend of mine once suggested, Americans have had cars so long they no longer feel they have to prove anything behind the wheel. But even in the United States, cars kill approximately forty-two thousand people a year (more than guns do), and traffic accidents are the leading cause of death for people aged two to twenty-four. In Europe, fifty-five thousand people die every year in traffic accidents (while three times that many are permanently disabled) tallying to 150 deaths a day.

Picture the uproar if the airplane were killing this many people: a jet would be falling from the skies of Europe every single day. A second jet would be crashing daily in the United States. But precisely because the car's carnage occurs on a daily basis, it is not treated as news. Cars kill 270 people a day in the United States and Europe combined, and the fatalities pass all but unremarked, as if they were an inescapable fact of life.

Things that seem unavoidable or too difficult to fix or stop often get ignored. despite desparity

The car is a killer, and yet the human animal cannot resist it. When General Motors executive John Smith said that people everywhere want cars, it was not just self-serving propaganda. I saw the proof in nearly every country I visited.

Bangkok was a flagrant example of how, even under the worst circumstances, the car has a magnetic appeal for humans. Despite the ungodly traffic jams, Bangkok residents continued to buy motor vehicles. When I was there, five hundred cars and seven hundred motorbikes were being added to the city's streets every day. True, Bangkok is an extreme case—its traffic may be the worst in the world—but much the same thing is happening in nearly every big Asian city.

With my own eyes, I saw the terrible congestion in Bangkok, Hong Kong, Beijing, Guangzhou, Chongqing, and Xi'an. News reports and policy studies have indicated that Seoul, Jakarta, Shanghai, Manila, Kuala Lumpur, Bombay, Delhi, and Karachi, among other cities, are no less congested. Most also endure miserable air quality. Though lead is extremely toxic and known to damage children's mental development, most gasoline sold in Asia is leaded. Thailand did not begin offering unleaded gas until 1991 nor outlaw leaded gas until 1996.

The UN World Health Organization (WHO) has recommended that levels of total suspended particulate (TSP) matter (soot and dust, in lay terms) not exceed 60 to 90 micrograms per cubic meter per day. Tokyo, with its level at 50, satisfies that standard, and Hong Kong, with 82, is at the high end of acceptability, but none of the other cities just mentioned even come close. Just across the river from Hong Kong, Guangzhou measures 300. Bangkok registers 171; Shanghai, 225; and Beijing, 363. Cars are not the only source of this pollution—power plants and factories are the other big contributors. But motor vehicles are the chief culprit in high-traffic cities like Bangkok, and their predominance may well increase in years to come. Asia, after all, is regarded as *the* big growth market by the global auto industry. Sales are expected to double between 1995 and 2000.

The American social critic Lewis Mumford warned in 1963 that granting cars unlimited access to the city invited the city's destruction. Today, his prediction rings true not just in Asia but around the world. Traffic jams cost the United States $100 billion a year in lost productivity, not to mention the emotional stress on the drivers, who, in the words of one pop song, are "crammed like lemmings into shiny metal boxes, contestants in a suicidal race." Even European cities with good mass transit systems have been nearly ruined by cars. Paris and London still rank among the world's great metropolises, but their outdoor charms have been sadly diminished by the near-constant noise and exhaust of cars on their streets.

"The unrelenting growth of transport has become possibly the greatest environmental threat facing the UK," the Royal Commission on Environmental Pollution warned in a much-publicized report in 1994. The nation's auto-dominated transport system was "unsustainable," warned the Commission. When I visited London a few months later, in May 1995, it was easy to see why the commission was so alarmed. I took a run one morning along the Grand Union Canal, which snakes across north central London toward Battlebridge Basin. It was a muggy day, and even along the water the air smelled metallic and heavy. Only when I climbed a grassy hill with a view back downtown, however, could I grasp just how polluted the air actually was. Layers of gray and yellow smog throbbed in the midmorning heat, leaving the distant skyline a murky blur. Each of the half-dozen taxi drivers I questioned complained that air pollution had gotten much worse in recent years, especially during winter. "I come home with a headache every day now," said one man who had been driving taxis for a decade. By 1996, London was exceeding WHO air quality levels at least once a week, and even the conservative government of Prime Minister John Major was urging Britons to reduce their driving. Traffic jams were nastier, too, thanks to the 70 percent increase in traffic volume that London had experienced over the preceding twenty years.

Traffic conditions had worsened even more rapidly in Paris, where congestion had increased 200 percent in the same period. When I visited Paris in May 1991 near the start of my global expedition, the contrast with the Paris I remembered from ten years earlier was shocking. Not only were there far more cars on the streets, but the cars themselves were a different breed. No longer tiny, utilitarian two-seaters, Paris's cars now tended to be large, luxurious sedans that would not have looked out of place on the streets of Dallas or Chicago. The effect of the city's automobile explosion was to give the place a much more American look and feel, while increasing the city's tension level and diminishing its attractiveness. After all, outdoor life is basic to Paris's charm. But sidewalk cafés lose some of their allure when one inhales lungfuls of exhaust with one's cappuccino, and architectural glories are hard to appreciate amid the smog and racket produced by an endless swarm of automobiles. One silver lining was the sunsets, which a friend and I made a point of watching every evening from a footbridge over the Seine. The density of pollutants could make Parisian sunsets spectacularly vivid—on the evenings when they were not blotted out altogether.

"I love this city, but the cars are making it unlivable," said Jean-Francis Held, a journalist and author who has been the chief editor of two of France's leading newsmagazines, *Le Nouvel observateur* and *L'Événement du Jeudi.* Paris residents are famous for leaving town en masse for the month of August, but Held, a vigorous, outgoing man in his fifties, told me he so regretted what automobiles had done to Paris that August was now one of his favorite times in the city. "No traffic," he explained.

Held, it should be noted, by no means hated cars; he had raced them semiprofessionally for a time in his youth and had written extensively about automobile culture. But cars, he believed, must be kept in their place. He found it exasperating that so much of Paris's traffic was caused by wasteful personal indulgence. "Today, everyone feels he must have a car, and use it, even though you don't need one in Paris at all," he complained. "The Metro is excellent here. It gets you across town much faster than a car."

And yet, as in Bangkok, people in Paris continued to buy and drive more cars. By 1994, traffic congestion had increased by a third over 1991 levels. Authorities announced a system of pollution warnings that urged residents not to drive at times of heavy smog, but compliance was voluntary, and, besides, the smog warning levels were twice as high as European norms. Even those levels were breached during the summer of 1995. In August, after a period of hot and windless days, the Environment Ministry told Parisians not to use their cars "except in absolute emergencies." Environmentalists chose more direct tactics. Wearing face masks against the fetid air and hoisting bicycles over their heads, they waded into traffic on key boulevards and physically brought it to a halt. But the underlying problem persisted. One day in October 1997, Paris's air pollution became so serious that authorities closed the city to half the cars that normally entered it, though only for the day, a move that gave new meaning to the term "half measure."

Meanwhile, European cities that lacked decent mass transit, such as

Athens and Rome, were nearly as besieged by motor traffic as Bangkok was. The week I reached Athens, in October 1991, air pollution hit record levels, causing the government to ban private cars from the city center for two days in a row, while 650 people sought medical treatment. Did that episode frighten residents into reducing their driving? No, no more than the deaths of eight hundred Athenians during the summer smog emergencies of 1987 had done.

Topography makes Athens especially vulnerable to smog—like Los Angeles, the city is in a basin, surrounded on three sides by treeless mountains that trap pollution—and the abundance of motor vehicles turns this vulnerability into a hazard. The city government had tried to restrict traffic flow with a system based on license plate numbers—cars whose plates ended in an even number were forbidden to enter the center one day, cars with odd-numbered plates were forbidden the next. But desperate drivers and enterprising mechanics made a mockery of the restrictions. Just outside the city limits were dozens of roadside shops where drivers could rent properly numbered plates for the day, installation included.

On my third day in Athens, I rode by funicular to the top of St. George Hill, which should have commanded a spectacular view of the Athens basin. On my left and right, sandy-brown mountains stretched in parallel chains down to the sea. Straight ahead and far below lay the Parthenon, the enduring symbol of ancient Greece. But its magnificence was hard to admire, for one could barely see it. The smog in Athens was said to be worst in summer, but here it was, the second week of October, and a puffy gray haze covered the city like a poisonous blanket, shrouding in blurry outline the cradle of one of the world's great civilizations. Thus the car was not only tainting the present but devouring the past and threatening the future. After all, how many tourists would visit Athens if they could not see its principle attractions?

A companion pointed out that I was lucky to have arrived before it was too late. "By the year 2000, you probably won't be able to see the Acropolis at all from up here," she said. That may be true. In 1992 and 1993, smog emergencies sent hundreds of Athenians to the hospital. In the summer of 1994, the city government launched a new crackdown, doubling fines on cars that illegally entered the center. It was not enough.

The following summer, the government again had to ban cars—all of them this time—from the city center during smog alerts, a spectacle that was repeated in 1996 and again in 1997.

The good news is that Athens is building a subway system, scheduled for operation in 1999. But subways are no cure-all, as Athens's Mediterranean neighbor, Rome, demonstrates. With two cars for every three citizens, Italy is the most car-saturated country in all Europe. Traffic speeds in Rome average seven miles an hour, barely more than a brisk walk. For the white-gloved *vigili urbani* who try to impose discipline at Rome's most clogged intersections, fainting on the job is not uncommon. So-called acoustic smog is also at dangerously high levels, with daytime noise averaging 80 to 90 decibels (80 decibels is as loud as a vacuum cleaner; at 85, a level high enough to cause ear damage, a person must shout to be heard).

I passed through Rome a number of times during my travels, including the week in November 1994 when the city government announced its policy of carless Thursday evenings. From 3:00 until 9:00 P.M. every Thursday until Christmas, all cars lacking catalytic converters would be prohibited from entering the city center.

For some reason, the first carless Thursday actually took place on a Wednesday (November 23), and I spent that afternoon walking across Rome to observe its effects firsthand. I began in the northeast, at Villa Ada park, and ended up across the Tiber River above Trastevere, the oldest part of Rome. Along the way, I passed through Villa Borghese park, down the Spanish Steps, along the Corso, through Piazza Venezia, and past the Forum and the Colosseum before hiking back along the river to cross the Garibaldi Bridge and climb Gianicolo Hill south of the Vatican. It was a pleasant, sunny day, ideal for walking, and it took just a few blocks to establish what the entirety of my walk confirmed: the ban did reduce traffic volume and the attendant noise and exhaust, but only enough to render Rome the kind of city that, say, Munich is on a daily basis. There was still plenty of traffic about, and not because Romans were disobeying the law. It was simply that the ban applied only to cars that lacked catalytic converters; that is, cars bought before 1991. Newer cars still circulated, as did taxis, medical vehicles, and the hulking orange public buses whose engines roared and smoked like portable blast furnaces. Nor was there a respite from the penetrating whine of *motorini*, the ubiquitous motorbikes whose two-wheeled status exempted them from the ban. While it was certainly easier to breathe than usual, it was striking how loud and busy Rome still was.

Yet even this partial reduction in traffic allowed a fresh appreciation of Rome's visual splendor. Normally, traffic is a smelly, blaring, distracting eyesore that dominates Roman street life; that day, it had receded to the background. Near the end of the day, I was walking along the Tiber as the sinking sun bathed the city in rich autumn light. Despite the sycamore trees overhead and the occasional glimpses one had of St. Peter's Cathedral, this was not usually a place for strolling; the avenues that extended along both sides of the Tiber were usually among the most thickly traveled in all Rome. But today the traffic was sparse enough that a riverside *passeggiata* was quite agreeable. To the right were the remains of the Theater of Marcellus, built by Augustus Caesar in 13 B.C.; to the left, the Ponte Rotto, ancient Rome's first arched stone bridge. Farther upriver, the ochre wall of an elegant palace glowed in warm repose beneath the setting sun, a perfect marriage of Mediterranean color and light. It was nice to be reminded of how beautiful Rome could be when given the chance.

Even Amsterdam, the bicycle capital of Europe, has been smitten by auto infatuation. Bicycles still slightly outnumber people in Holland, and lots of Amsterdammers still get around on two wheels. At traffic lights, whole squadrons of them wait patiently in the morning mist—high school and university students, of course, but also professional men and women in stylish raincoats with briefcases strapped over their back wheels. The light changes, and they push off en masse, pedaling calmly and methodically, their faces glazed into the expressionless mask of commuters the world over. Amsterdam also boasts an efficient system of trams and buses, as well as good train connections for more extended travel. Nevertheless, more and more people are opting for private cars, with the result that the city is becoming annoyingly congested.

"People who live outside the city, who in normal times would need half an hour to drive to their jobs, are typically spending one and a half to two hours in rush-hour traffic jams every day," said Astrid, a thirtyish office manager with whom I shared a table at breakfast one morning. Sitting beside Astrid was her coworker Ton, an aeronautical engineer and

part-time punk rock singer. Aged twenty-three, he had just completed
the year of volunteer service required of young men who refuse military
induction on conscientious objector grounds. "A few years ago, all my
friends were arguing against using cars," he said with a wry smile. "Now
they are applying for driver's licenses and trying to buy secondhand cars."

The hunger for automobiles seems even stronger in cities like Prague
and Berlin, where people have long been carless not for philosophical reasons but because of economic and political restrictions. I first visited
Berlin in June 1991, just days after it was voted the national capital again.
A police officer I interviewed beneath the Brandenburg Gate, which used
to divide the city, said that East Berliners in particular were cheered by
the vote because they thought it would bring much-needed jobs. A couple afternoons spent wandering through East Berlin revealed why that
mattered so much. Shops offered poor selections of shoddy merchandise;
stately old buildings were crumbling from neglect; streets were dark and
deserted. What few cars one saw were wheezing old clunkers.

When I returned to Berlin in February 1995, though, things looked
very different. The federal government had underwritten a construction
and renovation boom that had brightened the downtown considerably.
East Berliners now had access to essentially the same goods "Westies" did.
Wages in the East still lagged well behind skyrocketing prices, yet somehow many Easterners now had cars. The streets of neighborhoods like
Prenzlauerberg were crammed with them—not the elegant Mercedeses
and BMWs found in West Berlin, to be sure, but serviceable sedans that
must have put their new owners well into debt. I witnessed a similar burst
of car purchasing during two visits to Prague in 1991 and 1994; automobile ownership among Czechs increased by a third between 1990 and 1995,
bringing the total to two million cars in a country of ten million people.

The same pattern is found the world over: an automobile is one of
the first things lower-income people buy once they come within striking
distance of the middle class. Indeed, there is no starker symbol of the
chasm between the world's haves and its have-nots than the automobile.
Eighty percent of the world's five hundred million cars belong to 20 percent of its people. In the United States there is one car for every 2 people;
in Nigeria there is one for every 151 people, and in India one for every 367.

Psychology is part of the reason newly affluent people buy cars.

Nothing certifies that someone has "made it" like having a car does, and as auto advertising around the world relentlessly proclaims, the higher one's social status, the fancier one's car has to be. At the same time, for the world's poor majority, desiring a car is no mere mind game. As riders of *matatus* in eastern Africa (and of similar vehicles around the world) can attest, having access to a motor vehicle constitutes a dramatic improvement in one's standard of living. When the only other option is to walk, often with a punishingly heavy load on one's back (a reality that countless peasants I saw in Kenya, Uganda, and southern Sudan lived every day of their lives) the sound of the engine can be a beautiful thing.

During my global journey I did not forget that my home country was central to the environmental predicament, and nothing better signified this truth than the automobile. In America, everyone drives. Americans spend so much time on the road that they all but live in their cars, some of which (e.g., sport utility vehicles) are larger than the houses of many people in the Third World. If places like Bangkok and Berlin illustrate John Smith's claim that people everywhere want cars, then the wealthy, car-saturated United States offers the purest example of a corollary truth: once a human has become used to having a car, it is almost impossible for him or her to give it up.

When pollsters asked Americans which modern inventions they could least live without, the car surpassed both the lightbulb and the telephone. Most elected officials could have guessed as much. The car is one of the true untouchables in American politics. Even the gasoline shortages of the 1970s, when motorists waited in line for hours to fill their tanks, produced no weakening of the nation's devotion to the automobile. The car is central to the American dream and to many Americans' sense of personal identity. As much as the clothes they wear and the houses they live in, the cars they drive express who Americans think they are. A Mercedes says wealth; a Corvette, speed; a minivan, parenthood; and a sport utility vehicle, all of the above.

In the Third World, you're rich if you have a car; in the United States, you rank among the very poor if you do not have a car. And you are bound to stay that way, since many jobs can be reached only by car. Yet less than one-quarter of all American driving is work related; one-third of the nation's car trips are for social reasons (dinner, movies, and such), while another third are for errands. Between 1969 and 1995, the vehicle population of the United States grew six times as fast as the human population and twice as fast as the number of drivers. Six out of ten households own two or more cars; two out of ten own three or more.

Life without a car is literally unthinkable for most Americans, an assumption comically skewered by cartoonist Tom Toles. His cartoon opens with four people agreeing that the greenhouse effect threatens global catastrophe and that carbon dioxide production therefore has to be reduced. But when one person says, "The biggest problem is automobiles," a silence falls over the group. "Somehow," the narrator dryly observes, "the discussion always stops at this point." The solution the four finally hit upon is to tell South Americans to stop burning down their rainforests. "Yeah," one character says with relief, "the South Americans."

The tyranny of the car that prevails in the United States goes hand in hand with an enforced lack of alternatives. One of the most striking differences an American traveler discovers in Europe is how easy and pleasant it is to travel by train. Car use has risen sharply in Europe in recent years, but one can still get around quite well there without a car because of the thriving rail network. Trains leave punctually and often, prices are affordable, downtowns are connected to their airports by rail. The entire experience is efficient and civilized.

One especially enjoyable trip took me south from Paris to Marseilles via the TGV fast train, which covered the 350 miles in less than five hours, including stops. The next day, in nearby Aix-en-Provence, I boarded a local train for Briançon, a town at the base of the Alps, a hundred miles away. The old train poked northward through farmland brightened by patches of orange and red wildflowers for an hour and a half before the crags of one of the world's highest mountain chains came into view. The train tracks paralleled a foamy green river, and as we followed its twists and turns the Alps seemed to play hide-and-seek with us, first disappearing for long minutes while we rounded a bend, then suddenly looming above us in breathtaking grandeur. I soon found myself in Embrun, a compact, sleepy town whose cobblestoned main street was too narrow

[handwritten marginal note: most practical, what people want.]

for cars to traverse. Embrun rested on a bluff above an agricultural valley bisected by the same green river I'd seen from the train. On the far side of the valley, perhaps a mile away, the land rose into sculpted green hills dotted with dark wooden houses before steeply giving way to the massive spreading countenances of the Alps. The glistening snow-covered peaks presented a spectacular panorama, which I drank in from a park on the edge of Embrun's bluff before hiking down to the valley, where I saw a man and boy, assisted by two eager hounds, leading a flock of sheep to water. Life felt grand.

But only because the train delivered me to Embrun in the first place. In ninety-nine cases out of a hundred, a similarly out-of-the-way place in the United States would be impossible to reach without a private car. The United States had a vibrant railway system prior to World War II (until the 1920s, virtually all urban commuting was by rail), but the rise of the car strangled the life out of it. The decline of rail was due partly to the inherent appeal of cars, but political favoritism and corporate treachery—including outright sabotage and conspiracy by General Motors and other companies that stood to gain from a shift to an auto economy—played a decisive role. Indeed, the nation's mass transit system was systematically dismantled from the 1930s on, while government bodies looked the other way, or worse.

In what must rank among the great corporate crimes of the century, General Motors secretly joined with Standard Oil of California, Firestone Tire and Rubber, Phillips Petroleum, and Mack (Truck) Manufacturing in 1932 to form National City Lines, a phony front company whose modus operandi was brutally simple: buy up rail and trolley lines in cities across the land, then shut them down and tear out the tracks. National City Lines eventually closed approximately one hundred streetcar systems in some forty-five cities, including New York, Baltimore, and Los Angeles. At the time, Los Angeles boasted one of the finest transit systems in the country. While its trolley and rail lines were being shut, a replacement network of so-called freeways—highways that were anything but free to the unwitting citizens whose taxes paid for them—was being built. It was a classic case of carrot and stick. As mass transit options narrowed, Los Angeles residents turned increasingly to private automobiles. Widespread auto use in turn propelled the city's geographic sprawl to where living

without a car became difficult indeed. The profits gained from such skul-
duggery were incalculable (the scam essentially secured monopoly status
for the automobile within the U.S. transportation system), but the punishment was next to nothing—when a federal court convicted GM and its
coconspirators of antitrust violations in 1949, the fine was a mere $5,000.

The government's promotion of the auto economy accelerated after
World War II. Federal housing subsidies enabled millions of returning veterans to buy tract homes in the newly expanding suburbs. Along with
cheap oil and rising personal incomes, the suburbanization of the United
States helped fuel an extraordinary economic boom, not least because it
seemed to make purchase of a second family car (for mom's errands and
shopping) a necessity. "And we didn't hesitate for a moment to give that
impression," Robert Lund, a GM vice president, later recalled with a
smile.

The most powerful boost the government gave the auto economy
came in 1956, and, once again, GM was in the thick of it. From his post as
chief of the Federal Highway Administration, Francis DuPont, whose
family was the single largest holder of GM stock, led the charge for congressional approval of a national highway system. Hailed by President
Eisenhower as "the largest public works program in the history of the
world," the highway project amounted to a massive taxpayer subsidy to
the car industry. The scores of billions of dollars in public spending were
justified on ludicrous national security grounds; the highways, it was
claimed, would allow rapid deployment of military vehicles against a Soviet ground attack and speedy evacuation of cities in the event of a nuclear strike. But politicians were also attracted by the program's vast
opportunities for patronage; a great deal of loyalty could be bought while
dispensing the contracts and jobs involved in building 42,700 miles of
highway.

What the National City Lines scam had done to cities like Los Angeles the interstate highway subsidy program now did to the nation as a
whole. As railroads were starved for funds and more and more countryside was covered with expressways, Americans responded by driving
more and taking mass transit less. By 1970, only 9 percent of all U.S. commuters used public transit; by 1990, the figure was 5 percent. Although
trains were six times as energy efficient as automobiles at transporting

commuters, federal spending on highways between 1958 and 1989 totaled $213 billion, nine times as much as was spent on railroads. Today, the subsidy to drivers in the United States—including police, traffic management, street maintenance, and the health costs of air pollution and accidents—is commonly estimated at $300 billion a year.

While Europe and Japan were rebuilding their war-ravaged mass transit systems in the 1950s and 1960s, the United States was consolidating its identity as a society of, by, and for the automobile. Thus the United States now finds itself locked into a future where continued reliance on the car seems unavoidable, literally set in concrete. The car became the basis of the U.S. transportation system when oil was cheap and secure, open space was plentiful, and many of the car's environmental hazards were unknown. Conditions have since changed, but society's infrastructure has not. The suburbs are here, and scores of millions of Americans rely on cars to get to their jobs. The existing infrastructure must therefore be maintained and policed, at enormous public cost. The transportation bill that Congress passed in 1998 continued Washington's pro-automobile bias, devoting $217 billion to highway construction and repair, compared with $41 billion for mass transit.

Although no country is more enthralled by the automobile than the United States, the rest of the world is gaining fast. In 1950, the United States had three-quarters of the world's cars; today, its share is only one-third. While the world's human population has doubled since 1950, the number of cars has grown by a factor of ten, to five hundred million. The ratio of cars to humans will become even more disproportionate in the years ahead as scores of millions of people in the emerging markets of Asia, Latin America, and Eastern Europe buy their first automobiles.

The car, it seems, is nothing less than addictive for human beings. Like cigarettes, cars are a source of seductive pleasure that eventually comes to enslave its users; no other possession goes from luxury to necessity faster. In Mexico City, where an estimated 2.5 million cars help make the air as dirty as anywhere on earth, the government tried in the mid-1990s to defend public health by requiring all cars to stay off the streets one day a week. But the program was emasculated when many well-to-do citizens simply bought a second car to evade the restriction. In a city where seven out of ten children already suffered impaired mental development from lead poisoning, such behavior bordered on the criminal. But addicts will be addicts.

"The car is the most beautiful toy in human history," the writer Jean-Francis Held told me in Paris. "But it must only be a toy. It is physically impossible for billions of people to use individual cars, now or in the future. And it is insane to treat cars as a universal necessity for daily transportation. We will choke to death!" Nevertheless, Held sounded optimistic. To ban cars outright is impossible for democratic governments, he said, but there are other options: "If the government is responsible, there are many ways to domesticate the car so that it is not so ecologically damaging: mandating increased efficiency, less polluting fuels, higher user fees, car pools. People must remain free to buy cars, but if you limit the use of the car, they will buy fewer of them. You want a BMW that goes 250 kilometers an hour? Okay, but you must still obey the speed limit. You want to own a dozen cars? Fine, but you still are not allowed to park them on the sidewalk."

Can the car be domesticated into something humans can live and breathe with? As Held said, many reforms are possible. Most of them fall into three categories: build better cars, buy fewer cars, use cars less. But whether humans will actually implement these reforms—and do so quickly and comprehensively enough to counter the momentum of a world careening toward one billion automobiles by 2020—is a different question, one that goes to the heart of the global political economy. Humanity's addiction to cars is not simply a matter of lifestyle, psychology, and infrastructure. It is also economic, and reinforced by long-standing political relationships that involve some of the most powerful commercial enterprises in history.

Indeed, the automobile is arguably the most economically important product of our time. Manufacturing cars is the biggest industry in the world; fueling them is the second biggest. The car accounts for approximately one of every seven jobs in the United States and millions more around the world. Stop producing cars and the global economy would collapse. Even encouraging people to buy fewer cars could weaken economies, and it would certainly provoke ferocious opposition from the

Large economic enterprises, no matter how damaging will continue business.

car lobby—the oil and auto companies that have accumulated immense wealth over decades of feeding the car addiction and that have no intention of stopping now.

The political and economic influence these corporations will wield over the future of the car is tremendous. It is not only because they design, produce, and market cars but also because the jobs and investments they can pump into a region or country gives them decisive political leverage. The authority of the corporations is not absolute, as their recent concession to start offering high-efficiency cars illustrates. But they can often block or delay changes they dislike; for example, they *could* have been building such cars long ago. All this can only give pause to anyone familiar with the industry's history.

After all, these are the same corporations that did not want to put safety glass and seat belts in cars. In an episode nearly as unsavory as the National City Lines conspiracy against streetcars, the Big Three U.S. automakers—GM, Ford, and Chrysler—had to be forced in the 1950s and 1960s by consumer activists like Ralph Nader and conscientious public officials in New York and California to install not just seat belts but eventually shoulder belts, collapsible steering wheels, and numerous other basic safety features. Industry resisted these reforms through a dual-track defense it would employ often in years to come against safety and environmental reforms alike: first deny that any problem exists (cars don't cause accidents, the industry blustered, drivers do) and then claim that the reforms would make cars prohibitively expensive. In the case of seat belts, the industry went a step further. After the New York State legislature in 1964 forced automakers to make seat belts a standard feature, the industry had the nerve to issue public statements claiming it had favored belts all along.

The automakers have displayed comparable public spiritedness repeatedly over the past forty years. When California tried to require tailpipe exhaust controls in the 1960s, when the federal government tried to impose such controls nationwide and bolster them with fuel efficiency standards in the 1970s, when environmentalists urged that high-efficiency cars be produced in the 1980s and 1990s, the industry's answer was always the same: it can't be done, and we will exert all possible influence to make sure it isn't done. The industry's intransigence has sometimes been em-

barrassing. The claim that high-efficiency cars are impractical was publicly mocked when Greenpeace activists "kidnapped" a Renault Vesta-2
prototype in 1993 and brought it to the International Car Show for spec-
tators to test drive. The Vesta-2 had a maximum speed of 140 miles an
hour but got 107 miles per gallon in highway driving. Toyota, GM, Ford,
and Volkswagen had produced similar prototypes but, like Renault, had
not put them on the market, claiming they were uneconomical.

Ambushes aside, the industry gradually grew more clever about its
public relations. Above all, it learned to be obstructionist mainly behind
closed doors, while presenting itself publicly as an honorable corporate
citizen. The sanitizing strength of modern propaganda even proved capable of making a hero out of Lee Iacocca, the executive known for turning around the failing Chrysler Corporation in the 1980s but not for
green-lighting the infamous Ford Pinto in 1970, despite crash tests showing that the Pinto exploded when struck from behind. Iacocca denied
personal knowledge of the crash tests.

But as I mentioned earlier, the power of the automakers is not without limits. Cars sold in the United States are much cleaner today than
they were twenty-five years ago; tailpipe emissions of carbon monoxide
and other smog-causing chemicals have fallen sharply. There is still a
long way to go—car exhaust causes 30,000 deaths and 120,000 premature
deaths every year. But the historical record shows that reform of automobiles is possible, if government is vigilant and market forces are well
harnessed. When high standards are established and corporations are
forced to meet them, progress is made. Tailpipe emissions were sharply
reduced beginning in the 1970s because Washington demanded it, despite
Detroit's howls of protest. Electric cars are on sale in the United States because California, a market too big to ignore, decreed that by 1998 2 percent of all vehicles sold in the state had to be "zero-emissions" cars.
(Pressure from industry later led that deadline to be rescinded, though
the requirement of 10 percent zero-emissions cars by 2003 remains.)

Assertive government is rare these days, however, and not just in
the United States. Governments have increasingly come to accept the corporate argument that setting tough environmental standards—an
approach corporate spokespersons absurdly deride with the communist-
tinged label of "command and control" regulation—is counterproductive.

The preferred alternative is a voluntary approach that stresses words like "cooperation" and "flexibility" and just happens to impose no obligations on the corporations in question. In the United States, the Clinton administration has not simply embraced volunteerism but subsidized it. In 1993, Clinton and the chief executive officers of Ford, Chrysler, and GM announced the formation of the Partnership for a New Generation of Vehicles (PNGV), a research and development effort aimed at developing, by 2010, a midsize sedan that would get eighty miles a gallon without sacrificing safety, comfort, or affordability. But the automakers did not commit to producing such vehicles, causing environmentalists to complain that the PNGV was merely an excuse for not tightening fuel efficiency standards. Fuel efficiency has stagnated at about twenty-four miles per gallon in the United States since the mid-1970s. It has stagnated in Europe and Japan as well, albeit at thirty to thirty-five miles per gallon.

Yet raising fuel efficiency is essential to fighting climate change. Better filters can reduce the amount of smog a given car produces to practically zero, but until the laws of chemistry are repealed, a gasoline car will always produce carbon dioxide, about nineteen pounds of it for every gallon of gas burned. That leaves two options (beyond the obvious one of using cars less): switch cars to a different fuel, such as hydrogen or electricity, or increase gasoline fuel efficiency, very sharply.

In global terms, if gasoline-powered cars were to become twice as efficient by 2020 as they are today, it would exactly offset the doubling in the number of cars expected by then. In effect, this would maintain the greenhouse status quo, despite a huge increase in humanity's auto use. But the status quo is far from good enough, given the IPCC's recommendation that greenhouse gas emissions be cut by 50 to 70 percent. For these reductions to be achieved, fuel efficiency must increase by a factor of five, from today's 25 to 30 miles a gallon up to 125 to 150 miles a gallon—a breathtaking leap, considering how little efficiency has improved recently.

Could electric cars close the gap? Not in their current form. Electric cars help reduce smog because they emit no tailpipe exhaust; California classifies them as zero-emissions vehicles for that reason. But the dirty little secret of electric cars is that their exhaust is merely displaced; it is released from a few power plants rather than from thousands of auto tailpipes. Advocates of electric cars say this makes the exhaust easier to control, which is true as far as smog is concerned. Unfortunately, the best controls in the world do nothing against carbon dioxide; the only solution is to burn less fossil fuel in the first place. Since most of the world's electricity is generated by burning coal, electric cars end up producing approximately 90 percent as much greenhouse gas as gasoline-powered cars do. Maneuvered into this intellectual corner, advocates of electric cars say that coal is on its way out as the prime electricity fuel; solar, wind, and natural gas will soon give us greenhouse-friendly electricity. That rebuttal sounds nice, but for the moment at least, it reflects cheerleading more than demonstrable fact.

One solution is to build hybrid cars. A hybrid car combines the best features of gasoline and electric vehicles. It uses gasoline to free itself from the range limits and crushing weight of conventional electric cars; it relies on an electric motor and flywheel to avoid the enormous amounts of energy that gasoline cars waste during braking and deceleration. Amory Lovins, an energy expert and director of research at the Rocky Mountain Institute in Colorado, predicts that hybrids will get 150 to 400 miles a gallon or more, if they are built with high-strength plastics and composite materials rather than iron and steel. Composites are stronger and much lighter than steel; when combined with aerodynamically slippery design, they produce a car that, according to Lovins, is "safer, sportier, more comfortable and quiet[er]" than ordinary cars, while costing about the same, producing one hundred times less air pollution, and using five to ten times less fuel. Lovins bases his projections on a "re-invention of the car" that he and his Rocky Mountain Institute colleagues began in 1991. He claims that more than twenty firms (including traditional automakers and others) are pursuing the concept with the institute. Officials I interviewed at Chrysler and GM scoffed that Lovins was vastly overestimating what hybrid technology could accomplish while underestimating how much such cars would cost. Sound familiar?

A second alternative to standard electric cars is fuel cell electric cars. Fuel cells produce electricity by mixing hydrogen and oxygen. There is no combustion, only a chemical reaction whose "exhaust" is mainly water vapor. In its purest form, the process produces no noxious byproducts to cause smog and no CO_2 to spur global warming. But this assumes the fuel cell uses pure hydrogen gas, which is difficult to store and can be dangerous if a tank leaks. If the hydrogen instead is derived from methanol or gasoline, the process will emit carbon dioxide, but in substantially smaller amounts than standard electricity production does. The upshot is a car that produces virtually no smog and much less carbon dioxide, even as it delivers twice the gas mileage.

After years of stonewalling, the world's automakers are at last preparing to bring some of these technologies to market. Compelled by California's zero-emissions mandate, General Motors began offering the EV1 electric car in December 1996. The move that truly shook the industry, however, came in October 1997. At the Tokyo Motor Show, Toyota unveiled several low-emissions, high-efficiency cars and announced plans to have them in showrooms within a matter of weeks. By December, Japanese shoppers could purchase a Prius, a small car that got seventy miles to the gallon, for the very competitive price of $17,000. The race was on. Also in December, Ford announced it would join Daimler-Benz of Germany in producing fuel cell electric cars; Daimler-Benz envisioned mass-producing one hundred thousand such vehicles a year by 2005. In January 1998, General Motors exhibited prototypes that would get sixty to eighty miles per gallon and said it hoped to have a hybrid car ready for production by 2001 and a fuel cell electric car by 2004. Chrysler, too, joined the procession, promising to develop a fuel cell vehicle whose hydrogen would be extracted from gasoline purchased at conventional service stations.

How many people will actually buy high-efficiency vehicles remains to be seen, but the appearance of such cars in the marketplace is a big step forward. The credit belongs in no small part to the marketplace itself. The Big Three U.S. automakers remembered how Japanese and European automakers had captured U.S. market share in the 1970s by selling fuel-efficient cars after the oil shock, and they knew they could not afford to be left behind a second time. GM's John Smith admitted as much. It was Toyota's maneuvers at the Tokyo Motor Show, he said, that convinced him GM had to be a leader in high-efficiency, low-emissions cars.

Yet market forces alone do not explain industry's abrupt about-face. No one was impertinent enough to mention it at the time, but the industry's sudden embrace of high-efficiency cars implicitly confirmed what activists had long charged: such cars could have been produced years earlier. After all, new car models do not appear in showrooms overnight—the process of planning, designing, testing, and manufacturing them often takes five years or more. Manufacturers had been producing prototypes of four- and five-passenger cars that got between 67 and 138 miles per gallon as far back as the mid-1980s. If market forces alone could have put such cars in showrooms, they would have done so.

was no consumer demand

What changed in the 1990s was the scientific and political environment in which automakers operated. Above all, the greenhouse effect emerged as an overriding environmental concern. As consensus grew among scientists, citizens, politicians, and such key industries as insurance that global warming was a grave problem that needed to be dealt with, it became clear that significant changes in government policy and commercial behavior would be required, and probably sooner rather than later. In a follow-up to the Earth Summit of 1992, Japan was preparing to host an international meeting in Kyoto in 1997 at which nations would agree to binding reductions in greenhouse gas emissions. The U.S. government was proposing only a return to 1990 emission levels by 2012, but the Europeans were urging cuts of 15 percent. In any case, the run-up to Kyoto signaled that governments were about to order policy shifts of some sort and that the twenty-first-century marketplace would value environmentally friendly products more highly. It only made good business sense for the world's automakers to prepare for these new conditions. If Japanese and European carmakers recognized the approaching policy shift sooner than their American competitors did, it was perhaps because the public dialogue about global warming was far more clear-eyed in Europe and Japan than in the United States, where the discussion still focused on whether the problem actually existed.

How much the new auto technologies end up changing the global warming equation depends on which technologies triumph and when. At the moment, roughly half of all new motor vehicle sales in the United States are of trucks and sport utility vehicles. The fuel efficiency of these vehicles is very poor—sometimes barely ten miles per gallon—but their profit margins are spectacular, so manufacturers are far more interested in promoting them than high-efficiency vehicles, thus retarding progress. In any case, even a confirmed technological optimist like Amory Lovins cautions that high-efficiency cars cannot solve what he considers the basic problem posed by the car: "too much driving by too many people in too many cars." Hybrids could even make such problems worse, he warns, by making driving a relatively environmentally harmless activity with essentially free fuel, thereby encouraging people to use cars more. In a world where eight million Los Angelenos and a billion Chinese would be driving hybrid cars, Lovins told me, "we wouldn't run out of oil or air, but we'd surely run out of roads and patience."

This returns us to the idea of encouraging humans to use cars less, a task that becomes all the more urgent if the development of hybrids does not proceed as rapidly as Lovins expects. Recall Jean-Francis Held's caveat "if the government is responsible, there are many ways to domesticate the car." Since the car's dominance depends on massive public subsidies, the most powerful step that governments could take would be to redirect those subsidies to trains and other environmentally superior forms of transit. Related policy options include limiting auto access to city centers, halting new road construction, upgrading mass transit, discouraging suburban sprawl to make cars less necessary in the first place, and initiating public education programs to urge less driving.

There are heartening developments. Germany plans to invest more in its rail system than its roads over the next ten years. Activists in Great Britain have blocked numerous national highway projects. Cities as diverse as Portland, Oregon, in the United States and Curitiba in Brazil have demonstrated that intelligent planning of mass transit and land use can produce pleasant, prosperous cities oriented around people rather than cars. Cars sold in California are now the cleanest internal combustion vehicles in the world. The European Union in 1998 finally outlawed leaded gas entirely and mandated tough new air pollution regulations. But these and other examples remain isolated cases, overwhelmed by an avalanche of public policy and mass behavior heading in the opposite direction. For example, California's cleaner engines have made Los Angeles's air the cleanest it has been in four decades, but that air remains the most polluted in the United States. Around the world, most governments still lopsidedly favor cars and highways over trains and mass transit, even when their own experts are advising them differently. "I see people being quite reactive," Carter Brandon, a senior economist in the Asia division at the World Bank, told me. "All the mega-cities in India are failing to deal adequately with the problem of traffic. Yet the bank hasn't seen fit to invest in mass transit [in Asia] because they don't see it being able to pay for itself. I find that a dumb argument. The subway systems in New York or Paris or London don't pay for themselves either, but that doesn't mean they're not good for the cities."

Given the imposing political muscle of the automobile lobby, most governments will oppose the lobby only when emboldened by firm, sustained citizen pressure. This raises the other big question about reducing auto use: how willing are people to accept even minor cut backs in their own driving? The answer, so far, seems to be not much. Earlier, I likened humanity's attachment to cars to an addiction to ciga rettes, but weaning humans away from cars will be difficult precisely because of the way the two addictions differ: driving a car offers not just fleeting sensual pleasure but real material advantages. Of course, the drawbacks of cars will become increasingly apparent as more time passes; traffic jams in particular are bound to get much worse. But what I witnessed in Bangkok, Athens, Los Angeles, and most other places I visited makes me doubt that even killer congestion will convince most people to drive less; in the minds of most drivers, traffic jams are always the fault of *other* people.

Twentieth-century humans have much for which to thank the automobile. Just as my young Ugandan friend praised the dam on the Nile for the electricity it provided, so the car "has done much good for us," giving us mobility at previously unimaginable speeds and stimulating overall economic growth. But it may be time for humans to retire the car before it retires us. At the very least, the car must be fundamentally transformed, becoming many times more efficient than it is at present, if humans are to coexist with it. And even that solution reflects a technocratic mind-set that assumes the answer to technology's failures is more technology, a perspective whose shortcomings have been made clear time and again in human history.

Meanwhile, the human appetite for cars is deeply rooted and far from satisfied; all over the world, it seems we can't say no to cars. Everyone who can afford a car has one, and so do many people who can't. Owners don't just take their cars for granted, they can't imagine how they ever lived without them. In Swiss writer Friedrich Durrenmatt's telling phrase, the car has become "a high-quality artificial limb," irreplaceable and inseparable from modern humans.

I would take Durrenmatt's analogy a step further. Think back to the evolution of finches on the Galápagos Islands and how, from one generation to the next, they took on such features as longer or deeper beaks. Does not humanity's tightening attachment to the automobile over recent generations (especially in the United States, where having a driver's license has become virtually a requirement of citizenship) bear some resemblance to the finches' evolutionary process? Of course, the finches were responding to weather conditions beyond their control, while humans choose cars precisely in order to gain control over their environment, so the trigger mechanisms are different. But the results might not be.

Just as Galápagos finches with large beaks flourished until floods made it impossible for them to find enough seeds to eat, so humans who depend on cars could find themselves in peril if a similar interruption in the oil supply makes driving problematic. (This is no remote possibility. *Scientific American* reported in 1998 that technical obstacles to oil production would bring the era of cheap, abundant oil to an end within the next decade.) Moreover, while a finch's beak size affected only its individual fate, the use of cars by even one-fifth of humanity throws into question the environmental prospects of the entire species, if only because of the dangers of global climate change.

Acquiring the artificial limb known as the automobile may turn out to be the most important evolutionary adaptation twentieth-century humans have made. Whether they will survive the environmental side effects cannot yet be predicted, but only the reckless would disregard the many threatening clouds on the horizon. The question will be answered by the same eternal forces that determined which Galápagos finches lived and died: the interaction of natural selection with the vagaries of nature and the behavior of the species in question. Droughts and floods were what did in the finches unlucky enough to have the wrong beak sizes. If global climate change continues to intensify, humans will experience plenty of droughts and floods themselves. But surely this is coincidence, a mere cosmic joke at our expense.

To the Nuclear Lighthouse

> Nuclear weapons are irrational devices. They
> were rationalized and accepted as a desperate
> measure in the face of circumstances that were
> unimaginable. Now as the world evolves rapidly,
> I think that the vast majority of people on the
> face of the earth will endorse the proposition
> that such weapons have no place among us.
> —GEN. GEORGE LEE BUTLER,
> former commander, Strategic Air Command

As an example of the environmental contradictions of twentieth century industrial "progress," the city of Leningrad is hard to beat. For decades, Leningrad has ranked as one of the world's great cultural centers, boasting some of the finest art, theater, dance, and architecture to be found anywhere. But when I reached Leningrad in July 1991, I quickly learned that its water was absolutely unsafe to drink, thanks to a witch's brew of human and industrial waste that poured constantly into the Neva River. Much of the industrial waste was military-related; more than 70 percent of the factories in the Leningrad area belonged to the Soviet military, according to Alexei Yablokov, deputy chair of the Supreme Soviet's environmental committee and one of the nation's leading environmentalists. "The military is like a law unto itself," Yablokov told me. "It

No one to blame to

can pollute rivers and hold underground nuclear tests and nobody is held responsible because the rest of the government cannot discipline the military."

I arrived in Leningrad six weeks before the military-led coup against General Secretary Mikhail Gorbachev that spelled the end of the Soviet Union. Days before my arrival, the citizens had voted to restore their city's original name of St. Petersburg, but the margin was close: 56 percent in favor, 44 percent against. Many senior citizens had clung to the old name in honor of loved ones who died defending Leningrad against Hitler's army during the ghastly nine-hundred-day siege of World War II, when an estimated two million Russians perished after being reduced to eating glue and sawdust. Advocates of the name change, on the other hand, saw the new name as a blow against the still ruling Communist Party and a reassertion of the glories of Peter the Great, the czar who had made his namesake city Russia's window onto Europe in the early 1700s.

The physical appearance of the city seemed as divided as the opinions of its citizens. Leningrad resembled nothing so much as a classic Rolls-Royce that had deteriorated into a rusty, dented, filthy shadow of its former self. The grandeur of Peter's Winter Palace along the Neva River, the magnificent holdings of its Hermitage Museum, the stately stone buildings and canals that recalled Paris and Amsterdam all harkened back to the prosperous St. Petersburg of old. I was lucky enough to arrive in high summer, the season of white nights, when even at midnight there was enough light to read a newspaper outdoors. One night at about 11:30 I strolled down the main street, Nevsky Prospekt. The light was soft and luminescent, bright enough to appreciate the lines and proportions of the palatial buildings on both sides of the street yet dark enough to obscure the gaping holes, crumbling facades, and lack of paint that marred their elegance. For one enchanted evening, I felt I had been transported back to prerevolutionary St. Petersburg.

Yet the shabbiness and hard times of present-day life were inescapable. I was met at the train station by Vlad, a Russian photographer I had known in San Francisco who would be serving as my interpreter here. His friend Alex had access to a run-down, old jalopy (the windows no longer rolled down, the seats had lost their springs), and as we drove across town the city itself seemed in no better shape than the car. Buildings were caked with so many years of dust it was hard to tell what their original colors might have been. The streets were pocked with huge potholes and lined on either side by weeds and waist high grass. Vlad took me to his parents' place, where I would be taking over his old room for a few nights.

His parents were factory workers who lived in a nine-story apartment building in the north of the city. The downstairs entrance was a plywood door that opened into an unlit vestibule that smelled powerfully of mildew and urine. The elevator was suffused with the same odor, but Vlad apparently no longer noticed it. During the very slow ascent of the elevator, it suddenly stopped with such a crash I was sure it had broken and left us stranded. But again, Vlad, Alex, and their friend Leonid were nonchalant. Vlad simply heaved open the elevator and we stepped into another dark, dank hallway, down which lay the door to the apartment.

Inside was cozier, thanks to a couple lovely old pieces of furniture and the great warmth of Vlad's folks, especially his father, who informed me with impassioned hand gestures that under no circumstances was I to drink water from the tap. This I knew already—water pollution was the main environmental story I planned to investigate in Leningrad. Nevertheless, when he filled a glass from the tap, it was sobering to see the water's greasy texture and smell its metallic scent. Vlad's father emphasized that I should drink only from the green pitcher in the icebox, which contained water that had been boiled for ten minutes. He added that the building had been completely without water from five until nine o'clock that evening and, worse, that there would be no hot water at all until September 1, two months away. Also, the phone was out again. "Welcome to Soviet Union," Vlad said to me with a smirk. "Is interesting country."

On our way downtown the next morning, Leonid and I stopped into a neighborhood food store. Dust coated the front door and windows so thickly one could not see through them. Inside, most shelves were empty; the only product available in any quantity was bread. In one corner were half a dozen five-kilogram bags of potatoes, one of which Leonid purchased. That afternoon, while walking the city, we passed through a private vegetable market with ample, fine-looking produce, but prices were five times the normal rate and Leonid was too proud to let me buy him anything. A few blocks later, we found ourselves on a street where peddlers stretched on for an entire block. Mostly desperate pensioners, they had on offer a pathetic range of items: pencils, an old hallway mirror, some empty bottles, a pair of very worn lady's shoes. Capitalism had only begun to arrive in Russia, and already the bottom was falling out for the lower classes.

I spent a week in Leningrad interviewing environmentalists, city council members, engineers, scientists, high party officials, and average citizens about the state of Leningrad's water and its larger ecological situation. Everyone knew that the city's water was unsafe; the media had widely reported it. But except for the boss of the local Pepsi Cola bottling plant, who told me he regularly took colas home for his family, most people drank the water anyway. "What else can I do?" asked Dimitri, a twenty-year-old student of English with curly dark hair and a bright, ambitious intelligence. "It's the only water we have. I try to boil it first, but that's not always possible." And even boiling the water eliminated only bacteria, not industrial toxins.

Leningrad drew its water from the Neva River, which was fed by Lake Ladoga, approximately fifty miles to the north. Ladoga was the largest lake in Europe. In olden days its purity was so renowned that sea captains would insist on stowing Ladoga water aboard before long journeys. Now, however, the lake was ringed with scores of paper mills and other factories that discharged vast amounts of heavy metals, acids, and chlorine. The Neva was further polluted while passing through Leningrad by the city's approximately two thousand factories, only 10 percent of which treated their waste before discharge. Human waste from the hundreds of thousands of households in Leningrad also poured into the Neva, generally without benefit of prior treatment. Drinking water was treated before being distributed to homes and offices, but with limited effect. The Neva still contained concentrations of *olgino* (a stomach bacteria) that were ten thousand times higher than the legal limit, according to a study by the city council.

Compounding all these problems, a massive dam was being built across the Gulf of Finland twelve kilometers from downtown Leningrad, supposedly to protect the city from floods but also to provide a ring road for auto traffic. A colossal boondoggle of centralized planning that originated during the Brezhnev era, the dam would interfere with the Neva's traditional self-cleaning method of exchanging water with the gulf. Environmentalists pointed out that the dam would act like a cork in a bottle, stopping up the Neva with its pollutants and rendering the gulf a fetid swamp.

Despite its poisonous water and highly contaminated air, Leningrad did not rank among the ten most polluted cities in the Soviet Union. Competition for that honor was stiff in a country where two-thirds of the drinking water did not meet health standards, air pollution in over one hundred cities exceeded legal limits by a factor of ten, a chemically saturated river somewhere in the country burst into flames once a month, and 20 percent of the population (about forty million people) lived in areas that scientists had labeled zones of ecological "conflict," "crisis," or "catastrophe."

After a week in Leningrad I took the train to Moscow, where I stayed with a friend of Vlad's named Kiril. A twenty-five-year-old former prison inmate who thought nothing of beating up his downstairs neighbor to get him off their shared telephone line, Kiril happened to be the grandson of the man who served as Josef Stalin's ambassador to the United States immediately after World War II. One day, when I tried to present Kiril with the U.S. $20 he said he needed to pay for a license to marry his live-in girlfriend, his pride was so wounded that he angrily turned on me: "Take back the money, Mark, or you will become my enemy!" Still, he generally liked me, and he even agreed to let me use the English typewriter he had inherited from his grandfather, an ancient Royal manual whose *y* and *z* keys were transposed but otherwise functioned well enough. Typing up my notes one morning while Kiril and a buddy were out muscling in on a land deal, I wondered about the messages that had been typed on this machine decades before by the ambassador, an official who must have been privy to some of the most sensitive aspects of U.S.–Soviet relations at the dawn of the Cold War. If only typewriters could talk!

Part of the reason Kiril treated me decently, I realized, was that I represented a possible entrée to the United States, a country he yearned to visit. He was ashamed of the Soviet Union, a place he ridiculed as backwards, ugly, and poor. Alas, this was not an entirely unfair characterization. This was my first visit to the USSR, and I was frankly astonished. Combine the appalling environmental degradation with a stagnant economy, poor living standards, and the shoddy technology everywhere on display, and it was difficult to regard the Soviet Union as much of a superpower. Throughout my five weeks of travel there, I often found myself thinking, "This is the place we were supposed to be so afraid of during the Cold War?"

Yet Americans had legitimate reason to fear the Soviet Union during the Cold War, for Soviet leaders, like American leaders, had their hands on the most deadly technology of the twentieth century. If the automobile was the most economically important technology of the century, nuclear fission was the most important technology, period, because it raised the question of whether there would *be* any human life beyond this century. Unlocking the atom's secrets was arguably the single most fateful step *Homo sapiens sapiens* had taken in their two-million-year pursuit of technological mastery over the natural world. By discovering how to produce nuclear reactions, humans were exploiting the very forces that generated sunshine and made life on earth possible in the first place. These were powers that earlier humans had ascribed to gods, a point not lost on the atomic bomb's chief designer, physicist Robert Oppenheimer. At the moment the first test bomb (or "gadget," as its creators called it) exploded above the New Mexico desert on July 16, 1945, a line of Hindu scripture flashed through Oppenheimer's mind: "I am become Death, the shatterer of worlds."

Within a month, Oppenheimer's creation had incinerated two Japanese cities and killed hundreds of thousands of civilians. But Hiroshima and Nagasaki were by no means the only cities devastated by the technology that, in Einstein's famous phrase, changed everything except our way of thinking. As I was soon to see, on the western edge of Siberia was another city that carried the scars of its nuclear past—a city called Chelyabinsk.

The express train from Moscow took thirty-six hours to plod across the featureless Russian plain to Chelyabinsk, a dusty industrial city with a million inhabitants. I awoke a few hours before arrival, when the train finally began lumbering up the shallow inclines of the southern Ural Mountains. Green, stony hillsides dotted with graceful white birch trees broke the visual monotony for the first time since Moscow, but the relief was temporary. Chelyabinsk lay just over the rise, at the edge of the vast steppes that stretched on to the Pacific.

Fifty miles north of the city was an industrial complex whose Cold War code name was Mayak. Translated, Mayak means "lighthouse"—an ironic name, considering the place had not existed on Soviet maps for more than forty years. (Chelyabinsk itself was still officially closed to outsiders when I got there in July 1991.) Covering almost eighty square miles, the Mayak complex had recently been called "the most polluted spot on earth" by a team of visiting foreign scientists, a judgment Mayak officials did not dispute. Built by forced labor shortly after World War II, Mayak had been the Soviet Union's primary nuclear weapons production facility from 1946 until November 1990, when the last of its five plutonium reactors was shut down. Mayak was, in short, the heart of the Soviet nuclear production apparatus.

As such, Mayak was the site of perhaps the biggest nuclear catastrophe in history after the Hiroshima and Nagasaki bombings. There had been three nuclear disasters at Mayak whose damages were comparable to, and probably worse than, the reactor meltdown in 1986 that made Chernobyl a household name around the world. The difference with the Mayak disasters was that they never became media events. On the contrary, they were kept secret—not only from the outside world but from the Russian people, including hundreds of thousands of local residents who were exposed to massive amounts of radiation. In a striking case of Cold War duplicity and doublethink, the news from Mayak was suppressed by both the KGB and the CIA, each of which apparently feared an informed populace as much as it feared the enemy arsenal. (The CIA learned about the accidents in the course of normal intelligence gathering but declined to publicize them.) Thus, when I reached Chelyabinsk in 1991, the three Mayak nuclear disasters still remained largely unknown to all but a handful of international nuclear policy experts.

Astonishingly enough, the first Mayak disaster was not an accident at all but the result of deliberate policy. From 1949, when the Mayak complex produced the Soviet Union's first nuclear weapon, until 1956, Mayak officials poured their nuclear waste directly into the nearby Techa River. Tens of thousands of people living downstream received average doses of radiation four times greater than those subsequently received at Chernobyl. For the twenty-eight thousand people most acutely exposed, average individual doses were fifty-seven times greater than at Chernobyl. Nevertheless, only seventy-five hundred people were ever evacuated from their homes, and people were not forbidden to use the river water until 1953, four years after the contamination began.

The second, and most terrible, Mayak disaster took place on September 29, 1957, when a nuclear waste dump exploded, spewing seventy to eighty metric tons of waste into the sky. The waste facility had been constructed in 1953 as an alternative to more river dumping. When its cooling system malfunctioned, the waste began to dry out and heat up, eventually reaching the unearthly temperature of 350 degrees Celsius. The resulting explosion was equivalent to seventy to one hundred tons of TNT—enough to blast a thick concrete lid off the tanks and hurl it twenty-five meters away. The total amount of ejected radioactivity measured twenty million curies—ten times more than had already been dumped in the Techa River. Ninety percent of the radioactivity fell immediately back to earth, but the remaining two million curies formed a plume half a mile high that spread across the Chelyabinsk region, severely contaminating air, water, and soil. All the pine trees in a twenty-square-kilometer area died over the next eighteen months. Approximately 272,000 people were exposed to average doses of 0.7 rems of radiation, the same amount that 750,000 Chernobyl victims would experience in 1986.

The third Mayak disaster occurred in 1967, and again nuclear waste was the culprit. In 1951, after Mayak officials realized they could no longer dump waste in the Techa River but before they built the storage facility that would explode in 1957, they began pouring waste into Lake Karachay, a natural lake within the Mayak complex; since Karachay had no outlets, this measure, it was assumed, would keep the waste from contaminating the regional water system. However, in 1967, a cyclone swept across the drought-exposed shores of Lake Karachay and whirled its deadly silt high into the air and across the surrounding landscape. Five million curies of radioactivity were dispersed over fifteen thousand square miles; nearly half a million people were affected.

My guide in Chelyabinsk was Natalia Miranova, a tenacious, red-haired woman in her forties who had fought to win medical treatment and protection from further danger for her neighbors; her efforts had recently won her election as the people's deputy to the regional Supreme Soviet. I had met Natalia at an academic conference in the United States a few months before; now, she had kindly come to meet Vlad and me at the Chelyabinsk train station. Joining her was Valodya Ishkvatov, whose crinkly eyes and flat, honey-colored face reminded me that I was now on the Asian side of the Soviet Union. Valodya served as our driver over the coming days, and like many other locals, he had been personally affected by the Mayak disasters.

When the Mayak waste dump exploded in 1957, Valodya was a boy of eight, living along the Techa River with his parents and three brothers and sisters. His father worked at a large orchard next to the river where pears and apples were grown. Valodya's family was not evacuated until a year after the explosion, and even after evacuation his father kept working in the orchard; every day, he walked the five miles from the nearby village where the family had been forced to take shelter in a farmer's barn. The orchard was kept in production, and its fruit sold throughout the Soviet Union, until 1964, when the government ordered the trees to be burned.

Valodya, who now looked perhaps fifteen years older than his actual age of forty-one, said that all six members of his family were among the sixty-six thousand victims of the Mayak disasters upon whom the government had kept health records. But he had no confidence in those records. Many times that number of people had actually been irradiated, he pointed out, and he added that his own family's case illustrated how the official records substantially underestimated the number of victims.

"Neither I nor my older brother and sister were willing to undergo the spinal operation that is used to test for radiation sickness," he told me through Vlad. "When it's done improperly, you never walk again, and none of us trusted the local doctors enough. So in our records each of us is listed as not having radiation sickness, even though each of us has had all kinds of symptoms. I myself had to have a growth [the size of a golf ball] removed from my neck a couple of years ago." His mother and baby brother suffered still worse fortune. The brother was cursed at birth with an overlarge head and shrunken chest, and since then he had been plagued ceaselessly by severe migraines and arthritis. The mother died at sixty after suffering for years from a variety of ailments associated with radiation sickness, including a weakened heart, high blood pressure, and fatigue.

The complete death toll of the Mayak nuclear disasters will never be known, and bookkeeping errors are but the most banal reason why. More far-reaching and despicable was the outright deceit practiced by Soviet authorities. Until 1989, Chelyabinsk health officials were prohibited from even acknowledging the existence of radiation sickness, much less admitting that it had been killing local people for forty years. Instead, they had to diagnose patients as suffering from ABC disease, a code name handed down from the Ministry of Health in Moscow that carried the grotesque translation "weakened vegetative syndrome."

"It was unpleasant, but I had to conform," said Dr. Mira Kossenko when I asked how she felt about lying to her patients for so many years. A slim, serious woman who now headed the clinical department at the Institute of Biophysics in Chelyabinsk, Kossenko added, "When they asked me what was wrong with them, I simply told them there was something wrong with their blood. It was a complex moral situation. We did our best to treat the people who came to us. But talking about radiation sickness was considered to be revealing a state secret, and I would have served seven years in prison."

Instead, other innocents were conscripted. On my first day in Chelyabinsk, Natalia took me to the local children's hospital. Nearly all of the approximately thirty children on the leukemia ward were bald, thanks to the radiation therapy that, in a perverse twist, was now being applied in a last-gasp attempt to save their stricken bodies. The kids ranged in age from fifteen down to one. At night, their mothers slept beside them on cots. When a dozen of the mothers gathered in the playroom late in the afternoon to speak with me, the mother of one sad-faced, heavyset girl could not stop sobbing. Her daughter, who looked about ten, reached over and stroked her mother's arm to comfort her. This unleashed a deep, aching wail from the mother that drove her from the room. The mother's peers looked on with sympathy, dread, and a couple forced, painful smiles. They knew what the children did not: the doctors expected 75 percent of these children to be dead within five years, and some of them much sooner than that.

Not all these children's cancers were necessarily the result of atomic contamination, of course. But the doctors had concluded that some, and perhaps many, were. A study that Mikhail Gorbachev had ordered of the ecological situation in Chelyabinsk (which I will refer to as the Gorbachev report) concluded the same. Nevertheless, for three years in a row Moscow had told the doctors that there were no funds to buy even such basic medical equipment as blood cell separators, which would cut the leukemia ward's death rate significantly.

"The problem is that all the assistance goes to Chernobyl," Dr. Leva Zhukovsky told me. "Our mothers understand better than anyone the pain of the mothers in Chernobyl whose children have leukemia. We only want the public to know that our children are dying, too."

we get ourselves into situations we cannot fix. we have the technology to create disasters but we don't have the tech to fix them

Soviet authorities tried to cover up the Chernobyl accident as well. Just as in Chelyabinsk, doctors were ordered not to diagnose Chernobyl patients as suffering from radiation exposure; the total amount of radiation Chernobyl unleashed was also considered classified. The deception began during the very first moments of the accident, when engineers who reported that reactor number four had suffered a catastrophic meltdown were ignored by the Chernobyl plant's director, who went on to assure Moscow that the situation was under control. (The engineers who investigated the blue-hot remains of the reactor were rewarded for their honesty with lethal doses of radiation that killed them within ten weeks.) The blast had released fifty tons of radioactivity into the sky. Although the fallout drifted across western Europe and eventually was detected as far away as Point Reyes, California, the bulk of it contaminated fifty thousand square miles of prime farmland in Belarus, Ukraine, and Russia. Five million people were exposed to dangerous levels of radiation, but only 135,000 were evacuated from a so-called exclusion zone that extended in a thirty-kilometer radius from the plant. Food grown in the irradiated areas continued to be consumed as well.

In Moscow, some top officials apparently wanted to stonewall the outside world, but that was impossible. Radiation monitors in Sweden and elsewhere had registered alarming readings after the blast, and foreign governments were demanding answers. Gorbachev, who had announced his policy of glasnost only three months earlier, now had to live up to it; sixteen days after the accident, he went on television to admit that "a misfortune . . . has befallen us" at Chernobyl. But even

Gorbachev's remarks grossly understated the crisis; he was relying on reports from officials at the site and throughout the Soviet system who were continuing to censor information. Andre Pralnikov, a journalist I met in Moscow who had infiltrated the Chernobyl site days after the accident, told me that cleanup specialists told him that there were "at least three Hiroshimas" worth of radioactivity inside the ruined plant. That estimate turned out to be much too low; Chernobyl actually released about two hundred times as much radiation as Hiroshima and Nagasaki combined. Pralnikov did not include such information in the reports he smuggled out of Chernobyl, however, "because it would have been no use. My editor [at *Izvestia*] would have taken it out, because he would have known that, if he didn't, the official censor above him would have."

Although they could not entirely suppress the truth about Chernobyl, Soviet authorities did succeed in sowing long-lasting confusion about the accident's true scope and consequences. They had help, especially from the pronuclear International Atomic Energy Agency (IAEA) of the United Nations. The IAEA's report on Chernobyl, published in 1991, declared that the accident had not caused any physical health problems for the local population, only psychological ones. Critics pointed out that the IAEA had reached this astonishing conclusion on the basis of information provided almost entirely by the Soviet government; among its most egregious errors, the agency had simply ignored the two groups of people exposed to the largest doses of radiation—the tens of thousands of people who had lived within the plant's thirty-kilometer "exclusion zone" and the eight hundred thousand so-called liquidators who cleaned up after the accident.

Nevertheless, the same argument—the danger of radiation is all in people's heads—resurfaced five years later on the front page of the *New York Times* in a story by reporter Michael Specter titled "10 Years Later, Through Fear, Chernobyl Still Kills in Belarus." One had to wonder whether Specter and his editors read their own newspaper. After all, the *Times* had recently published two stories reporting that the World Health Organization, after a more independent and rigorous investigation than the IAEA's, had reached very different conclusions about Chernobyl's effects. The WHO found that thyroid cancers among children in Belarus had become 285 times more common in the years after the accident. Illnesses of all kinds had increased 30 percent, and 375,000 people remained displaced or homeless. Nor was it clear how much additional suffering lay ahead, for no one was bothering to monitor the health of the eight hundred thousand workers and soldiers who, as liquidators, had endured the most concentrated exposures to radiation. Moreover, many kinds of cancer, such as leukemia, can take years to manifest; the very high incidence of thyroid cancer (a rapid metastasizer) suggested that the ultimate number of deadly cancers could likewise be very high. Dr. John Gofman, professor emeritus of biophysics at the University of California at Berkeley, estimated that Chernobyl would eventually cause between 50,000 and 250,000 cancer deaths in the former Soviet Union, plus an equal number in the rest of the world.

So it was no wonder that the world paid a lot of attention to Chernobyl. By contrast, while Chelyabinsk had also suffered grievously, its travails were too little known to attract much international concern, a fact that exasperated some of its defenders. "Please do not compare Chelyabinsk with Chernobyl, because [Chelyabinsk] is a much different and far worse problem," pleaded Alexander Penyagin, the people's deputy who represented the Chelyabinsk region in the national Supreme Soviet in Moscow, where he showed me a draft of the Gorbachev report. "Because of glasnost, nobody could sweep Chernobyl under the rug. But the disasters at Chelyabinsk were top secret until just two years ago [1989]. They continued for many more years and they released far more radioactivity than Chernobyl did."

Penyagin went on to assert that the Mayak disasters were one hundred times worse than Chernobyl, hyperbole that seemed beside the point. There was quite enough suffering to go around in both Chelyabinsk and Chernobyl without having to pit one against the other. According to the Gorbachev report, Chelyabinsk was the cancer capital of the entire Soviet Union—no small achievement, given the dreadful state of Russia's environment and public health. Again, the Mayak disasters were not the only reason. Chelyabinsk was also a major agricultural area with extensive chemical use, as well as an industrial nerve center with a long military history. The region had produced weapons for Russian leaders since the time of the czars, and it played an especially critical role during World War II. In 1941, after the Germans overran the Soviet

Union's western border, the Soviets transferred their entire metallurgy industry—factory by factory, machine by machine—from Ukraine to the relative safety (behind the Urals) of Chelyabinsk. But the factories had not been improved upon since, and the environmental consequences were devastating.

Since Chelyabinsk factories had no air purification filters, they had released some 391,000 tons of pollutants in 1990, giving Chelyabinsk some of the most polluted air in the Soviet Union. Drinking water also contained "very high levels of pollution," the Gorbachev report stated—five to twenty times as much iron and forty to sixty times as much copper as it should have. It is hard to be healthy under such conditions, and the people in Chelyabinsk were not. You could see it in their faces—drawn, pasty, permanently fatigued—and you could track it in the health statistics. Both morbidity and mortality rates jumped during the 1980s. Growth in diseases of the blood circulation system increased 31 percent. Bronchial asthma increased by 43 percent; congenital anomalies, by 23 percent; and gastrointestinal tract illnesses, by 35 percent.

Even the bland bureaucratic language of the Gorbachev report could not wholly mask the severity of the crisis in Chelyabinsk: "An extremely unfavorable ecological situation has developed, which is made worse by the lack of proper medical services. An especially critical situation exists in the zones of ecological tension" (extraordinary phrase!) "where about 80 percent of the population reside."

I toured one of these zones of ecological tension on my second day in Chelyabinsk. We headed out early in Natalia's creaking, rust-holed Lada, Valodya at the wheel. Leaving the city behind, we passed spacious green cow pastures, golden cornfields swaying in the breeze, and countless stands of the shining white birch trees that dominate the Ural landscape. Once we turned off the main highway, we had to share the road with horse-drawn hay carts and slow down to avoid geese crossing our path. After about an hour, we pulled up to the seven-thousand-hectare Neva state farm, which produced meat, milk, potatoes, and feed corn. We were greeted by Nikolai Chvelev, a squat, energetic man with a farmer's ruddy complexion and an eyewitness's memories of what it felt like to live along the Techa River in the 1950s.

Nikolai Chvelev was now the top man at the Neva farm, but in 1954 he was a twenty-one-year-old soldier who had fallen in love with a local girl while stationed in Chelyabinsk. He told me he and his fellow recruits were brought to the Techa one day in 1954 and told not to swim in it. But because they were not told why the river was off limits, and because there did not look to be anything wrong with the water, they sometimes disobeyed these orders, especially on hot summer days. Local people also continued using river water for drinking and cooking. Not until the second disaster, the waste dump explosion in 1957, did people begin to suspect that something was seriously wrong.

"A lot of rumors circulated, but people were afraid to ask too many questions. At the time, you could get thrown in jail for that," Chvelev recalled. "My brother-in-law worked for the police, and he told me privately what little he knew: that some kind of accident had happened at the complex. Only then did we stop swimming in the river. For two weeks after the explosion, the water in the river was black, but after that it became clear again. The dairy where my wife worked was kept in operation until 1959. She and her coworkers didn't know what 'radiation' meant, and they weren't told anything by the authorities, so they saw no reason to stop producing milk, except on the days when the water was black."

After a hearty lunch of beef, noodles, and vodka (!) in the Neva communal dining hall, Chvelev and I hopped in his truck and headed for a far corner of the farm, where the Techa flowed past. It was a glorious summer afternoon, breezy and warm, and as we bounced across sun-splashed fields brimming with pink and white wildflowers, all notions of ecological disaster seemed impossibly remote. Radioactivity, after all, cannot be seen, felt, smelled, tasted, or heard; although I knew better, it seemed inconceivable that something so deadly could be lurking amid so much natural beauty.

Pulling to a halt a quarter mile from the river, Chvelev estimated that we were about eighteen miles downstream from the site where nuclear waste was originally dumped in the Techa. At the truck, the dosimeter Natalia Miranova had brought registered approximately 25 microrontgens, close to the normal background level of 20 micro-rontgens. I hiked ahead with Vlad to scout photo opportunities while Natalia changed her shoes for the walk to the river. A couple of minutes later, I heard her behind me calling out numbers from the dosimeter. As Vlad translated them, they rose steadily the closer Natalia got to the water: "61, 64, 67, 64, 73." By the time she stood a few feet from water's edge, the dosimeter read 98. She reached down and placed it at the very edge of the water and again we watched the readings climb—first to 120, then 140, 160, and finally to a peak of 221.

That day, ten times the normal background radiation level seemed very high to me. But the next day I traveled farther downstream, to the village of Muslyumova, and suddenly a reading of 221 seemed tame indeed.

Muslyumova lay twenty-two miles downstream from the Mayak complex as the crow flies, fifty miles as the fish swims. On our way there, Valodya pulled the car to a stop about half a mile short of the village, next to a pasture where cows were lazily munching grass in the sun and white geese were splashing in and out of the river. At water's edge, the dosimeter read 445—twenty times the normal background level.

The road into Muslyumova was little more than a dirt path. We stopped on a bluff and looked across the Techa, one hundred feet below, to rows of low, wide houses on the other side of the river. A barbed wire fence, its rusted strands disintegrated into a tangled mess, was strung along the edge of the bluff, a pathetic remnant of the official effort to dissuade residents from going near the river. Peering through the fence, I saw a boy of nine or ten wading into the river from the far bank, fishing rod in hand.

Natalia and I walked down to the water's edge. Here along the river's flood plain, nearly all the dosimeter readings were very high—500s and 600s. Natalia held the device over a piece of dried cow dung. The meter shot up to 850, a reflection of the fact that radioactivity becomes more concentrated as it passes through the food chain.

We returned to the top of the bluff, where I was surrounded by a group of about twenty local residents, mainly women whose small children darted behind their mothers' skirts to stare at the man with the notebook who was, they said, the first journalist ever to visit their village. A young woman of about twenty-five emerged as the group's spokesperson. She delivered spirited answers in a calm but authoritative voice and even spoke a little English; it was no surprise to learn she was one of the village schoolteachers. She claimed that the people of Muslyumova were told how dangerous the river was only a year and a half ago and that this warning came not from local authorities, who continued to insist that the villagers could safely remain where they were, but from local environmentalists and a team of visiting foreign scientists. I inquired about the children's health.

"They often have a hard time holding their pencils," she replied. "They tire very easily and complain of pain in their joints. During the spring and autumn, every fourth or fifth child suffers from chronic nose bleeds. Almost all of them have low red blood cell counts."

A man behind me broke in to say that the village had voted recently in favor of evacuation, but the government had refused to help. Tests last spring, he added, had shown that half the local livestock had leukemia, since they still drank from the river. The teacher explained that a well had been dug two years ago to supply the villagers with drinking water, but it did not provide enough for the livestock as well.

"What makes you think the well water is any safer than the river?" I asked.

She shrugged. "We don't really know. The local sanitary station checks the water, and they tell us it's normal, but they never give us the actual readings, so we don't have much faith in them. Still, we must hope that it is a little safer. What other choice do we have?"

As deadly as the Techa River was, it did not pose the greatest immediate environmental danger in Chelyabinsk. There was a place inside the Mayak complex where an adult male could die from radiation in less time than it takes to read a morning newspaper. If he stood on the shore of Lake Karachay, next to the pipe that had poured hundreds of millions of gallons of nuclear waste into the lake since 1953, he would encounter a "radiation exposure rate [of] 600 rontgens per hour, sufficient to provide a lethal dose within an hour," according to Thomas Cochran, a senior scientist at the Natural Resources Defense Council in Washington, D.C. Cochran had been part of the team of foreign scientists invited to the Mayak complex in 1989, and it was he who wrote that it was "the most polluted spot on earth." Now, in 1991, Lake Karachay remained in perilous condition; indeed, a repeat of the 1967 accident, or worse, could happen at any minute.

Since 1951, Lake Karachay had accumulated an awesome 120 million curies worth of radioactivity and absorbed nearly one hundred times more strontium 90 and cesium 137 than was released at Chernobyl. Moreover, although Mayak officials had assumed that waste dumped in the lake would be isolated from the regional water system, this had not proven to be the case. Dr. Cochran's team documented that 93 percent of the radioactivity in the lake had filtered down into the soil beneath, and 60 percent of it had reached the underlying water table. From there, it had already migrated half a mile away from the lake.

The danger was not only that Lake Karachay's radioactivity would infect the water table through groundwater migration; there was also a danger that the lake would be struck by another natural disaster like the cyclone that caused the 1967 incident. Despite years of decay, the radioactivity remaining in Lake Karachay amounted to seven Chernobyls' worth of strontium 90 and cesium 137. Thus, another cyclone could bring nightmarish results. And the risk was real. Driving back to town one afternoon, Natalia and I came upon an entire grove of flattened birch trees, their splintered trunks a stark testament to the power of local windstorms. In addition, running beneath the Mayak complex were a number of geological fault lines, which could flush irradiated water across hundreds of miles via underground channels during an earthquake.

I wanted to see Lake Karachay for myself, but the Mayak authorities refused to allow me inside the complex. Back in Moscow, Alexander Penyagin, the people's deputy for Chelyabinsk, had secured a promise during a long-distance phone conversation with the Mayak director, Mikhail Fitisov, that I would be able to visit the site. That promise evaporated, however, when Fitisov refused to return my calls in Chelyabinsk. Instead, I had to be satisfied with an interview with Eugene Ryzkhov, who had worked at Mayak for thirty-five years, mainly as an engineer, before joining the public affairs department a few years ago.

Ryzkhov lived up to the stereotype of press flaks the world over: unctuous smile, slippery claims, unswerving devotion to the official line. He blamed the shutdown of Mayak's five production reactors on the political fallout of the Chernobyl accident, denied that residents of the

Techa River area ran a significantly higher risk of getting cancer, and ridiculed as "crazy" the suggestion that contaminated villages like Muslyumova be evacuated. At the end of our interview, I asked him whether, back in the 1950s and 1960s, he and his Mayak colleagues had ever tried to warn the public of the terrible dangers he and I had spent the last two hours discussing.

"No," he said. "We were raised not to do that. We worked at a secret military enterprise. Besides, I was only a rank-and-file engineer."

"Do you regret not having warned people?"

"I think my colleagues and I have our regrets that we were unable to prevent the disasters. But life has taught us a great lesson. And for the last twenty years, the overall radiation level here hasn't gotten worse, it's gotten better."

"But do you regret the deception?"

"It wasn't deception. We obeyed the rules of the system in which we lived. . . . It was impossible to do otherwise without being a dissident. My work in containing the problems was objectively useful to people. I did my duty. I have nothing to regret."

The doctors at the Institute for Biophysics in Chelyabinsk were more troubled by their participation in the Mayak cover-up. Perhaps seeking to exorcise their guilt, Mira Kossenko and her colleagues Marina Degteva and N. A. Petushova had authored a study that is now regarded as one of the definitive analyses of the health effects of the Mayak nuclear disasters. After guiding me through the intricacies of their research, Kossenko and Degteva took me on a tour of their hospital, where patients were housed four to a room on plain single beds. There were none of the tubes, monitors, and other apparatuses that clutter Western hospital rooms, and my escorts were clearly embarrassed to show me lab equipment that, even to my untrained eyes, looked remarkably out of date. Dr. Kossenko explained that the hospital had always been underfunded. "When this [hospital] was built [in 1957], the authorities were reluctant to set up a large facility because they thought local people might draw conclusions about what kind of complex Mayak really was. And of course this was a military secret. . . . I'm only able to talk about this now because the regime of secrecy was lifted two years ago."

The recitation of this history led me to suggest to the two doctors that, until 1989, their institute had engaged in rampant medical deception of the very people it was supposed to be healing. I worried for a moment that my bluntness had insulted them. Dr. Degteva shifted some papers on her desk, then lifted her gaze and looked me in the eye. "*Da,*" she murmured gloomily, before adding, in English, "just like Hanford."

Located in a remote area of Washington State, in the northwest corner of the United States, the Hanford Nuclear Reservation was constructed during World War II to build the world's first nuclear weapon. And Dr. Degteva was right. The parallels between what happened in Chelyabinsk over the years and what happened at Hanford are nothing short of eerie.

In Hanford, too, the immediate postwar climate fostered a mentality in which production took precedence over safety. And so, in 1945, Hanford officials released 340,000 curies worth of radioactive gases into the atmosphere, without warning the local populace, apparently because it was the simplest way to get rid of the waste. Later, Hanford officials also elected to pour nuclear waste directly into the nearest waterway. As a result, the mighty Columbia became the most polluted river in the United States. Soil was also contaminated. According to the Brookings Institution study, *Atomic Audit,* "between 1946 and 1966, in excess of 120 million gallons of liquid wastes were *intentionally* discharged from the Hanford tanks directly to the ground." [Emphasis in original.]

Nor were Hanford officials any more forthcoming about the risks of their secret actions than their counterparts at Mayak were. The 1945 venting of gases was not made public until 1986, when environmentalists near Hanford forced the release of nineteen thousand pages of official documents. By then, the Hanford complex had also discharged approximately eight trillion liters of low-level liquid radioactive waste directly into the soil, making the complex arguably the most polluted site in the United States.

The Hanford experience was not an aberration. The U.S. government knowingly understated the health and ecological risks of nuclear weapons production throughout the Cold War. In the 1950s, the government assured the public that nuclear testing in the Pacific posed no more health dangers than a chest X ray. The crew of a Japanese fishing boat found out differently when they encountered a radioactive cloud from one of the tests and immediately fell ill; one crew member died before reaching home port two weeks later. Residents of Utah and Nevada were likewise told "not to worry if their Geiger counters went crazy" during a detonation at the Nevada test site, advice that led many people to go outside and watch the blast and receive dangerously high doses of radiation. When ranchers sued after their sheep began dropping dead, the Atomic Energy Commission went so far as to lie to a judge and pressure witnesses not to testify in the case.

The government also used citizens as guinea pigs to test the effects of nuclear weapons. In an episode that U.S. Secretary of Energy Hazel O'Leary, in 1993, likened to the Nazi experiments of World War II, approximately 250 experiments were conducted in the United States between 1944 and 1973 on an estimated one hundred thousand individuals, including hospital patients who were injected with plutonium and pregnant women who were given radioactive pills. Thousands of American soldiers were ordered to march through the mushroom clouds of atomic test blasts. Not until 1988 did the U.S. Congress grant these "atomic veterans" compensation for the resulting health damages. In 1990, compensation was also offered to those who had lived downwind from the Nevada test site, as well as to uranium miners, who toiled in what analysts called probably "the deadliest part of nuclear weapons production."

In short, the U.S. and Soviet experiences with atomic energy during the Cold War were often more alike than not. Secrecy and cutting corners in one country were used to justify secrecy and cutting corners in the other. Each nuclear establishment acted like a state within a state whose officials sometimes seemed to have more in common with their adversaries than with their fellow countrymen and women. The two nuclear establishments even used the same vocabularies, as when U.S. military planners recommended creating "reeducation" programs to "correct" the thinking of American citizens worried about nuclear fallout.

The starkest example of this shared mind-set was the cloak of secrecy the CIA and KGB draped over the second Mayak disaster, the 1957 waste dump explosion. For twenty years, the two intelligence agencies declined to inform the rest of the world about an accident of cataclysmic scope and consequence, even though this meant, for the CIA, foregoing the chance to score points in its propaganda campaign against the "Red Menace."

Secrecy and sacrificing innocents are part of most wars, but they took on new meaning during the Cold War because of the unique nature of nuclear technology. During World War II, a lapse in security might have endangered a given infantry unit or tactical maneuver. But during the Cold War, any such lapse could endanger the entire country, for the arms race led both superpowers to accumulate enough firepower to destroy each other many times over. As Soviet premier Nikita Khrushchev remarked in 1962, any additional weapons would "only make the rubble bounce." Because the range and speed of nuclear weapons made defending against them impossible, each side based its security on a stated willingness to answer any attack with a crushing counterblow—the doctrine of "mutual assured destruction" that was as terrifying as its MAD acronym suggested. The two countries, and the larger world, were constantly poised minutes from doomsday. Further complicating matters, Great Britain (in 1952), France (in 1960), and China (in 1964) also exploded nuclear weapons, followed in the 1970s by India, Israel, and South Africa. The nuclear genie was out of the bottle.

Robert Oppenheimer compared the taming of the atom to man's fall from grace in the Garden of Eden. The creators of the bomb had "known sin," wrote Oppenheimer, "and this is a knowledge which they cannot lose." Acquiring such unprecedented power was cause for celebration in other quarters, though. When the defeat of Hitler erased the original rationale for developing nuclear weapons, Einstein urged that the Manhattan Project be halted; the White House ignored him. Although officials in the Japanese government had signaled a readiness to surrender, the American military insisted on using the bomb not just once, in Hiroshima, as a sort of demonstration blow, but twice, needlessly leveling Nagasaki and killing additional tens of thousands of civilians. After the war, some of the scientists who had helped develop the bomb urged that it be placed under international control, but they too were dismissed by Washington officials who saw a nuclear monopoly as a means of projecting American power around the world. President Harry Truman, the man who decided to drop both the Hiroshima and Nagasaki bombs, apparently assumed the U.S. nuclear monopoly would last forever. Meanwhile, commercial interests looked forward to exploiting atomic energy to produce electricity they promised would be "too cheap to meter."

Talk about playing the sorcerer's apprentice! Few of the early nuclear champions realized how little they actually knew about this technology, which, after all, had been developed very rapidly under wartime conditions. Their ignorance proved costly; the old rule of thumb about the commensurate costs and benefits of a revolutionary new technology was soon validated once again. Nuclear fission represented the greatest power humans had ever tapped, but the associated costs and challenges were no less monumental. The embrace of atomic energy not only threatened the end of human civilization, it condemned humanity to environmental and health injuries that would take decades if not centuries to heal, and it saddled us with waste disposal responsibilities that for all intents and purposes will last forever.

The damages at the Hanford and Mayak facilities only begin to tell the story. There are hundreds of sites around the world where aspects of nuclear weapons production have been undertaken, from uranium mining to plutonium reprocessing to weapons testing to waste storage. At virtually all these sites, the soil and water have been polluted and human health compromised, often severely. The most reliable and comprehensive account of the damage is found in *Nuclear Wastelands*, a handbook compiled under the auspices of International Physicians for the Prevention of Nuclear War, whose work was awarded the Nobel Peace Prize in 1985. "A kind of secret low-intensity radioactive warfare has been waged against unsuspecting populations," the book argues. "Destruction before detonation is the hallmark of nuclear weapons production."

The precise number of casualties attributable to nuclear weapons production is impossible to determine, not least because the official secrecy, deception, and disregard of individuals' welfare practiced at Mayak and Hanford has long been the norm at nuclear facilities the world over. Even in the United States, whose laws give citizens considerable freedom to uncover official wrongdoing, much remains unknown. For example, only in 1997 did government health officials announce that nuclear blasts at the Nevada test site near Las Vegas in the 1950s had caused between ten thousand and seventy-five thousand thyroid cancers, mainly among children. Many of these cancers might have been avoided had the military located its test site on the Atlantic coast, where prevailing winds would have swept the fallout over the ocean. But the Nevada site was chosen because it was more convenient and secret—the military already owned it.

The United States and the other nuclear states conducted approximately nineteen hundred nuclear weapons tests between 1945 and 1990, an average of one test per week. The 518 tests that were conducted in the atmosphere are calculated to cause 2.4 million cancer deaths worldwide beginning in 1949. (Nearly half a million of those deaths would occur before the year 2000.) The Limited Test Ban Treaty of 1963 reduced the damage by moving tests underground (a precaution France and China ignored until 1974 and 1980, respectively). But even underground tests posed enormous hazards. After all, nuclear test sites amounted to unlicensed nuclear waste dumps, and poor ones at that, since their ability to contain radioactivity had been compromised by the force of the blasts themselves.

One nuclear disaster zone is found at the top of the Kola Peninsula in northwestern Russia, near the border with Norway and Finland. The harbors of Kola were home to the Soviet Union's Northern Fleet during the Cold War and now contain enough radioactive materials to rival Chelyabinsk. For decades, Soviet authorities treated the sea surrounding Kola as a waste dump, casting used submarine reactors, spent fuel, and other nuclear debris into one of the world's richest fishing areas. By 1991, when the dumping stopped, the waters contained two-thirds of all the nuclear waste ever dumped into the world's oceans. Seventy nuclear submarines on Kola still await decommissioning, each containing large amounts of enriched uranium. Thousands of spent fuel rods in corroded containers likewise continue to leak radioactive pollution into the sea.

But at least no nuclear warheads were compromised at the Kola Peninsula (as far as we know). In 1986, a Soviet submarine sank off the coast of Bermuda while carrying sixteen nuclear missiles and thirty-four warheads. "Those warheads are still sitting on the bottom of the ocean, as are approximately fifty more warheads from other Soviet accidents all over the world's oceans," said William Arkin, a nuclear weapons expert with *The Bulletin of the Atomic Scientists,* in a 1997 interview. Noting that what happened in the Soviet Union before 1986 "is still a deeply buried state secret," Arkin added that "there are undoubtedly scores more of such accidents that we don't know about. Since the United States Department of Defense acknowledges that there were thirty-two nuclear-weapons-related accidents in the United States alone prior to 1980 ('accidents' being a technical term, rather than the thousands of 'incidents' that have occurred), one can only imagine how many accidents the Russians have experienced."

Future accidents, either in Russia or the United States, can by no means be ruled out. An internal report by the U.S. Energy Department's nuclear safety director warned in 1993 that there was a "high" likelihood of disaster at U.S. nuclear weapons plants because of deteriorating equipment, poor management, and worker sabotage. Government experts first admitted in 1985 that tanks containing fifty-seven million gallons of nuclear waste at the Hanford complex were not only leaking but could well explode, just as Mayak's waste tank had in 1957. In Hanford's case, the danger arose from an accumulation of hydrogen gas in the tanks. Officials have since begun "bleeding" hydrogen from the tanks on a regular basis, which they claim reduces the likelihood of explosion to "low." Outside experts are less sanguine. "There continues to be some risk of explosions—the Department of Energy says it's low, but the truth is, we don't know enough to quantify it—both at Hanford and at the Savannah River nuclear production facility in South Carolina, especially now that Savannah River has begun reprocessing again," said Arjun Makhijani, president of the Institute for Energy and Environmental Research and one of the editors of the *Nuclear Wastelands* study. Meanwhile, the storage tanks at Hanford have continued to leak, with radioactive material contaminating an acquifer of ground water 230 feet below the complex.

Cleaning up the environmental mess left behind by nuclear weapons production will be a long, difficult, extremely expensive job. In the United States alone, the Department of Energy expects the job to take another seventy-five years and cost $200 billion. A study by Stephen I. Schwartz of the Brookings Institution estimates the cost at $365.1 billion. Because the crisis is so much more dire in the former Soviet Union—more ecosystems were more severely polluted there—the job will cost much more and take far longer to complete there. "It is hard to quantify how much the Soviet cleanup will cost," Makhijani said, "partly because nobody has a good idea about what to do there. How do you dispose of the one hundred million curies of radioactivity in Lake Karachay? There are no good answers."

As the episodes recounted in this chapter show, nuclear waste has been an Achilles' heel since the earliest days of nuclear production. It was always considered tomorrow's problem, something to deal with after more urgent tasks—producing a bomb, catching up in the arms race, introducing nuclear-generated electricity—had been achieved. Finding a solution to the waste disposal problem is complicated by the fact that many radionuclides—most important, plutonium—are not only extremely toxic but have half-lives of thousands of years; thus, they must be isolated from ecosystems and human contact for a period of time equal to the known length of human civilization. And the amount of waste that must be managed is huge and growing. To date, humans have produced some seventy thousand nuclear weapons. The consequent waste products include an estimated four hundred thousand metric tons of depleted uranium, three billion curies of high-level plutonium-related waste, and one hundred to two hundred million metric tons of uranium mill refuse. Ironically, these figures will increase as post–Cold War disarmament proceeds and more and more nuclear weapons are dismantled. Of course, civilian power reactors also produce nuclear waste. In fact, 95 percent of the waste now in existence (if measured by radioactivity rather than mere volume) came from commercial nuclear power stations.

Devising an acceptable method of waste storage has been a major logistical problem for both the civilian and military wings of the nuclear industry. In the United States, the government finally approved an underground repository for military waste in 1998. Despite opposition from environmentalists, the Energy Department plans to bury some five million cubic feet of radioactive debris 2,150 feet below the desert near Carlsbad, New Mexico, in a complex called the Waste Isolation Pilot Project. But no such remedy is in sight for the waste generated by nuclear power stations. As a result, some electric utility companies may have to shut down their nuclear reactors earlier than expected because they are running out of room to store the waste they produce. For years, the civilian industry has been pushing the federal government simply to declare victory over the problem and open a permanent nuclear waste disposal facility. The industry has pinned its hopes on a government project to build a storage facility deep within Yucca Mountain in Nevada. Unfortunately, it turns out that Yucca Mountain sits above thirty-two active geological faults. Government scientists have also discovered that rainwater leaks from the top of the mountain into its core, raising the danger that any nuclear waste stored there would eventually reach the larger ecosystem. Such uncertainties have delayed the opening of a permanent disposal facility, originally scheduled for 1998, until well into the twenty-first century.

Ten-thousand-year time spans may humble those impressed by the fragility of human institutions and the limits of our vision, but nuclear industry officials are undaunted; they have long been confident that they can isolate their waste products from the environment for as long as necessary. After the accident at the Three Mile Island nuclear power plant in Pennsylvania in 1979, I interviewed scores of these men for my book *Nuclear Inc.: The Men and Money Behind Nuclear Energy*. Even then, the lack of a solution to the waste problem was threatening the industry's prospects, and I frequently asked its leaders how they planned to overcome the obstacle. John West, vice president of reactor manufacturer Combustion Engineering's nuclear division, told me that the real trouble was not that there was no solution to nuclear waste but that there were too many solutions and the dithering federal government, as usual, could not make up its mind which one was the best.

"I have a vulgar analogy," West confided. "It's kind of like you have a blond, a brunette, and a redhead, real glamorous gals all lined up for action, and you can't decide which one you'd like to go to bed with. They're all good."

And if the industry's certainty about nuclear waste storage turned out to be wrong, so what? "To me, it's the craziest thing," another top executive told me, referring to the many governors, legislators, and average citizens who had declared their states off limits to nuclear dumping in the late 1970s. "Neither they nor their descendants are going to be there at the time when anything could conceivably go wrong. If you do a halfway decent job of disposing of [nuclear waste], it's at least a few hundred years before anything could go wrong, and they won't even be there then."

And the nuclear industry wonders why people don't trust it.

Yet, there is cause for hope. This book would have struck a decidedly gloomier tone were it not for the remarkable breakthroughs in nuclear weapons negotiations that have occurred in recent years—not just the cutbacks agreed to by the United States and the former Soviet Union in the late 1980s and early 1990s but also the Comprehensive Test Ban Treaty signed by 158 countries in 1996. Since nuclear weapons cannot be reliably deployed without first testing them, an effective test ban would block new weapons development and a return to the arms race. There is even talk of abolishing nuclear weapons altogether, a cause being championed by some of the very men who used to have their fingers on the nuclear trigger, including Gen. Charles A. Horner, leader of the North American Aerospace Defense Command, which protects the United States and Canada from nuclear attack. "The nuclear weapon is obsolete," said Horner in 1994. "I want to get rid of them all."

Nuclear war is the ultimate environmental danger, the single greatest threat to continued habitation of the planet, and for decades it looked like humanity was heading for it as surely as a heat-seeking missile locked on to its target. The Cuban Missile Crisis of 1962 was the most serious flashpoint, but there were at least eleven other occasions when the United States threatened to use nuclear weapons, including during the Vietnam war in 1969 and 1972 and the Soviet invasion of Afghanistan in 1979. (How many times the Soviets and other nuclear states made similar threats is not publicly known.) And the superpowers were not the only loose cannons. In May 1990, war nearly broke out between India and Pakistan during a quarrel over a disputed border region. U.S. officials monitoring the conflict termed it an even more dangerous episode than the Cuban Missile Crisis.

The risk of global nuclear war seemed especially great in the early 1980s. Both superpowers were feverishly expanding arsenals that already bulged with overkill capacity. In the United States, Reagan administration officials spoke openly about fighting and winning a nuclear war. "With enough shovels" to dig their bomb shelters, Americans would survive a nuclear war, one official declared in a remark that horrified the vaunted Reagan public relations apparatus. Reagan himself said he could imagine a nuclear war being limited to Europe. He also "joked" about bombing Russia and demonstrated a frightening ignorance of basic nuclear facts. In 1982, he told a press conference that nuclear missiles launched from submarines were not that dangerous because they could always be recalled after launch. Belligerent rhetoric emanated from the

Soviets as well, and among strategists on both sides brinksmanship was the order of the day. When the Soviets insisted on deploying SS-20 missiles, the United States and its NATO allies contended that this justified their own deployment of new cruise and Pershing missiles. It was bad enough that both superpowers' arsenals boasted ever larger numbers of ever more powerful weapons. But the more worrisome trend was the sharp decline in battlefield reaction time. Cruise and Pershing missiles could strike targets within six minutes of launch. That left opposing military commanders precious little time to decide whether what they saw on radar screens was a genuine attack that had to be countered or a mere computer error to be ignored.

With the hair trigger stretched so taut and U.S.–Soviet relations so embittered, the outbreak of nuclear war, whether by accident or design, was a real and present danger. Political activists capitalized on popular fear to catalyze massive public opposition to the arms race. Millions of people marched through the streets of Western European capitals in the spring and summer of 1981 to demand a nuclear-free Europe. In June 1982, nearly a million demonstrators filled New York's Central Park as part of a national movement to "freeze" the arms race.

But there was no substantive shift in official policies until Mikhail Gorbachev became Soviet general secretary in March 1985 and began making one unilateral concession after another. That July, he announced an eighteen-month moratorium on Soviet nuclear weapons testing, which he later extended three times. In January 1986, the Soviet leader announced a plan to rid the world of nuclear weapons entirely by 2000. The Reagan administration repeatedly rejected Gorbachev's initiatives, calling them "nothing but propaganda." Yet Reagan had spoken in general terms of his wish to eliminate nuclear weapons, a desire Gorbachev seized upon during their talks in Reykjavík in 1986. Gorbachev revived his proposal for nuclear disarmament in Reykjavík, but Reagan rejected it, since it would have prohibited development of Reagan's beloved Strategic Defense Initiative—space weapons.

No such foot-dragging was possible, however, when Gorbachev later accepted the Americans' long-standing call for the withdrawal of all Euromissiles from Europe. The result was the Intermediate-Range Nuclear Forces (INF) Treaty of 1987. The treaty obliged the Soviets to eliminate far more weapons than the Americans would, but Gorbachev accepted the imbalance, explaining that "parity" had no meaning in an age of massive overkill. Indeed, insisting on strict numerical balance between the superpowers' arsenals only blocked progress toward disarmament. This was revolutionary new thinking, and it ended up transforming the world.

If the INF Treaty marked humanity's first step back from the nuclear abyss, the second came in 1991, when the two superpowers signed the Strategic Arms Reduction Treaty (START). By then, the fall of the Berlin Wall, Gorbachev's calm acceptance of the end of the Warsaw Pact military alliance, and his demobilization of hundreds of thousands of Soviet ground troops had proven to all but the most intransigent Cold Warriors that a fundamentally new era had begun. START was lauded as the first nuclear treaty that would actually reduce arsenals rather than manage their continued growth. There was less to START than met the eye, however. During the nine years it had taken to negotiate the treaty, the superpowers' arsenals had grown by about the same number of weapons as START would eliminate. START thus turned out to be a classic case of running to stand still; it merely maintained the nuclear status quo.

The momentum of the arms race was not decisively reversed until the START II Treaty was signed in 1993. Under the terms of START II, each nation had to reduce long-range nuclear warheads to approximately thirty-five hundred—about one-third as many as each had had in the late 1980s—by 2003. For the first time in more than four decades, the United States and the former Soviet Union would be dismantling more nuclear weapons than they were building.

To reverse the drill for death known as the nuclear arms race was a genuinely historic achievement, a stirring victory for humanity and its future. There were many lessons in the victory, but perhaps the most basic was never to give up hope. At certain moments in history, despite the darkest of outlooks, conditions can change enormously, and all but overnight—a heartening reminder as one ponders how to defuse the more gradual environmental crises that cloud humans' future.

No sooner had the nuclear hair trigger relaxed, however, than a bland complacency overcame many citizens and policymakers. Few seemed to appreciate how close they had come to catastrophe, how lucky humanity had been to dodge the nuclear bullet, or how much further they still had to go to secure a truly safe world. After all, what if Gorbachev had not come to power and launched his unilateral initiatives? Amid the self-congratulations and relief that followed the Cold War, the nuclear arms race and the era it had defined were regarded as a sort of bad dream, best forgotten amid the cheer and possibility of the new morning.

Some of the men who labored inside the nuclear system knew better. George Lee Butler, a retired U.S. Air Force general who directed the Strategic Air Command from 1992 to 1994, warned that humanity was in danger of lurching "backward into the dark world we so narrowly escaped without thermonuclear holocaust." Butler was one of sixty top military officials from around the world who released a statement in December 1996 urging a redoubled commitment to nuclear disarmament and abolition. The statement received virtually no coverage by the mainstream news media, at least in the United States—a symptom of the very complacency the former warriors were trying to puncture.

The generals may not have known it at the time, but the world had come distressingly close to an accidental nuclear war just two years earlier, long after the end of the Cold War had supposedly turned Americans and Russians into trusting friends. On January 25, 1995, the Russian military confused a scientific research rocket launched from the Norwegian Arctic island of Andøya for a NATO missile, even though Russia had been warned in advance of the launch. Calling the episode "the most serious in the history of nuclear weapons," Peter Pry, a former CIA officer, revealed that the Russian military "made all the preparations for starting a nuclear war except making the decision to launch." President Yeltsin went so far as to consult the secret codes used to order a nuclear strike before the situation was clarified.

"No major change in the U.S.–Russian nuclear equation has occurred—not in war-planning, not in daily alert practices, not in strategic arms control, and maybe not even in core attitudes," nuclear weapons expert Bruce Blair said in *The Gift of Time,* a study of the prospects for nuclear abolition written by Jonathan Schell. Schell, whose 1982 book on the dangers of the nuclear arms race, *The Fate of the Earth,* had sparked great public debate and anti-nuclear activism, argued in *The Gift of Time* (1998) that nuclear abolition was an idea whose time had come, not least because the current nuclear situation remained so dangerous. "No issue that could justify even the smallest of conventional wars divides the two former super-enemies," wrote Schell, "yet each nation's nuclear arsenal is still poised on a hair trigger to blow the other to kingdom come several times over, and no plan that would fundamentally alter this state of affairs is even on the drawing boards."

The size of the remaining arsenals makes accidental war a grave hazard. Presidents Yeltsin and Clinton agreed in principle in 1997 that a START III treaty would reduce long-range missiles to no more than twenty-five hundred apiece by 2007, but there was a big catch. START III could not be negotiated until the Russian parliament ratified START II, which was no sure thing. Even before NATO's decision in 1997 to expand eastward into Poland, Hungary, and the Czech Republic, a move despised by Russians of virtually all political stripes, ratification of START II had been blocked by communist and nationalist deputies who resented what they saw as America's high-handed ways in the aftermath of the Cold War. But the Americans had little room to criticize Russian foot-dragging; the U.S. Senate had not gotten around to ratifying START II until 1996. And even under START II, both Russia and the United States would still wield more than enough firepower to obliterate human civilization.

The political and economic turbulence in Russia gave rise to perhaps the most harrowing nuclear hazard of the post–Cold War world: the so-called loose nukes problem. How to keep nuclear weapons and materials from falling into the "wrong" hands had been a concern since the dawn of the nuclear age. Now, with so many weapons being dismantled at the very time state authority was crumbling in the former Soviet Union, vast amounts of warheads and weapons-grade material were suddenly becoming vulnerable to theft or diversion onto the black market.

It takes about fifteen pounds of plutonium to make a Hiroshima-strength bomb. There were four hundred thousand pounds of plutonium lying around in the former Soviet Union in 1991, plus 2.4 million pounds of enriched uranium. Often this material was poorly guarded. In 1996, U.S. government investigators were able to wander into the Kurchatov Institute in Moscow without even showing identification. There, they discovered a cache of nuclear materials guarded by a single unarmed policeman. In a land where organized crime now made a mockery of the rule of law, such laxness courted disaster. In June 1997, U.S. agents posing as drug traffickers arrested two Lithuanian smugglers in Miami who promised to sell them surface-to-air missiles and tactical nuclear weapons, and the smugglers proved they could deliver on the deal. In his 1996 book, *Avoiding Nuclear Anarchy,* Harvard professor Graham Allison documented six cases in which stolen nuclear materials were smuggled out of Russia before foreign security forces intercepted them. How many smugglers have gone undetected is, of course, not known.

Imagine the horror if the terrorists behind the Oklahoma City bombing of 1995 (or the recurring bombings in the Middle East, Northern Ireland, and other global hot spots) had relied on nuclear rather than conventional explosives. Professor Allison is convinced such a tragedy is only a matter of time, and he is not alone. Theodore Taylor, a former nuclear weapons designer at the U.S. Los Alamos National Laboratory, told the *Bulletin of the Atomic Scientists* in 1995 that the proper setting of its minutes-to-nuclear-midnight clock depended on what definition of "doomsday" one used. If doomsday meant a nuclear World War III, Taylor was optimistic enough to choose 11:30 P.M. If, however, doomsday meant "the time remaining before nuclear terrorists kill more than 100,000 people, I would set it at two minutes to midnight."

The collapse of the Russian economy magnifies the peril, for it is not just theft of nuclear materials by outsiders that poses a danger; an inside job, by workers impoverished and disgruntled after months without wages, would have the same effect. In the years since the dissolution of the Soviet Union, more and more of its people have been reduced to penury and desperation. The shock transition to capitalism implemented by President Yeltsin at the insistence of the International Monetary Fund and Western governments has caused industrial production to fall nearly 50 percent, while driving a quarter of the population below the official poverty line. "A large twentieth-century middle class is being transformed into nineteenth-century subsistence farmers, who must grow on tiny garden plots [the food] they need to survive but can no longer afford to buy," reported Russian specialists Katrina vanden Heuvel and Stephen F. Cohen in August 1997.

Some Russians responded to their situation with sardonic humor: "We know now that everything the Party told us about communism was false. And everything they told us about capitalism was true." Others, less resilient perhaps, succumbed to drink, despair, and suicide. Which returns us to Chelyabinsk. In October 1996, Vladimir Nechai, the director of the Chelyabinsk-70 research laboratory, killed himself after leaving behind a note expressing shame at the government's failure to pay the salaries of his staff. The world could scarcely ask for a more penetrating wake-up call on the vulnerability of nuclear materials in Russia. Nechai's suicide made news in the West for a day or two, then dropped from sight.

Humanity has made extraordinary progress against the threat of nuclear destruction in recent years. The reversal of the U.S.–Soviet arms race—so unexpected for so many years—is the most encouraging political development of our time. But there is still a long way to go before we are out of the nuclear woods.

The greatest immediate danger arises from the hair-trigger status of the approximately ten thousand nuclear warheads that remain in each of the American and Russian arsenals. To lower the risk of accidental war, Russia and the United States could revamp their nuclear doctrines by taking their forces off "launch on warning" alert status and by making "no first strike" pledges. They could also shrink their arsenals to as close to zero as possible, and redouble efforts to prevent the proliferation of nuclear weapons to other nations. Although the United States has begun to work with Russia to improve security at some Russian nuclear facilities, insufficient funding has delayed completion of the project until 2006 at the earliest. In a world increasingly fractured by violent ethnic and religious disputes and by state and individual terrorism, the availability of large amounts of nuclear-weapons-making material is a recipe for catastrophe. Numerous attempts to steal plutonium and sell nuclear weapons on the black market have already been made; they need succeed only once.

Above all, Russia and the United States must make clear by both word and deed their commitment to the eventual abolition of nuclear weapons. This applies especially to the United States, the world's sole remaining superpower. Unfortunately, American policymakers show few signs of embracing this goal. Although President Clinton has signed the Comprehensive Test Ban Treaty, Congress has yet to ratify it. More disturbing, the Department of Energy has continued to conduct so-called "sub-critical" nuclear tests, which the department claims are allowed under the test ban treaty because the resulting nuclear reactions are contained short of full explosion. Numerous other countries and independent experts dispute that claim. They point out that sub-critical tests enable a nation to continue designing new nuclear weapons—precisely what the test ban treaty is intended to prevent. If that is not the American goal, they ask, why is the United States planning to spend $45 billion over the next ten years on its Stockpile Stewardship Program, assembling a vast collection of lasers, supercomputers, and other equipment designed to develop nuclear weapons?

India in particular has accused the United States of hypocrisy for urging nuclear have-not nations to remain that way while the United States refuses to agree to a specific timetable for nuclear disarmament. This was India's rationale for not signing the test ban treaty, and it was trotted out again in May 1998, when India carried out five underground nuclear weapons tests, including one test of a "thermonuclear," or hydrogen, bomb. Within days, India's neighbor and rival Pakistan responded with three nuclear tests of its own, raising tensions in Asia to a frightening pitch. The United States and other nuclear weapons states responded with economic sanctions and calls for India and Pakistan to foreswear further weapons development. But such pressures seemed doomed to ineffectuality as long as the nuclear weapons states insisted on maintaining their own arsenals. The Indian and Pakistani tests changed the world; no longer could the traditional weapons states decide among themselves who else could join the nuclear club. From now on, it seemed the only way to limit the club's membership was to close it down altogether.

The nuclear era of human history is only beginning. If traveling in Africa teaches one to look to the deep past before gauging our species's environmental prospects, traveling to Chelyabinsk shows that imagining the distant future—the world our offspring may one day inherit—is no less important. The Cold War may be over, but the plutonium it left behind, and the knowledge of how to use it, will last forever. Nearly 80 percent of the 2.4 million cancer deaths expected to result from atmospheric nuclear testing will occur after the year 2000. Vast swaths of land, including the areas around the Chernobyl and Mayak nuclear complexes, will be uninhabitable for centuries. An enormous quantity of nuclear waste will remain so radioactive it must be insulated from human contact for millennia. Contrary to the assurances of nuclear power executives, humans have yet to discover whether this can be accomplished. Meanwhile, existing stocks of plutonium remain susceptible to weapons diversion, yet both civilian and military plutonium production continues.

The nuclear danger will persist and intensify until the so-called plutonium economy of nuclear weapons and power production is dismantled, according to Arjun Makhijani. "We do not need to spend lots of money" on this task, Makhijani told me. "It's a matter of billions of dollars a year, compared to the $4 trillion we have already spent to build nuclear weapons. Four steps are needed: detach nuclear warheads from their delivery systems to minimize the risk of accidental nuclear war; vitrify [encase in glass or ceramic] all existing plutonium to make it less usable in weapons; convert all highly enriched uranium into low-grade uranium, for the same reason; and stop producing plutonium, everywhere."

The last step may be the most controversial, for it has important economic implications. It means not only closing weapons production plants but also prohibiting plutonium use in civilian power reactors; such a move would limit the role of nuclear energy as a major energy source in the twenty-first century. France closed its Super-Phoenix plutonium breeder reactor, which accounted for 50 percent of the world's breeder capacity, in 1998. Japan, another great champion of civilian plutonium use, is rethinking its program after serious accidents at the Monju breeder plant in 1995 and the Tokai reprocessing facility in 1997. Nevertheless, the global nuclear industry has not given up its dream of a plutonium-fueled future, and it still has plenty of political muscle, as its post-Chernobyl behavior demonstrates.

It seems incredible, but there are fifteen nuclear power reactors still in operation in the former Soviet bloc that are at least as dangerous as the reactor that exploded at Chernobyl in 1986. The plants lack not only emergency core cooling systems, which can head off a reactor meltdown, but containment vessels, which, in the event of a meltdown, are supposed to keep any fallout from reaching the atmosphere. These and many other inadequacies led the United States and other Western nations to press for these reactors to be shut down; in 1990, economic aid was pledged toward that end. But Russia refused to shut the plants, claiming that their electricity was irreplaceable. The West relented to Russia's insistence, which apparently stemmed more from the ambitions for the Russian nuclear program than any shortages of electricity. The unsafe plants could, in fact, be closed without disrupting local economies, according to William Chandler of the U.S. Department of Energy: "Demand for electricity in Russia has fallen since 1990 by twice as much as all the electricity that Russia's unsafe nuclear reactors now produce," he said in 1997.

But instead of pressing to close the unsafe plants, Western governments are underwriting training programs and other marginal efforts to upgrade their safety, as much as that is possible without adding emergency cooling systems and containment vessels, a technical and financial impossibility. The winners under this deal are the Russian nuclear ministry, along with Westinghouse, Mitsubishi, Électricité de France, and the other nuclear companies gobbling up the Western subsidies. The losers— big surprise—are the people who live near these plants, which, as Chernobyl demonstrated, means people throughout the European continent.

Did Chernobyl teach us nothing? Even in the aftermath of the worst nuclear power accident in history, commercial pressures are allowed to mock public safety. Meanwhile, Chelyabinsk is offered but a pittance to clean up its mess and forestall the accident-in-waiting at Lake Karachay. In effect, it is being written off, by both the West and the Russian government, as a sacrifice zone, a place too polluted to ever be salvaged.

Some people in Chelyabinsk suspected all along that this would be their fate. I remember saying good-bye to Valodya, the Tartar who had cheerfully driven Natalia Mironova and me around Chelyabinsk. Valodya had been fascinated by the idea of my global journey—to travel around the world as one's job seemed incredible to him. Hands clasped, we were saying our farewells when Valodya asked earnestly whether I would ever return to Chelyabinsk. I tried to deflect the question, noting that I had many other countries I had to visit first. Yes, he persisted, but if you could come back, would you want to? I didn't want to lie, but I couldn't tell the truth. I looked at the ground in silence. Sensing my embarrassment, Valodya clapped me on the shoulder and said, "It's okay, Mark. We know not to expect things in Chelyabinsk. But you are always welcome here."

Five

"Is Your Stomach
Too Full?"

> "Master," I cried, "who are these people
> By black air oppressed?"
> —DANTE, *The Inferno*

Of all the things you need when traveling around the world, luck may be the most decisive, and meeting Zhenbing was certainly lucky. I must confess, it didn't feel like luck when my original interpreter in China abruptly abandoned the project thirty-six hours after I reached Beijing. But then Zhenbing appeared, introduced by a mutual friend, and I soon realized fate had done me a favor. A thirty-year-old economics instructor at one of Beijing's major universities, Zhenbing had a ready laugh, high cheekbones in a long, angular face, and glasses that did not quite hide the slightly off-center gaze of his right eye. He had a few weeks free before next semester's classes began, and the idea of traveling around within his native land appealed to his adventurous spirit. His English was strong (he had recently returned from two years of graduate study at one of Europe's most prestigious academic institutions), and he was an experienced interpreter. Nor was he troubled by my plan to operate below the government's radar screen, traveling as a tourist to evade the media minders who monitored all permanently stationed foreign reporters in China.

Zhenbing's greatest strength as an interpreter was that he could talk to anybody, from lowly peasants to high party leaders. He was just one of those people everyone likes on the spot. Part of his secret was that he was extremely funny and laughed a lot, especially at himself, a talent he had plenty of opportunities to display, for he had without question the worst sense of direction of anyone I met in all my travels. This is generally not a talent one looks for in interpreters, and it wasn't long before I took over all our navigating. But Zhenbing was a good sport, and to amuse ourselves at the end of the day, we would sometimes compare ideas for how to get back to the hotel or train station. At these moments Zhenbing would look around intently, furrow his brow in concentration, and finally point decisively in one direction or another. But his eyebrows, raised in bravely hopeful fashion, betrayed his inner uncertainty, and when he learned he had chosen wrongly once again, he would burst into uproarious laughter.

Above all, Zhenbing was a great, sometimes endless, talker and an enthusiastic storyteller. I can't count the number of times he struck up a conversation with the person next to him in a bus station or train compartment, and soon it wasn't just the two of them but a whole crowd of previously silent strangers chattering back and forth like long-lost relatives. Zhenbing was invariably at the center of things, with the crowd hanging on his every word—laughing, interjecting questions and commentary, and generally having a fine old time. It was a bit like watching the Pied Piper in action, except that Zhenbing's charm captivated grownups and children alike—and gave me access to countless candid conversations that I never would have heard otherwise.

Another advantage of traveling with Zhenbing was that he saw life the way a peasant did, which is to say, the way the vast majority of Chinese did. Historically, 90 percent of China's population had lived in the countryside, and even after the great urban migrations sparked by the economic reforms of the 1980s, the figure remained over 70 percent. Born in 1966, Zhenbing had grown up very poor, passing his first twenty years in a small village northwest of Beijing near the border with Inner

Mongolia. The second of three brothers, he was fourteen years old before he got his first pair of shoes. His family inhabited a mud straw hut and <u>was too poor to buy coal in winter; for heat, they burned straw</u>. They did this in a climate as cold as Boston's or Berlin's, with winter temperatures that dipped 30 degrees below zero. "Often the straw was not enough, so the inside wall of the hut became white with icy waterdrops, like frozen snow," Zhenbing recalled. "In my village, when a girl was preparing to marry, the first thing the parents checked was, will the back wall of the would-be son-in-law be white or not? If not white, they approved the marriage, because that meant his family was wealthy enough to keep the house warm."

Zhenbing was an invaluable reality check for an outsider like myself investigating China's environmental crisis. He never made an issue of his background, perhaps because it was such a common one in China. But in ways large and small he helped me understand that, while there were plenty of things the Chinese masses might not like about their existence, by far their biggest complaint was being miserably poor, and they would put up with a great deal of aesthetic or environmental unpleasantness to escape that poverty.

As recently as 1950, the average life expectancy in China was thirty-nine years, a level not seen in Europe since the Industrial Revolution. And many Chinese of the 1990s still had firsthand memories of suffering through the greatest man-made disaster of the twentieth century, the famine caused by Mao Zedong's Great Leap Forward campaign. As Jasper Becker documented in his powerful book, *Hungry Ghosts,* Mao's famine killed some thirty million people between 1959 and 1961—more than Hitler's and Stalin's death tolls combined—and brought misery, starvation, and even cannibalism to virtually all parts of rural China.

Chinese life spans averaged about seventy years by the mid-1990s, yet hundreds of millions still lived in desperate poverty. In one village Zhenbing and I visited in Sichuan province, on an early January day when my feet were barely comfortable inside polar-insulated hiking boots, I watched a grim-faced peasant woman seated on a rocky river bank, her bare feet dangling in the frigid water as she washed her family's clothes. On the other side of the village, three young children amused themselves with the only toy they had, a plastic water bottle filled with pebbles, which they pulled around like a wagon on a string. Behind them, a bare-footed man stamped around on a pile of loose, moist coal, looking like an eighteenth-century European peasant crushing grapes for wine. In fact, he was manufacturing—by foot, as it were—the briquettes of fuel whose carcinogenic combustion would provide what little heat he and his neighbors enjoyed in their windowless mud huts.

Back in Beijing, Zhenbing agreed after some coaxing one day to show me his dormitory room at the university. It was on the third floor of a long, barnlike building. Out front, scores of bicycles slouched against one another like tired children. From the moment we pushed through the scarred wooden doors on the building's ground floor, we were surrounded by the smell of toilets, stale air, and general uncleanliness. The concrete stairs were filthy, the walls splattered with all manner of stains and dirt—nothing had been painted for decades. Though it was two o'-clock in the afternoon, it was very dark inside the dorm. Following Zhenbing up to the second floor, I saw that the only light came from dim, naked bulbs dangling from the ceiling at intervals of fifty feet or so. The dusty, shadowy corridors were like obstacle courses, overflowing with boxes, cooking gear, desks, and whatever other belongings could not fit inside the rooms.

The odor was rather worse upstairs, because the toilets led directly off the corridor, and their doors were wide open. Along one wall of the men's room was a stand-up urinal that did not seem to flush. Along the opposite wall were six squat toilets that were open in front—"There is never privacy here," Zhenbing complained—and separated only by concrete partitions. There was no toilet paper, only a hole in the concrete floor and a small red lever with which to flush away waste. One recent visitor had neglected that task, leaving a deposit of frighteningly large proportion to fester and stink. There were no bath or shower facilities; all washing was done in a room down the hall, where two rows of concrete tubs offered cold water only. Mirrors, counter space, towel racks, and similar luxuries were nonexistent.

Inside Zhenbing's room the smell was also unpleasant, for which he quickly apologized. The walls were grimy and flaking from mildew. When I tried to open the one tiny window, I found it had been broken and the missing piece literally papered over with a sheet of notebook paper.

The room was six paces wide and four paces long, yet it somehow contained two cast-iron bed frames, a metal bookcase, two wooden desks, a coat stand, and a set of rickety shelves. The light was again a single overhead bulb, and the beds were as hard as stone: a wooden plank topped by a bamboo mat and thin cotton quilt.

The place felt shabby and grim and spirit-crushing, but Zhenbing pointed out that it was in fact a relative privilege. As a junior faculty member, he had this room to himself. Ph.D. students would live three together in the same space, while master's students were packed four together and undergraduates six to eight. When I asked whether these humble conditions ranked above those of the rural majority in China, Zhenbing replied, with some bitterness, "Oh, yes, definitely. You must remember, the government is afraid of students getting upset, so they treat us relatively well. We have access to electricity, to running water, to central heating. Paradise!"

Zhenbing was riled up now, and as we headed off campus for our next appointment his frustration burst forth in a tirade that alternately attacked and defended the government. "This is why economic development is the most important goal for China," he said. "It is more important than environment, or human rights, or the other issues the Western media and governments complain about. You may think it is propaganda, but most Chinese support the government for building the Three Gorges dam [a controversial project under construction on the Yangtze River]. That is no business of yours! You may complain about our contributions to global warming or the ozone hole, because those issues affect everyone on earth. But how much pollution we make, how many trees we cut or dams we build is nobody's business but ours."

Among his many other talents, Zhenbing also possessed world-class sleeping abilities. The man could fall asleep anywhere, anytime, for short periods or long, then wake up alert, cheerful, and eager to chat. One day, on a train from Shenyang to Beijing, he slept for more than eight hours seated directly beneath a loudspeaker that blared tinny Chinese pop music, inside a compartment where passengers were crammed knee-to-knee and shoulder-to-shoulder, while the aisle overflowed with people pushing and shoving toward the rear to refill tea thermoses or visit the reeking toilets. Zhenbing told me he learned his sleeping skills in college, by necessity, while sharing a room with seven other students. (Endless college card games also made Zhenbing a good enough gambler that years later, in the West, he supplemented his income by visiting casinos and regularly beating the house, but that's another story.)

My most vivid memory of our train ride to Beijing, though, was of all the spitting that went on. I was wedged between a sniffling peasant girl on one side and her older brother or cousin on the other. Zhenbing and I had "hard-seat" tickets—the lowest class, and the only ones available. I was a curiosity to these peasants, and to show friendship the young man offered me a few of the sunflower seeds he and his family of seven were munching. Upon finishing his own seeds, he washed them down with a swig of tea and then, with a deep hocking sound, summoned from his throat a prodigious gob of phlegm, which he casually spat onto the floor in front of us. Out of the corner of my eye, I watched as he then reached out his foot and rubbed the spit into the floor, as if stamping out a cigarette. It was 8:15 in the morning, there were fourteen more hours to Beijing, and there was lots more spittle loosed throughout that packed compartment before we got there.

Everyone seemed to spit in China—on the sidewalk, in the classroom, on the train, in restaurants, wherever. Middle-aged housewives, rowdy teenagers, toothless old men, beautiful young women—the habit was universal. The communists had tried to eradicate spitting when they came to power in 1949; it was one of their first exhortations to the masses. They failed. Spitting lived on in the 1990s because it was a habit of peasant life, and the vast majority of Chinese were still peasants or only one generation removed. The habit itself apparently derived from the basic conditions of peasant life, which included rampant lung infections and other respiratory diseases. These, in turn, resulted from a historical fact with enormous environmental implications: for centuries, Chinese peasants had lived with very little heat in wintertime, so they were frequently sick. They burned wood—if they were lucky—or straw and cropstalks, as Zhenbing's family did. In the 1990s, Chinese peasants still relied on such "biomass" fuels for 70 percent of their energy consumption (and found themselves short of fuel between three and six months out of the year).

Coal therefore represented a great advance for the Chinese people; it kept a body much warmer. But it did so at terrible cost. Coal smoke was the most important element in the air pollution that was killing at least 1.9 million people a year in China, according to the World Bank. "Pollution is one reason chronic obstructive pulmonary disease—emphysema and chronic bronchitis—has become the leading cause of death in China, with a mortality rate five times that in the United States," reported the bank. Outdoor air pollution was second only to cigarette smoking as a cause of lung cancer in China's cities, where lung cancers had increased 18.5 percent since 1988. Coal smoke was also the main component of the indoor pollution from home stoves that caused most rural lung cancers, especially among women.

Like the automobile, nuclear fission, and so many other technologies in human history, coal burning in China was both a blessing and a curse. Coal had brought China into the industrial era and enabled the Chinese masses to be warm in winter for the first time in history. At the same time, it had poisoned the air and water, not to mention people's lungs, beyond description. The huge amounts of coal burned in China (especially in the north, which relied on coal for winter heat), along with the primitive technologies often employed, gave large northern cities such as Beijing and Shenyang some of the filthiest air on earth. The levels of total suspended particulates (TSP) routinely climbed as high as 400 and 500, even 800 in winter—four to nine times greater than World Health Organization guidelines.

I had breathed plenty of bad air in my travels, but none like Beijing's; it made Bangkok's air seem merely unpleasant. According to He Kebin, vice chairman of the Department of Environmental Engineering at Tsinghua University, approximately 75 percent of Beijing's air pollution in winter was caused by burning coal, the fuel that heated and powered the city. I had been warned in advance about Beijing's air. One journalist told me about a Western diplomat friend, an avid jogger who refused to give up his habit during a two-year stint in the capital. When the diplomat finally returned to his home country and underwent a routine medical checkup, his physician told him he absolutely had to quit smoking. Need I add that cigarettes had never passed this man's lips?

Nevertheless, when I arrived in Beijing, I had to wonder if all that talk had been mere journalistic exaggeration. My first morning in town,

I bounced out of bed and eagerly headed outside for my first walk in the People's Republic. The temperature that December day was a bracing nineteen degrees, but the real surprise was the brilliantly blue and sunny sky. There were no signs of pollution anywhere. How could this be?

It was just before eight o'clock, and the four-lane boulevard outside my hotel was crowded with commuters, a stream of humanity so dense and fast-moving that at first I could only stand back and watch. As wheezing buses rumbled to a halt at the side of the boulevard, bunches of pedestrians would dash shouting and laughing into the throng already waiting at the bus stop in hopes of squeezing their way aboard the crushingly packed vehicle. A few people traveled by taxi or car, but the vast majority was on bicycles, usually the stolid black Chinese model called Flying Pigeons. There were also lots of three-wheeled cargo bikes. They had long, drooping gear chains hanging inches off the ground and, in back, wooden flatbeds that carried everything from bulging sacks of fruit and vegetables to freshly skinned sides of pork, to couches, toilets, televisions, and small mountains of crushed cardboard boxes destined for recycling. I was especially intrigued by bikes carrying the coal briquettes locals called "honeycombs" (because of the holes drilled in the briquettes to encourage cleaner burning). Round, black, the size of small coffee cakes, the honeycombs were stacked by the hundreds into squat pyramids and sold off the carts for burning in the home stoves of the poor. Honeycombs were said to be the cause of much of China's air pollution, but where was that pollution?

It took me forty-five minutes to walk a square city block that morning, partly because the wind was so fierce that I sometimes had to duck into doorways to escape it. One sidewalk was occupied by dozens of street vendors, each swathed in countless layers of clothing and hunched behind frozen piles of nuts, grains, eggs, fruits, or shriveled white cabbages. Whatever romantic expectations I had had of Beijing's architecture were quickly extinguished by the pervasive drabness of the place. The buildings were square, ugly concrete boxes, and Beijing, like most of China, was the most litter-strewn place I had ever visited. Everywhere I looked there was trash—plastic bags, peanut shells, cigarette boxes, food cartons, construction site refuse—and the gusting winds sent it blasting along the streets like pellets from a shotgun.

I returned to my hotel chilled, exhilarated, and bewildered, and I spent the rest of the day plagued by a fear familiar to all journalists: had I come in search of a nonexistent story? Where was all the air pollution?

That afternoon, a government press aide proudly explained that the government had moved all the city's heavy industry out of the down-town area. It sounded plausible, and I subsequently learned that some factories had indeed been relocated. But it turned out that Beijing was, in fact, rarely graced by blue skies in winter, except immediately after Siberian winds had roared through and flushed away all the smog. By chance, such winds had struck the night I arrived and continued blowing throughout the following day. As a taxi driver told me that second evening, the only reason Beijing's air did not look dirtier was that "it's very windy today. If there were no wind, you'd notice [the pollution] very strongly."

Sure enough, the winds calmed the next day, a Friday, and over the following ten days I witnessed the sickening descent of Beijing into a city of murk and gloom.

At noon on Saturday, after just twelve hours of still air, I took a bus across town to a luncheon interview, traveling the main west–east boulevard through Tiananmen Square. Straight above me, the sky was still blue, but in the distance a fuzzy, pale gray layer of smog already frosted the skyline. When I came back outside four hours later, the layer had nearly doubled in thickness, its blurry density giving the sky an otherworldly aspect as it melted into a sunset of vivid pinks and yellows. The pollution accumulated with each passing day, and by Thursday I was used to waking up to a dull, gray-white haze that rested on the city skyline like a lid on a wok.

The haze would grow palpably worse through the course of a day, as countless thousands of boilers were fired up and internal combustion engines spewed exhaust. On Thursday, for example, Zhenbing and I were riding south on the second ring road that skirts Purple Bamboo Park on its way around the western edge of town. It was about 4:30 in the after-noon, and we were at the head of a long, flat stretch of highway. Our eyes should have been drawn to the Chinese national television (CCTV) tower, by far the tallest structure in the city, which lay directly in front of us, about four miles away. But by this hour the smog had become so thick that what had been a basically sunny day at noon now looked overcast and dark. Only because I knew the CCTV tower had to be up ahead could I faintly make out its needle-nosed outline against the sky. To my right, striding down the sidewalk, I saw a tall, young woman in a smart black overcoat. Behind her, two young girls wearing bright red and yellow athletic suits pedaled bicycles. Farther back, a wizened old fellow in a blue Mao cap was bent over an upturned bike, trying to repair it. I looked back toward the CCTV tower and realized that to anyone at the tower, the air around here must look just as bad as the air up there appeared to me, and I swallowed hard at how much poison was entering our lungs.

On Friday morning, I took a taxi to the National People's Congress on Tiananmen Square. Passing through the larger intersections of Beijing, I looked both ways down the cross streets but could see no farther than half a block; beyond that, an impenetrable gray mass concealed everything. When I reached Tiananmen Square, at 8:45, the sun hung white and barely visible above the southern gate to the Imperial Palace, like a weak lightbulb in a barroom full of cigarette smoke. Gazing north, past Mao's mausoleum and the site of the 1989 massacre, I could not see the far end of the square, much less the Forbidden City beyond it. The pedestrians crossing the square were like spectral figures—half ghost, half flesh—as they disappeared into the gritty mist.

It had now been a week since the Siberian winds had cleansed Beijing, and I craved a respite from the increasingly filthy air. On Sunday I took a public bus to the Great Wall of Badaling, seventy-five kilometers north of the capital. Though genuinely foggy at the wall, the fresh air felt wonderful descending into my lungs, and northern winds even brought a cheering patch of blue sky by midafternoon. But not to Beijing. Back in the city, I stepped off the bus into a grimy dusk. Beneath my feet, whorls of coal dust spiraled across the sidewalk like black snow flurries.

How did people stand it? The bicycle delivery drivers who strained to pedal their overloaded carts uphill amid air so polluted it nearly glowed—how did they do it? Zhenbing brushed aside such questions. "We are used to it," he said. "I have lived here for years, so my body has gotten used to this air."

I heard the same from countless other locals. Foul air was simply a fact of life in Beijing, they believed, an inevitable result of the sharp increase in motor vehicles, office buildings, neon lights, private shops, karaoke restaurants, and other forms of economic activity that had breathed new life into the city in recent years. Zhenbing did not enjoy breathing such air—he wasn't stupid. But to someone who had learned in childhood how trying life without coal could be, he saw pollution as the lesser of two evils. Observing Zhenbing's stoicism, I was reminded yet again of the young Ugandan at the park above the Nile, who defended damming the river because it meant electricity for his people. Zhenbing regarded extra pollution as a trade-off he was ready, even eager, to make in return for warmer apartments, busier factories, fewer power shortages, and a higher standard of living. Who could blame him?

Despite its terrible health effects, China's dependence on coal is bound to continue, if only because coal is one of the few natural resources China has in any abundance. No other country produces or consumes as much coal as China does. When I visited Datong, an ugly, low-slung city in Shanxi province known as China's coal mining capital, residents proudly told me time and again, "Datong sits on a sea of coal!" The Number 9 Coal Mine, ten miles from downtown Datong, produced ten thousand tons of coal a day, and there were fifteen mines of equal capacity in the near vicinity, plus hundreds of smaller privately owned ones. All day long, the roads around Datong were filled with sky-blue cargo trucks brimming with chunks of coal. I saw coal transported by virtually every other means imaginable, too: trains, smaller trucks, and wooden carts pulled by three-wheeled motorcycles and even by donkeys. The roadsides were crowded with members of the poorer classes, who used shovels, brooms, and bare hands to scoop fallen ore into baskets they balanced across their shoulders for the walk home. Although Datong's streets were almost literally paved with coal, I did not appreciate the sheer volume involved until the morning Zhenbing and I left town. Rolling south through Shanxi province, our train passed a number of huge, looming mountains of coal. The first mountain was easily seven stories tall and extended along the tracks for nearly two miles. On top of the mountain I saw a yellow bulldozer, shoving around the black ore and looking as tiny and inconsequential as a child's toy on a sandpile.

Coal accounts for three-quarters of China's total energy consumption, which is the other reason China's coal dependence is likely to last for decades: virtually the entire national infrastructure runs on coal. The exception is the transportation system, which is based on petroleum and human muscle. But three-quarters of electricity production and virtually all the factories and heating are coal-powered. To replace or upgrade this infrastructure—the hundreds of electric power stations, the hundreds of thousands of boilers in factories and apartment buildings, the millions of honeycomb stoves—is an essential but necessarily long-term project, especially because Chinese of all classes are impatient to have economic progress *today*.

After all, this is a country where the overwhelming majority of people does not have a refrigerator, electricity shortages are constant, and one hundred million peasants live without any electricity whatsoever. "Electricity shortages exist in all the big cities, except Beijing," Lang Siwei, director of the Air Conditioning Institute of the Chinese Academy of Building Research, told me. "Shortages are very severe among Yangtze River cities, especially small cities, which often have no electricity during the daytime hours of peak demand. The only reason Beijing is spared such shortages is that electricity gets diverted from other cities to keep the capital running."

"In 90 percent of the villages there is electricity, but it can be cut off at anytime because of the pervasive shortages in the system," said Zhou Dadi, deputy director general of the State Planning Commission's Energy Research Institute. "Most rural families have electric light, and a few might also have a television and a small refrigerator. But not more than that, because the weak fuses and supply lines can't handle it."

One reason for the shortages, of course, is that demand for refrigerators and other consumer appliances has skyrocketed in recent years. Only 7 percent of urban households had a refrigerator in 1985; by 1994, 62 percent did. During the same period, possession of color televisions increased from 17 to 86 percent of households; clothes washers, from 48 to 87 percent; and air conditioners, from zero to 5 percent. In the countryside, where more people lived but there had been much less development, the steepest increases were of electric fans and black-and-white televisions, which increased from approximately 10 percent of households in 1985 to 81 and 62 percent, respectively, in 1994. Clothes washers, color TVs, and especially refrigerators remained rare in rural areas—approximately 15 percent of households had one or more of these items—but here too the trend lines were climbing sharply.

Electricity is the fastest-growing part of China's energy demand. Its growth averaged 9 percent a year from 1984 to 1994 and was projected to continue at no less than 7 percent a year through 2000 and beyond. That means that China's total electrical generating capacity has to double every decade to keep pace. Thus, declared Lang Siwei, "electricity shortages are certain in China for the next five years. Whether they continue after that will depend on two things: whether we can improve our energy efficiency sufficiently, and whether we can build all the new power plants we have planned." According to Pan Baozheng, a senior engineer with the State Science and Technology Commission, China was planning to build thirty to sixty new electric power plants every year for the foreseeable future. The plants would have capacities of 300 to 600 megawatts each, and 75 percent of them would be coal-fired. China is "the biggest market for electric power plants in the world," Pan told me with pride.

All this economic growth has led experts to predict that China's total coal production, which reached 1.3 billion metric tons in 1996, would double if not triple by 2020 at the latest. Much will depend on how well China implements the energy efficiency and other infrastructure reforms mentioned earlier, according to William Chandler, director of the Advanced International Studies Unit at the Battelle, Pacific Northwest Laboratories of the U.S. Department of Energy. "Our scenarios predict either a doubling of coal consumption with efficiency reforms, or a tripling without them," said Chandler, who has been regularly visiting China and collaborating with its energy planners for over a decade. These increases would occur "by 2015 if one assumes a continuation of rapid economic growth in China, by 2020 with moderate growth," Chandler added.

Besides adding to China's problems with filthy skies and battered lungs, the additional coal burning would threaten the rest of the world by greatly increasing China's production of acid rain and greenhouse gases. In geographic contrast to the TSP problem, acid rain is most pronounced in the southern half of China, because the sulfur content of southern coal is higher and southern soils are naturally more acidic. As with TSP, sulfur dioxide levels in much of China are many times greater than the World Health Organization guideline of 40 to 60 micrograms per cubic meter annually. Among China's most afflicted cities were the coal center of Taiyuan, where the average level in 1996 was 230. Worst of all was Chongqing. Its 1996 average reading was 320, but in winter it climbed above 600!

Acid rain affects 29 percent of the land area in China and causes $5 billion worth of damage every year. Emissions of sulfur dioxide—the source of acid rain—corrode buildings, encourage respiratory disease, and damage forests, lakes, and agriculture. In the Chongqing area, 24 percent of the 1993 vegetable crop was damaged by acid rain, with cereal crops suffering equivalent losses. In a country that is straining to feed itself, and where the rural majority is so poor it relies on wood and other biomass for 70 percent of its fuel, such losses are especially costly.

Acid rain is "quite a serious problem" for China, said Wang Wenxing, a senior advisor at the Chinese Research Academy of Environmental Sciences, "but in the future it will be much worse," because its effects can take decades to manifest. "In the United States," added Wang, "you can see the effects of your earlier industrialization today. In China, the industrialization is under way now, so the effects will be growing later." Since sulfur dioxide emissions can travel many hundreds of miles before falling to earth as rain, China's coal burning has dramatically worsened the acid rain problem in both Japan and South Korea; China is responsible for half of Japan's acid rain and 80 percent of South Korea's. In Japan, as in Europe and the United States, electric utility companies were forced in the 1970s and 1980s to add "scrubbers" that removed sulfur dioxide from their coal plants' exhaust. Tokyo is now funding an extensive program of technology transfer aimed at outfitting Chinese plants with their own scrubbers.

No such technical fix is possible with the carbon dioxide emissions that enhance the greenhouse effect, however. Coal is the most potent carbon dioxide producer of all fossil fuels; it emits 25 percent more carbon dioxide per unit of energy use than petroleum does, and nearly twice as much as natural gas. Although China relies on petroleum for 17 percent of its energy consumption, its heavy reliance on coal is what makes it a greenhouse giant. By 1990, when China emitted 580 million tons of carbon dioxide a year, it had already surpassed the former Soviet Union as the world's second largest producer of greenhouse gases and trailed only the United States. With its immense coal reserves (over 165 billion tons), huge population, and booming economic modernization program, China will at least double and perhaps triple its greenhouse emissions by 2025, overtaking the United States and profoundly affecting the global struggle against climate change. According to the scientists of the IPCC, greenhouse gas emissions must decline by 50 to 70 percent to stabilize current concentrations of greenhouse gases. If China's carbon dioxide emissions grow as projected between 1990 and 2025, that growth alone would actually increase global emissions by 17 percent.

Without a radical shift in policies elsewhere in the world, such an increase would doom efforts to achieve the IPCC's target. It would accelerate the global warming already under way and plunge the world into potentially catastrophic territory. And China itself would by no means be immune. A study done by China's National Environmental Protection Agency (NEPA), the World Bank, and the UN Development Program concluded that a doubling of global carbon dioxide concentrations would have the following impacts on China: storms and typhoons would become more extreme and frequent; much of China's coastline, including the economic powerhouses of Shanghai and Guangdong province, would face severe flooding (an area the size of Portugal would be inundated and an estimated sixty-seven million people displaced); agricultural production would be affected most strongly of all, with increased drought and soil erosion lowering average yields of wheat, rice, and cotton; livestock and fish production would also decline. Again, for a country already straining to feed itself, such setbacks could be disastrous.

During our weeks of traveling together, Zhenbing and I fell into a running debate over which city had the dirtiest air in China. Was it Beijing? Zhenbing wouldn't consider it—mainly, I suspected, because he lived there. Was it Benxi, the Manchurian industrial city whose pollution was so thick that the city had vanished from satellite photos in the 1980s? Possibly. Though Benxi was now visible again from outer space, local officials admitted that its TSP levels remained very high. And my interviews with local residents suggested that what progress had been made in Benxi stemmed as much from widespread factory bankruptcies as from the government's vaunted cleanup campaign.

For a while I leaned toward the coal center of Datong. As bad as Beijing's air was, it was unusual to see smokestacks there belching pure black smoke; the pollution seemed more dispersed somehow. But in Datong, black emissions were routine and ubiquitous. Standing in the square in front of the train station, I counted three such thick black plumes within two hundred meters of me, and dozens more in the distance. But Datong was soon supplanted by Taiyuan, another major coal center that was located a day's train trip south of Datong. Notwithstanding its status as the provincial capital, Taiyuan likewise seemed to impose no controls over smokestack emissions. But it had a population of four million, nearly five times that of Datong, so its air was as soupy and gray as a foggy day in San Francisco, even though there was no natural fog within one hundred miles. Zhenbing insisted Taiyuan's air was the worst yet—"I feel choked to death!" he exclaimed as we hiked the mile and a half from the train station to central downtown—and it was hard to argue with him. Taiyuan had registered TSP concentrations in winter that were twenty times worse than the maximum daily standard in the United States, and its average readings in 1995 were a shocking 568.

Another formidable competitor was Xi'an, the ancient imperial capital known the world over for the massive collection of terra-cotta warriors buried outside of town. A splendid bell tower and massive city wall dating back to the Ming dynasty enhanced Xi'an's reputation as perhaps China's loveliest city. It gave me great sorrow to miss seeing these architectural treasures, but Xi'an's pollution made viewing them, or much of anything else, impossible. Even on a sunny day, the only sign of the orb itself was that one patch of sky was a bit brighter than the rest. As a test, I timed how long I could stare at that artifically veiled sun without hurting my eyes. After sixty seconds, I stopped counting.

But then Zhenbing and I arrived in Chongqing, the Yangtze River city where we encountered the chlorine waterfall and gas explosion described at the beginning of this book. It was in Chongqing that locals told me that, on the worst days of fog and smog, "if you stretch your hand out in front of your face, you cannot see your fingers." But during the bus ride in from the airport, Chongqing looked inviting enough. The mountain chain north of the city may have hindered air circulation, but it meant that Chongqing was governed by monsoon, rather than Siberian, weather systems. Instead of the parched, treeless landscape Zhenbing and I had seen everywhere up north, the gently sloping fields here were green and moist, dotted with trees and layered in narrow, well-tended terraces. Instead of subfreezing temperatures, the mercury was in the upper forties—almost tropical. As soon as we checked into our hotel, I gratefully shed my long underwear for the first time since arriving in Beijing three weeks earlier.

We spent the next morning at the Chongqing Environmental Institute, interviewing Peng Zhong Gui, an air pollution expert. Asked how Chongqing's air compared with that in northern Chinese cities, Peng was unequivocal: "The air pollution here is the worst in China. We are number one." Mr. Peng was not bragging. He had grown up in Chongqing, and he lamented what had happened to his hometown. The sulfur dioxide readings here were markedly higher than anywhere else in China, he said: an annual average of 320 micrograms per cubic meter, with daily spikes as high as 900. (WHO standards, of course, are 40 to 60.) Chongqing's particulate readings, while not the highest in China, were nevertheless above 300 micrograms per cubic meter—not far below those of Beijing. And because Chongqing burned little coal for heat, its TSP readings stayed relatively constant through the year, unlike Beijing's, which fell in summer. (Like all Chinese cities bordering on or south of the Yangtze River, Chongqing was discouraged from winter coal burning by the central government in order to save fuel. China may have had plenty of coal under the ground, but it had neither the infrastructure nor the money to mine and transport that coal to everyone who might want it.)

"Rates of lung disease and cancer in Chongqing are far higher than in other Chinese cities," Peng continued. "In 1989, there were fifty-four fatal cases of lung cancer for every one hundred thousand people; the Chinese average should have been ten cases." Peng's studies also revealed that workers' lungs didn't function nearly as well on smoggy days as on clear ones. Not that there were many of the latter. "In Chongqing we have only a few days a year when visibility is ten kilometers," Peng said. "Back in 1949, ten kilometers was the average visibility throughout the year. But since then, the average has dropped to three kilometers today, and it continues to fall year by year."

After the interview, Zhenbing and I walked back to our hotel, crossing the four-lane bridge that towered above the Jialing River. The bridge was half a kilometer long, but we could see only faint outlines of the office and apartment buildings at the far end of it. Even the river below us was scarcely visible through the dense gray mist. Halfway across the bridge, a wicked coughing fit came over me, and soon Zhenbing was hacking, too. At the end of the bridge, we had to leap aside when around the corner suddenly came two men, half hopping, half jogging beneath the weight of a massive carpet rolled around a bamboo pole slung across their shoulders. Zhenbing and I carried on through a shopping district whose streets were lined by yellow-nut trees. Anywhere from ten to thirty feet tall, the trees had leaves as long as a man's thumb and twice as wide; unfortunately, they were covered so thickly with dust that their natural green was rendered a sickly, puffy gray, like bread mold.

It was hard to tell how much of the muck in Chongqing was fog and how much of it smog—until we got lucky one day and the fog cleared long enough for the sun to make a very rare winter appearance. The sun I saw was so white it looked like the moon, except much smaller. The sky remained gray, but with lighter, yellowish tones, while visibility extended to perhaps three kilometers. The pollution that remained seemed not as bad as Taiyuan's—more like Xi'an's. I repeated my test of staring at the sun. I could not go on staring forever, as I had in Xi'an, but a ten-second gaze was no problem.

Yet Chongqing's air quality was arguably better than it had been in years. Average visibility may have been declining over time, as Peng Zhong Gui said, but on an annual basis both sulfur dioxide and TSP readings were improving. The first interview Zhenbing and I did in Chongqing was with Wang Gang, an elfin man with thick salt-and-pepper hair who had recently retired as the head of planning for the Chongqing Environmental Protection Bureau (EPB). Back in the early 1980s, Wang said, TSP levels were over 600 and sulfur dioxide levels were 450. That meant TSP levels had been cut by half over the past fifteen years and sulfur dioxide by nearly a third, despite solid economic growth—an improvement Wang attributed to the EPB's push to substitute natural gas for coal as the city's main cooking fuel. Some 75 percent of city households now relied on natural gas, said Wang, but current reserves would be exhausted in ten years, and he was not optimistic more would be found.

Air pollution remained the chief environmental problem in Chong-qing, Wang stressed, but the water pollution did not sound much better. Only 40 percent of the five hundred million cubic meters of industrial wastewater annually generated in the city were even rudimentarily treated before being discharged into the Yangtze and Jialing rivers. And virtually all the two hundred million cubic meters of household sewage annually produced went straight into the rivers without treatment, even though many people drew their drinking water from the rivers. (This was a commonplace in China, where only 4.5 percent of household sewage was treated before entering waterways.)

Wang Gang's words made a strong enough impression on me that I could hardly believe my ears a few hours later when Zhenbing suggested that we order fish for dinner at a local restaurant.

"Fish?" I asked. "Are you crazy? Don't you remember what Wang Gang told us about the pollution they dump in the Yangtze?"

"I am used to this kind of water," Zhenbing replied calmly. "The fish we eat in Beijing is not so different, I think. We do not have to eat fish if you do not want it, but for me it is no problem."

"Zhenbing," I remonstrated. "You may be used to it, but that doesn't mean it's good for you."

"I have been exposed to this kind of pollution for many years," Zhen-bing said. "For you, it is perhaps not a good idea to eat such fish. Your body is not used to the situation here in China. But for me, it is quite okay. My body has experience with such pollution. This is why I do not take your medicine. [Zhenbing had been coughing, and I had urged him to try some of my echinacea pills.] Ever since I was a boy, I have never taken such medicine because it weakens the body against the next cold. So I am rarely sick."

"But pollution doesn't work that way, Zhenbing."

He wasn't listening. He had launched into a condemnation of Dream Lady, a woman we had interviewed that afternoon. Zhenbing had nick-named her Dream Lady because she dreamed of moving to Beijing to es-cape Chongqing's dreadful air and water pollution, which, she said, made her sick nearly every day. She and her husband had visited Beijing in 1988, and she remembered it as a clean, cultured place with no pollution or traffic. When I gently disabused her of these outdated impressions, she was distraught: "But Beijing is my dream!" Zhenbing took no joy in

Dream Lady's disappointment, but he had little patience with her whining about Chongqing's pollution and the headaches it gave her.

"She will not live long," he said brusquely. "She is too soft. You have to train your body to tolerate such things. That way it stays strong enough to defend itself."

As we argued back and forth, I gradually came to realize an astonishing fact: Zhenbing, an extremely well-educated person by Chinese standards (less than 2 percent of eligible young people in China receive postsecondary education), was apparently ignorant of the basic biology of cancer. In keeping with the survival skills learned during his peasant childhood, he knew that a body could build up tolerance against the infections that cause colds. The problem was that he assumed the same principle held for the chemicals that poured out of China's factories. But then no one had ever informed him otherwise.

"I am used to it."

If I heard that phrase once during my six weeks in China, I heard it fifty times, and not just from Zhenbing. China's air pollution gave me a chest-burning cough simply because I was a foreigner; my hosts, on the other hand, were "used to it." On an overnight train to Shenyang, Zhenbing and I shared a compartment with a self-made Chinese capitalist who had struck out on his own fourteen years before and ended up a millionaire. Like his father and brother, this man had been a metalworker in Shenyang, earning, he recalled with ironic precision, 38.5 yuan a month. When the market reforms came, in 1982, he took his tools and expertise and set up private shop. After four years, he hired his first employee. Now, at the age of forty-one, he employed two hundred workers, owned five factories, and made ten million yuan in profits a year, about $1.25 million. He drove a diamond-blue Lincoln Town Car, wore a suede leather jacket, spent one million yuan a year (reluctantly) entertaining business clients in five-star restaurants, and bought private tutors for his thirteen-year-old daughter. He didn't really enjoy his work, the man told me; the constant chase after higher profits was just a way to keep score. Since he worked seven days a week, he added, he didn't have time to spend his money anyway.

After an hour of hearing about his life in Shenyang, I asked whether the pollution there was as bad as people said. He could not have been less interested in the topic; he was too busy with business to worry about such things, he said. But he was polite, and happy to be speaking with a foreigner, so when I raised the question a second time, he laughed and said, "If you think the air in Beijing was dirty, wait until you see Shenyang!"

"So why doesn't it worry you?" I asked.

"Because I grew up in Shenyang, and I am used to this kind of air," he explained.

"Well, what about your daughter? Don't you worry about her breathing such air?"

He flashed me a quizzical look, as if I must have been too slow to grasp his point the first time. "But she too was born in Shenyang, so she is also used to this air," he replied.

Exactly what being born in Shenyang could do to one's health was explained to me the next morning by Dr. Xu Zhaoyi, the chief of environmental epidemiology for Liaoning province. It was very cold in Shenyang that day, with temperatures in the teens and fierce winds, but Dr. Xu was gracious enough to wait on a nearby street corner for Zhenbing and me; a taxi, he said, would never find the side street where his public health station was located. A slender, dignified man with longish gray hair, Xu told me in excellent English as we walked back to his office that he had been working in Shenyang for thirty years. Soon, he would spend three months at Harvard University collaborating on a project on the health effects of pollution in China.

Inside the health station, we walked the three flights up to our meeting room; elevators were an extravagance rarely found in China's public buildings. On the upstairs landing we were joined by two young associates of Dr. Xu, who led us down a long, bare corridor to a meeting room. We stepped into a room furnished with little more than a conference table and some folding chairs, and before we even had time to sit down I realized that the room was completely without heat; the only difference from being outside was that we were out of the wind. One of the assistants tried to rectify matters by bustling over to the window and flipping some knobs on what looked like an antiquated air-conditioning unit, but he succeeded only in shooting frigid air into the room. He gave up, flashed a nervous smile, and sat down. I knew that public buildings south of the Yangtze were routinely left unheated to save fuel, but we were in

Manchuria, north of the Korean border. I also knew that workplaces in Beijing had gone unheated as recently as the 1970s, and even now thermostats in the capital were set so low that everyone wore long underwear beneath their work clothes. But it seemed clear that my hosts here in Shenyang were embarrassed not to have heat for a foreign guest, so in order not to make their loss of face any worse, I said nothing. For the next two hours, we all sat in our scarves and overcoats as we discussed the trio's findings, each of us acting as if this were perfectly normal. As Dr. Xu began the briefing, I wondered how many other freezing conference rooms this compassionate scholar had endured over the past thirty years.

Shenyang's air pollution had declined since a government cleanup program was initiated in 1981, said Dr. Xu, but the city's TSP levels nevertheless remained among the highest in the world. Dr. Xu and his colleagues had divided Shenyang into three pollution zones—low, medium, and high. Between 1986 and 1994, TSP readings in the low area averaged 366, equal to readings in Beijing. The medium area, however, was 478; and the high, 518. Juxtaposing these readings with morbidity statistics revealed that higher air pollution significantly increased death rates from strokes, chronic obstructive pulmonary disease, heart disease, children's pneumonia, and lung and esophagus cancer. "In Western countries, cigarettes account for 90 percent of the lung cancer risk," said Dr. Xu, "but in China it is 50 percent, because we have much more pollution." Of course, the weakest citizens fared the worst. When TSP levels stayed high for three consecutive days, said Dr. Xu, the city's daily death rate would increase by 10 percent as the old and the sick died.

After lunch, Zhenbing and I headed across town to visit the high pollution area, a neighborhood with lead battery, copper smelting, heavy machinery, and numerous other factories. On the way, we passed through the town square, where a ninety-foot-tall Cultural Revolution era statue of Mao, the largest in all China, was surrounded by English-language billboards hawking Shenyang's other big killer, cigarettes. We got lost twice (at this point, I was only beginning to discover Zhenbing's hopeless sense of direction), but eventually we made our way to what locals had nicknamed Unemployment Road, a narrow lane that curled off a major street and over the railway tracks. Here, peddlers wrapped in the usual countless layers of clothing stood stoically behind meager piles of goods: sunflower seeds, cheap gloves and shoes, cabbages frozen and mottled by the cold. The bridge over the tracks afforded a view of the western expanse of the city, where the plumeless spouts of dozens of smokestacks offered mute explanation for the peddlers' woeful circumstances. Many state-owned factories were not surviving the shift to a market economy; unemployment in Shenyang was said to exceed 30 percent.

Across the tracks and down a hill was a residential area of one-story brick buildings with rotted wooden doorways and roofs caked with soot. As we picked our way down the garbage-strewn, ice-slicked street, a steady stream of bicyclists and pedestrians came the other way and, suddenly, a gleaming black Audi sedan, beeping its horn to clear a path. We crossed a busy intersection and carried on past apartment buildings with bunches of Chinese onions hanging from windows like socks drying in the wind. After another half mile, we saw looming before us the heavy machinery factory Dr. Xu had singled out as one of the zone's most deadly polluters.

The sky was now melting into dusk, and the cold was piercing more deeply. Before Zhenbing and I could find anyone to approach for a chat, the street filled with hundreds of marching children. Chattering and laughing, they walked in rows six abreast, with brightly colored nylon book bags strapped across their backs. Beneath a streetlight on the corner, a couple of fruit peddlers were stamping their feet for warmth beside their pushcarts. Only when we came within speaking distance was it evident that the vendors were women. Both wore the olive-green greatcoats of the Chinese military, while white face masks covered their noses and mouths and shawl-like red scarves wound about their necks and ears, leaving only their eyes and the tops of their heads exposed. I bought a kilo of tangerines from one woman, who told me she lived just down the street. Many of her neighbors worked in the heavy-machinery factory, but they hadn't been paid for the past few months. Asked why, she replied simply, "The market economy."

Noticing puffs of gray smoke floating out of one of the factory's smokestacks, I said it seemed still to be operating.

"Oh, yes, people are still working," she said. "If they didn't, they could be fired. They are worried about keeping their pensions, collecting their back pay, and keeping their jobs when the factory starts making money again."

I asked whether her face mask was to protect her from the cold or the factory's pollution, and it was clear that the latter idea had never occurred to her. Removing her mask briefly, she squeezed her red, weather-burned cheeks and said with a bright, cheerful smile, "Try standing here in this cold all day! Without a mask, your skin will freeze in an hour." Laughing, she added, "Look at you, your cheeks are very red already."

By now a small crowd was gathering, drawn by the spectacle of my white face, and the atmosphere was welcoming and jovial. A short man with a droopy moustache and buck teeth was keen to know how tall I was, what size shoes I wore, when and where I was born. After a few minutes of such friendly talk, I asked again whether any of them worried about the plant's pollution. They impatiently waved away the subject, assuring me, "No, no, it's quite okay here," to return to what really intrigued them: whether I had been born in the Year of the Monkey or not.

Without spoiling the mood, I made one last attempt a few minutes later, telling them I had spoken just that morning with an eminent scientist who had said that the pollution from these factories was dangerous, and they should therefore be careful. But it was hopeless. The man with the moustache patted my arm and said with a smile, "This was good advice he gave you, because you do not come from around here. But we live here, and we are used to it."

The dilemma of these wonderfully hospitable but tragically ignorant workers in Shenyang was shared by millions of their counterparts, not only in China, but in Russia and many other countries struggling to join the capitalist world: a choice between unemployed misery on the street and jobs in factories that poisoned them. The workers' lack of concern about the pollution raining down on them reminded me of something I had been told back in Beijing by Ma Zhong, director of the Institute of Environmental Economics at Renmin University. Most Chinese, Ma said, even if made aware of the health effects of pollution, might still decide to take their chances if it meant more money in their pockets. "I saw a young man interviewed on television the other night," Ma told me. "They asked him, 'What is your number one dream in life?' 'Money,' he said. 'What is your number two dream?' 'Money.' 'Number three dream?' 'Money.' " Professor Ma was laughing by now, but his point was serious.

"Outsiders have to understand that kind of thinking if they want to influence China's environmental policies."

Most Chinese accepted the familiar view that economic growth required environmental damage, and they were quite ready to pay that price. "We have a saying in China," said one journalist who had tried to raise public awareness of the subject. " 'Is your stomach too full?' In other words, are you so well-off you can afford to complain about nothing? This phrase is used for Americans who talk about saving birds and monkeys while there are still many Chinese people who don't have enough food to eat."

The lack of environmental awareness in China may have even deeper roots, however. Denial of unpleasant realities had become part of the Chinese personality in recent decades, argued Orville Schell, the dean of the School of Journalism at the University of California at Berkeley and author of many books on China. "A society that has for decades had to ignore so many unjust and irrational things in order to just get along—the injustices of the gulag, families ruined during the Cultural Revolution, other kinds of government barbarity, the lack of a believable news media—is one in which the capacity to avoid recognizing all sorts of problems, including environmental ones, has become essential to survive," Schell told me.

The environmental movement in China thus had its work cut out for it—or it would, if it existed. The few individuals who dared work on the issue said that education was by necessity the top priority. Liang Conjie, the founder and president of Friends of Nature, one of the very few independent environmental groups in China, said his organization got permission to operate because it registered as a cultural, rather than political, group. He added that "Friends of Nature could never oppose the government directly, the way Greenpeace would. That will not work." Liang focused instead on raising public consciousness, particularly by prodding Chinese journalists to cover the environmental issue more attentively. With a mere four hundred members, Friends of Nature "cannot be involved in each of the thousands of environmental conflicts in China," said Liang. "But when people can be helped to understand these problems and how they can affect them, the government will have to respond."

It would be hard to overstate the power that state-run media exercises in China, so it is not surprising that Liang welcomed the increased media criticism of environmental problems that began appearing in 1996. That increase was orchestrated by the government itself, which had come to fear that environmental degradation was endangering overall economic growth. Chinese journalists were taken on official tours of environmental hot spots and encouraged to expose what they found. "To some local officials, a document issued in Beijing carries little urgency, but if you publicize the problem with stories on radio and television and print their names in the newspaper, suddenly they are in a hurry to fix the problem," grinned Yu Yuefeng, the staff director of the National People's Congress environment committee.

But there were definite limits on what the official media would say. "There has been more coverage, but it focuses mostly on what the government is doing to improve things, not on how bad the situation actually is," complained the journalist quoted earlier. "In my stories, I always have to begin with something positive—how NEPA has announced new policies to protect the air or raise water quality—not with how the pollution got there in the first place and what its exact effects are. So people don't know." Indeed, journalists were themselves often kept in the dark. This reporter, for example, was fluent in English and spent five days a week in a fully equipped newsroom in Beijing. Nevertheless, he was unaware of the biggest international news story of the day: the weeks of massive street demonstrations in Belgrade that were threatening to overthrow the government that had started the Bosnian war.

In any event, the official media was unlikely to raise a fuss about policies near and dear to Beijing, such as the Three Gorges dam project that had drawn so much criticism from environmental activists and governments around the world. In fact, uniformly positive media coverage helped explain why domestic opinion in China appeared to favor the project, just as Zhenbing had told me the day we visited his dorm room. Residents I interviewed in the Three Gorges area supported the dam by a ratio of approximately eight to one, mainly because they believed that the government's resettlement program would improve their standard of living. Just like the government said it would.

Coal is the inescapable fact at the center of the Chinese environmental crisis, and there is no easy (or even not so easy) remedy. Although China has taken many steps to mitigate the damages caused by burning coal, its options are limited by a scarcity of everything from water to alternative fuels to investment capital. For example, moving large factories out of city centers, as has been done in Beijing and Chongqing, does not reduce overall air pollution, but it does make the air breathed by city residents less debilitating. More effective, however, would be to "wash" the coal that factories and power plants use before it is burned. Washing, a chemical process routine in the United States and Europe, is especially important for the high-sulfur coal common in China. Nevertheless, only 20 percent of China's coal gets washed, partly because China is desperately short of water; its per capita supply is less than one-third of the world average. Lack of capital is equally problematic, according to Yu Yuefeng of the National People's Congress. "We know the coal should be washed first, but the investment needed to establish coal washing facilities is really enormous," said Yu. "We would need dozens of billions of yuan to achieve this. The Coal Ministry cannot afford it, and since most coal mines are also in poor economic shape, neither can they. So it will be a while before we can increase this 20 percent very much."

The same economic calculus undermines the prevention of acid rain. "Only a very small number of Chinese power plants have sulfur dioxide scrubbers installed, which is a really serious problem," said Wang Wenxing of the Chinese Research Academy of Environmental Sciences. "But China lacks capital. When setting up a power plant, 15 to 30 percent of the investment funds should be spent on environmental technologies like scrubbers. So China has a choice between building four plants that emit sulfur dioxide and other pollutants or three plants that do not."

Not all environmental technologies are so costly. Perhaps the most important are electrostatic precipitators, which remove particulates from coal smoke yet add only 1 percent to a power plant's capital cost. Such precipitators have been required on all large Chinese power plants since 1985; they reduce emissions of the largest particulates by 90 percent. Unfortunately, the most recent research indicates that precipitators are useless against the smallest, most harmful types of TSP. "The most dangerous particulates are those less than 2.5 micrometers in diameter," said Wei Fusheng, deputy director of China's National Environmental Monitoring Center, "and neither we nor anyone else in the world has the technology to eliminate them." Wei was among the scientists who served on a long-term research project sponsored by the U.S. Environmental Protection Agency that studied the health effects of air pollution in four Chinese cities: Guangdong, Wuhan, Chongqing, and Lanzhou. He explained that smaller particulates were the most dangerous for two reasons: "First, they get attached to heavy metals like oxidized mercury, lead, and other organic compounds that remain in the atmosphere for a long time. Second, because the particulates are so small, they are not blocked by our bodies' natural defense systems. So they can enter directly into the lungs and even the bloodstream."

Yet China's reliance on coal and its yearning for economic growth are so entrenched that even a health expert like Wei did not oppose doubling the nation's production of coal. Instead, he argued, "we must reduce the health effects of coal burning while we increase the production of coal." In addition to advocating the washing of coal, Wei suggested locating power plants at the mouth of coal mines. "In Beijing, the environmental carrying capacity of the city is reaching its maximum, so no more pollution should be allowed there," he said. "But we can set up a power plant [hundreds of miles away] in Inner Mongolia, next to the coal mine, and transmit the power from there to Beijing." This technical fix would create other problems, however. Where would the vast quantities of water required for the power plants' cooling towers be found in such an arid region? And not even a remote geographic location could neutralize the acid rain and greenhouse gases produced by coal-fired power plants.

So what is to be done? From an environmental and public health standpoint, the best idea is simply to use less coal in the first place. Many Chinese might regard this as a recipe for economic stagnation, but in fact the nation's recent history suggests otherwise. During the 1980s, as part of the economic reform program championed by Deng Xiaoping, China sharply reduced its rate of coal consumption by increasing its energy efficiency. Because the energy intensity of China's economy (i.e., the amount of energy used per unit of GNP produced) fell by 30 percent during the 1980s, the economy was able to grow an average of 9 percent a year even as energy consumption grew by only 5 percent a year.

This remarkable accomplishment was the result of deliberate government policy, implemented in response to the global oil crisis of 1979.

[handwritten margin note: trying to find another place to pollute instead of fixing the problem]

Energy planners realized that, in an era of rising oil prices and uncertain supply, China could not hope to produce or buy enough energy to achieve the quadrupling of the Chinese economy by 2000 that Deng Xiaoping wanted; the only option was to use energy more efficiently. Toward that end, economic restructuring shifted the emphasis of China's economy from heavy to light industry. Subsidies and price controls that encouraged waste by making energy appear cheaper than it was were reduced. Introduction of more advanced boilers, fans, and pumps in such core industries as electricity, steel, chemicals, concrete, and fertilizer further increased energy efficiency. More visible perhaps to the average Chinese was the phaseout of honeycomb home stoves—the least efficient, most polluting means of burning coal. This phaseout was pursued in concert with the demolition of vast tracts of *huotongs,* the traditional, one-story courtyard dwellings of urban China, where residents invariably used home stoves for cooking and heating. The *huotongs* were replaced by high-rise apartment buildings that relied on central heating from industrial-sized boilers.

As China prepares for the twenty-first century, efficiency remains the key to its energy future, the one bright spot in an otherwise gloomy picture. Because China began from such a low level of efficiency, there is still a long way to go—which, ironically, is good news for the environment. If the most efficient equipment and processes currently available on the world market were installed throughout China's energy system, the country's energy consumption could be cut by 40 to 50 percent, according to studies conducted by Zhou Dadi and his colleagues at the Beijing Energy Efficiency Center. For example, by introducing the most advanced technology for steelmaking and iron smelting at a single large factory in Baoshan, the government reduced the energy intensity of the entire steel industry by 10 percent. Meanwhile, experts at the Lawrence Berkeley Laboratories in California have been working with Chinese counterparts to develop refrigerators that would consume half as much energy as today's models. "The Chinese now manufacture more refrigerators than anybody in the world, ten million a year," said Mark Levine of Berkeley, adding that transforming the Chinese refrigerator industry "would make a major difference" to China and global greenhouse gas emissions. To encourage such reforms, the World Bank has been helping to create energy management corporations that would spread the word about efficiency opportunities within China. The corporations will approach factory managers with a simple message, said the bank's Robert Taylor: "Managing your factory in an energy efficient way will increase your bottom line. And if you don't believe it, here are some case studies that prove it."

Nevertheless, despite the enormous potential of efficiency improvements, the thrust of Chinese energy policy remains the expansion of supply. Although the government directs 6 to 10 percent of its energy budget to efficiency—much more than most other countries, especially poor ones—it invests ten times that much in expanding supply. The emerging private sector is similarly inclined.

Smitten with a capitalism they do not fully understand, many in government think that efficiency no longer needs government support now that prices better reflect economic realities. "Some people in China assume the market will take care of everything," said one top Western consultant. "They are holier than the pope!" In truth, however, since markets rarely internalize the true environmental costs of a given action, their price signals often discourage responsible behavior. For example, a provincial government in China could choose between investing U.S. $1,000 per kilowatt for the standard Chinese power plant, which operates at 30–35 percent efficiency, or U.S. $1,400 for an American-made plant that delivers both 45 percent efficiency and a 90 percent reduction in sulfur dioxide and nitrogen oxide emissions. Obviously, the American plant is environmentally superior, and it would probably save money in the long run through lower fuel bills as well. But most investors would choose the cheaper Chinese power plant because pervasive capital shortages force them to focus on the short term.

The shortage of capital is most decisive for the smallest end users, who happen to burn more than 50 percent of the coal consumed in China. Particularly in need of overhaul are the nation's estimated 430,000 industrial boilers, many of which were built in the 1950s on the basis of designs from the 1920s. However, "lack of capital means that up-front costs are far more important to the managers who make these investment decisions than lifetime costs are," said Jonathan Sinton, an analyst at the Lawrence Berkeley Laboratories who has interviewed many of these managers. "So instead of replacing what they have, they use the scarce capital to get a new piece of equipment, and they probably don't spend the extra money needed to get environmental features for that equipment."

What about supply-side alternatives to coal? Solar, wind, and nuclear energy each supply less than 1 percent of total electricity demand. Some planners envision a bigger future role for nuclear, but its high costs and safety problems argue against this. Chinese officials did announce a $2 billion reactor purchase from Russia in 1997, a tempting of fate if ever there was one. Not only was China buying the most accident-prone reactor technology in the world, the reactor would be sited on China's southern coast, amid perhaps the densest concentration of humans on earth. So, what about other options? Hydropower delivers 24 percent of China's electricity, and total installed capacity is expected to quadruple by 2020. However, because electricity makes up only 15 percent of China's total energy use, even hydropower is destined to remain a small share of the overall energy mix. To lower coal consumption an appreciable amount, the heating and industrial sectors have to be tackled. One option under discussion is to expand exploration for natural gas in western China and perhaps even import gas from Russia and Turkmenistan. The most optimistic forecasts hold that natural gas could cover 10 percent of China's total energy consumption by 2005—an important step in the right direction, but a relatively small one.

Coal therefore seems destined to remain a large and growing source of energy for China well into the twenty-first century, despite the ominous implications this carries for pollution in China and climate change for the planet as a whole. Bill Clinton has said that he told Chinese president Jiang Zemin in their first meeting, in November 1995, that the biggest security threat China posed to the United States was related not to nuclear weapons or trade agreements but the environment. Specifically, Clinton feared that China would copy America's bad example while developing its economy and end up causing terrible air pollution and global warming. Clinton proudly claimed that he could tell Jiang "had never thought about it just like that." No doubt. Jiang was probably wondering whether Clinton could possibly be serious. If ever there was a nonissue for China's leaders, global warming is it.

"Global warming is not on our agenda," a senior official of the Chongqing Environmental Protection Bureau said with a dismissive wave

Need replacements

of his hand when I asked about his agency's strategies to reduce carbon dioxide emissions. As if to underscore his contempt for the issue, the official added an assertion he had to know was false—"All the pollution produced in Chongqing is landing here in Chongqing, so it's not a global problem"—before he declared, "We can't start worrying about carbon dioxide until we solve the sulfur dioxide problem." The official and his counterparts throughout China consider sulfur dioxide more urgent because acid rain is landing on them and causing tangible damage today, while carbon dioxide emissions threaten merely potential, far-off, worldwide damage.

Short-sighted? Yes, but understandable. I arrived in China eager to investigate the climate change issue, but I almost forgot to raise the point during some interviews. When one is inhaling appallingly polluted air for weeks on end, one tends to focus the questions on *that*.

China has little patience with Western finger-pointing on the climate change issue, regarding it as a cynical means of constraining China's economic development. That is a paranoid view, but it contains a kernel of truth. For all its nuclear weapons, grand ambitions, and expensively dressed businesspeople wielding mobile phones, China remains a very poor country. On a per capita basis, it consumes barely 10 percent as much energy as the United States. It is the rich nations whose earlier industrialization has already condemned the world to climate change, argue the Chinese, so why should China be held back? Should not the right to emit greenhouse gases be shared equitably among the world's peoples? To the Chinese, global warming is a rich man's issue, and if he wants China to do something about it, he will have to pay for it. As one Western consultant with regular access to senior Chinese officials put it, "They know very well they can hold the world for ransom . . . and whenever they can extract concessions, they will."

"The Americans say China is the straw that breaks the camel's back on greenhouse gas emissions," commented Zhou Dadi, a self-described insider on China's climate change policies. "But we say, 'Why don't you take some of your heavy load off the camel first?' If the camel belongs to America, fine, we'll walk. But the camel does not belong to America. . . . China will insist on the per capita principle [of distributing emissions rights]. What else are we supposed to do? Go back to no heat in winter? Impossible."

[handwritten marginal note: didn't start dealing with it until after it was hurting them]

Zhou Dadi, I should make clear, was no party hack. He was fully aware of the prospects for global climate change and was doing what he could to prevent it. The traditional communist approach to energy planning, in both the Soviet Union and China, focused on expanding supply through mega-projects like the Three Gorges dam. By contrast, Zhou was an enthusiastic proponent of increasing energy efficiency, which he regarded as the smarter and cheaper alternative. As we talked in the lobby of a Beijing hotel one December afternoon, it became clear that Zhou also had personal reasons to reject the old ways. Like so many people of his generation (Zhou was fifty), he had lost some of the best years of his life to the Cultural Revolution. As a young man, he had studied physics at Tsinghua University, the country's premier science university, but upon graduation, he and his classmates were denied formal degrees, which were seen as signs of bourgeois hierarchy. Instead, Zhou was ordered to spend the next nine years in an electronics factory, repairing machinery. Not until the 1980s could he begin reclaiming his professional life by returning to Tsinghua to complete a master's degree in environmental engineering. Despite these hardships, Zhou's loyalty to his country (if not to the Communist Party) remained strong, and—again like many of his peers—the stolen years seemed only to convince him not to squander any more time now.

"China is not like Africa, you know, some remote place that's never been developed," Zhou told me, explaining China's goal of becoming what he called "a middle-class country" like France and Japan. "We used to be the most developed country in the world. Now, after many decades of turbulence, civil war, revolution, political instability, and other difficulties, we finally have the chance to develop the country again. And we will not lose that chance."

will do anything to Become developed

Six

How Population Matters

> The tendency to see in population growth an
> explanation for every calamity that afflicts poor
> people is now fairly well established in some
> circles, and the message that gets transmitted
> constantly is the opposite of the old picture
> postcard: "Wish you weren't here."
>
> —AMARTYA SEN,
> economist, Harvard University

Picture yourself in a boat on a river, with houses on stilts, and wispy blue skies. Shoeless brown children waved back from the shoreline, the equatorial sun in their eyes. Behind them the trees of the Amazon rainforest swayed in the afternoon breeze. Slender as lampposts, they crowded together, a wall of light purples and greens.

As Joao turned the boat toward the open river, the violet water beneath us met the green-brown current of the larger channel. For nearly ten minutes the colors flowed side by side, as distinct as oil and water, before the violet finally gave way without a trace. We were in northern Brazil, on the Uatumã River, one of the countless tributaries that drain the massive Amazon Basin. The nearest link to the modern world was twenty hours back up the Amazon, in Manaus, a ghost town of the nineteenth-century rubber boom that rainforest tourism had recently helped to rejuvenate. But we were heading downstream, to the isolated village of Urucará, to attend the party of the year.

Joao piloted this boat with his wife, Margarita, and today their kids were with them, too—eight girls and a boy, the oldest nineteen, the youngest a year and a half. A girl of about five beamed with pride at being in charge of the baby, but in an instant of inattention disaster nearly struck. Drawn by the foamy water rushing past the boat, the toddler made a sudden dash for the side. She was about to topple overboard when her mother swooped over to grab her. Calamity averted, Margarita folded her hands back in her lap and good-naturedly shrugged me a smile. Judging by the lines on her rounded brown face, she could not have been more than forty. She had given birth approximately every two years since she was a teenager, yet she was the picture of serenity, her beatific smile radiating a calm, joyous warmth as she watched over her brood.

Joao and Margarita operated this boat for the local Catholic mission, represented on board by Father Ron, a young Canadian with chestnut hair and twinkling eyes who liked the rock and roll of U2 and R.E.M. Ron had been the local priest for five years. But as he teased and giggled with Joao and Margarita's three teenage daughters in the front of the boat, the foursome looked more like a bunch of gossiping high school students than a priest with his charges. Ron called back to me that the girls wanted to know whether I was married. The nineteen-year-old was looking for a husband, and everyone agreed she and I would make a lovely couple. Everyone except her sister's friend, that is, who had been flashing me smoldering glances behind the others' backs from the moment we boarded the boat.

This had less to do with any attractiveness of mine than with the fact that sex was just in the air in Brazil, even when one was talking with a priest—sometimes even when a priest was doing the talking. When we landed in Urucará, I was the first one off the boat, and Ron made a crack in Portuguese that I couldn't follow but that made everyone else laugh. In English, he told me what he'd said: that I was in a hurry to go telephone my girlfriend, but not to worry, that left more girls for him! A bit risqué for a priest, I thought, but then sexual play and flirting were as common as breathing here, and even the devout had no trouble reconciling a love of God with a love of pleasure.

Soon it was the last hour before sunset. On the bluff above the river, a crowd milled about on Urucará's main street, waiting for the water parade that officially opened the festival of Sts. Peter and Paul. Down below, the boats that had transported people here from miles around were assembling themselves into a kind of ragged formation. Finally, the proud little flotilla was off. Horns tooting, masts festooned with brightly colored banners proclaiming the grace of Christ, the boats sailed forward into the final rays of a sun that turned their vessels yellowish gold and the river a sheet of sparkling silver. Up on the bluff, the crowd urged the boaters on with vigorous whistles and cheers.

As dusk gathered, the faithful lined up for a march across town, the little girls in white cotton dresses, the boys in blue shorts. Some members of the procession carried candles as they passed Urucará's ramshackle one-story buildings. Hoisted on two men's shoulders was a chipped plaster figure of the Virgin Mary. Ron had now changed into his vestments in preparation for Mass. I fell in step with him as the first firecrackers went off.

"Not what you expected?" Ron asked, reading my mind.

"Not exactly," I admitted as another burst of explosions sent a ripple of excitement through the procession. "I never knew church could be so exciting."

"You know, when I first got here, I thought this sort of celebration was kind of frivolous, a waste of scarce resources," said Ron. "I tried to get the people to have a less elaborate celebration and buy a few more books for the school instead. But they always refused. 'The celebration must be *muito animado,* Father,' they said—*very exciting.* This is the one night of the year when they can forget they are poor, forget the difficulties they face and just be glad to be alive. And of course," Ron added with a chuckle, "they get to talk about what fun they had for months afterwards."

The Mass turned out to be a kind of holy party. One of the luxuries rented for the evening was a scratchy loudspeaker system, which sat beneath a single strand of Christmas lights at the far end of an open field. Bordering the clearing were crude wooden stands offering sweets, fruit, meat, beer, and other refreshments, including *caipirinhas,* the ubiquitous Brazilian drink made from rum, sugar, and crushed lemons. The mood was very relaxed: worshipers would break away during the Mass to grab a bite or a drink, then drift back to rejoin the praising of the saints or the receiving of forgiveness. The younger children were delirious with happiness, cadging treats from the vendors and chasing after one another like frisky puppies, while their older siblings paired off for amorous strolls in the shadows.

The high point came at midnight, when the dancing started. A long shed beside the field had been converted into a disco of sorts. There wasn't much light inside, but the music—by a live band, another extravagance—was scorching, and the eroticism of the crowd was overt yet casual. The lambada was the dance of choice, and everyone was doing it, from earnest seven-year-olds and teenage show-offs to sweaty adults and sixty-year-old grandmothers. The grannies were amazing, their wrinkled faces impassive but their bottoms twitching furiously as they stutter-stepped across the floor at blinding speed. The beat was irresistible, and it kept the merry-making going until dawn.

Late the next morning (very late), I walked over to the low-slung, concrete-block building that contained Ron's office and the only telephone in Urucará. To my surprise, the phone was connected to the international network, so for the first time in many days I was able to retrieve messages from my answering service in the United States. One message was from a magazine editor who, when I called him back, gave me my second surprise of the day. Instead of asking me to write about the rainforest—the environmental issue of the day in Brazil and, in fact, the story I had come to the Amazon to investigate—the editor said he wanted an article about the environmental issue that no one was talking about: population growth.

Ever since biologist Paul Ehrlich had published his bestselling book, *The Population Bomb,* in 1968, population growth had been seen, in the United States at least, as an environmental threat of the first order. Concern had ebbed and flowed in the intervening decades, but now, in the 1990s, the environment was again a hot topic. The editor noted that Brazil had recently hosted the Earth Summit, but neither the politicians nor the journalists at the summit had said much about population. Why was that? Had the population bomb somehow been defused? Or were governments simply afraid to talk about the issue, scared off by the controversies it generated over abortion, women's rights, national sovereignty, and religious freedom? Intrigued, I told the editor about my recent boat ride with Joao and Margarita and their nine children. Now he insisted that I do the story, adding that I could tie it in to my rainforest investigation. I started pursuing the subject as soon as I hung up the phone.

Ron had invited me to join him for a leisurely bicycle ride around town so I could get the lay of the land. As we inflated the tires of two old black bikes, I said that I now understood why Ron's parishioners favored parties that were *muito animado*—last night had been magnificent. But as a Catholic priest, didn't Ron feel that the church's prohibitive teachings about sex and contraception were doomed to failure in Brazil, the only country I knew of where women who were not prostitutes whistled and called out to men on the street? Preaching abstinence seemed a total nonstarter here, and the "natural" methods of contraception tolerated by the church only resulted in families the size of Joao and Margarita's. As the briefest visit to a disco or beach made clear, Brazilians had their own rhythm method, and it had nothing to do with *preventing* pregnancies.

"Brazilians are very sexual people, there's no getting around that," Ron agreed with a smile. Ron was certainly no fuddy-duddy, and as a liberation theology priest he harbored no reflexive loyalty to Vatican dogma. What troubled him about his parishioners' unbridled approach to sex, he said, were the concrete effects it had on their lives. Liberation theology taught that the struggle for justice was central to Christ's teachings; helping people to achieve human dignity in the here and now was as important as helping them get to heaven later on. And escaping from poverty was hard enough, Ron argued, without the extra burdens of single motherhood, sexually transmitted diseases, and the like.

"The first thing a couple here does if there is mutual attraction is screw," Ron complained. "There's no affection, no courtship, just sex. So we have lots of unwed mothers, and lots of fathers who actually boast about having one child with this woman, another child with that woman, and so on. I would think that if I were a father, I'd want to support the children I had. But they end up having lots of kids they can't support."

Under the circumstances, Ron said, he had no qualms about endorsing birth control. "If they ask me, I tell them it is better to have three or four children you can support properly than to have many children who get sick all the time."

How many other priests around the world took that position, I asked. Fifty percent?

"It's hard to say, but it wouldn't surprise me," said Ron. "The other liberation theology priests probably say the same thing. On the other hand, the next prelacy down the river is run by priests from Italy, and they are very traditional—no birth control, no questioning of the status quo, none of that."

Yet even a priest as open-minded as Ron shrank from outright advocacy of birth control. When I asked how parishioners could obtain contraceptives, he replied, "Well, I'm celibate, so I don't really know how to advise them about the specifics of birth control. I do tell them to ask about it when the health workers visit."

I didn't press the point, but visits from health workers were not frequent in these parts, as Ron well knew. We pedaled slowly through the morning heat, and Ron fell silent except for an occasional hello to passing villagers. I was afraid I had embarrassed him, so I kept quiet, too. After a few minutes, the path dead-ended at a bluff above the river, where Ron braked to a halt. Staring into the distance, he seemed lost in thought until he turned to me and quietly asked, "Do you know what the hardest part of this job is? Burying children." At the end of the Amazon's dry season, in September, temperatures lingered above 50 degrees Celsius (122 degrees Fahrenheit) for weeks at a time, Ron said; during that hellish season, it was the children born to unwed mothers and extra-large families who fared the worst. "Those are the kids who get diarrhea and dehydrated, and many of them don't make it," he said. "Sometimes I bury three or four, even five, kids a week."

Absorbing Ron's words, I felt momentarily transported back to the bleak vistas of southern Sudan. There, the suffering was so acute, so distant from the daily ease of the United States and Europe, that I sometimes had to pinch myself to remember that we were all living on the same planet. Here in Urucará, it was now late June, the start of the dry season. Thus, in a few months, the exuberance of last night's celebration would be replaced for some local parents by the incomparable sorrow of putting their children's lifeless bodies in the ground. Adding to the cruelty of these deaths would be the fact that diarrhea and dehydration were easily treatable ailments; according to the World Bank, a health care package costing a mere eight dollars a person could wipe out such diseases throughout the developing world.

For so many children to be needlessly dying in Ron's prelacy was not just unspeakably sad, it was also another measure of the mammoth gulf separating the planet's privileged from its poor. The infant mortality rate in Brazil was fifty deaths per one thousand live births, more than five times higher than the nine deaths per one thousand that was average in industrial nations. An estimated forty thousand children died every day in the Third World, from malnutrition and disease. Meanwhile, disparities of income between the privileged and the poor were also growing wider. The world's top 20 percent of income earners made thirty times as much money as the bottom 20 percent did in 1960, according to the UN Development Program, but by 1991, they made sixty-one times as much. By 1994, the ratio was seventy-eight to one.

On my way to Brazil from Asia, I had stopped off in San Francisco for a few days to visit friends and recover my health (I had been hospitalized for a week in Bangkok after most of my white blood cells abruptly vanished one morning, the victims of contaminated water). My return to San Francisco marked my first complete circling of the world. After a year of travel, much of it in Africa and Asia, seeing my old hometown again was more than a little disorienting; I felt like a stranger in a familiar land. The sheer wealth of the place was staggering. With their leather jackets, designer eyeglasses, and stylish haircuts, many San Franciscans were *wearing* more money than African and Chinese peasants would earn in a lifetime. "Oh, look, peach shoes!" one prosperous shopper cooed to her companion outside a store window on Union Square. "Do you need peach shoes?"

At dinner with friends that night, I tried to explain why traveling among the Dinka and other impoverished peoples had touched me so deeply, why the existence of such misery on a planet of plenty seemed a moral rebuke to outsiders' comforts and complacency. Embarrassed perhaps, most of my friends hardly bothered to reply, while one vaguely left-of-center journalist almost seemed to take offense. "God, Mark, it's not our fault they're hungry," he complained. "Why don't they stop having so many kids?"

It was a simple question that nevertheless went to the core of the population issue and its complicated relationship with global poverty and environmental degradation. I wondered how to answer my friend. With the flippant observation that Third World villagers couldn't simply pop around to the corner pharmacy to pick up contraceptive gel or condoms—and even if they could, the price would probably be beyond their means? With the paradoxical fact that, despite such logistical obstacles, Third World birth rates had indeed been falling—dramatically, in some countries? Or with what doubtless would have been the reply of many Third World people I had met: that my friend's question was not merely simple-minded but arrogant—what right did people in wealthy countries have to blame the poor for their poverty, much less for humanity's environmental dilemma, when it was the rich countries' consumption patterns that were responsible for the vast majority of the world's resource depletion and ecosystem destruction?

"The average resident of an industrial country consumes 3 times as much fresh water, 10 times as much energy and 19 times as much aluminum as someone in a developing country," noted Alan Durning in his landmark book, *How Much Is Enough?* Industrial countries have about 25 percent of the world population but use about 80 percent of its energy. The United States alone contains 5 percent of the world's population but accounts for 22 percent of fossil fuel consumption, 24 percent of carbon dioxide emissions, and 33 percent of paper and plastic use. An average American uses 185 gallons of water a day for household tasks; the average Senegalese uses 8. An American consumes about fifty-three times more goods and services than a Chinese. Because similar disparities pertain to many commodities and activities, a baby born in the United States creates thirteen times as much environmental damage over the course of its lifetime as a baby born in Brazil, and thirty-five times as much as an Indian baby. My San Francisco friend had one child in diapers and a second on the way, thus giving him the Brazilian equivalent of twenty-six children—three times as many as Joao and Margarita had. Needless to say, however, he did not feel that he and his family were part of the global population problem.

"The United States is the most overpopulated country in the world," Paul Ehrlich has charged. And because its population is still growing, the environmental effects of its luxurious consumption patterns will increase yet further in the future. According to the President's Council on Sustainable Development, the United States will have to increase its energy efficiency 50 percent by 2050 merely to keep pace with population growth.

Under these circumstances, for the wealthy countries of the North to pressure the poorer nations of the South to limit their population growth strikes many in the South as a hypocritical attempt to evade moral responsibility and maintain the global status quo. Indeed, it was this perception that had fueled the South's resistance to making population part of the Earth Summit agenda. In the diplomatic phrasing of one Earth Summit briefing paper, "The fear has been that such discussion would shift the focus of debate to population growth in the South—and away from the North's contribution to global environmental degradation and its obligation to provide correctives."

But part of what makes population such a contentious issue is its many layers of complexity. There is no denying that the consumption patterns of wealthy countries cause much more environmental damage than the population size of poor countries. However, this does not make concern about rapid population growth an imperialist plot. Poor countries actually have self-interested reasons to limit their population growth.

Although Third World population growth is often portrayed, especially in the United States, as a bad and frightening thing, the so-called population explosion that began in the 1950s was in fact the result of undeniably good news: lower death rates and longer life spans for the human majority. In the apt phrase of Nicholas Eberstadt of the American Enterprise Institute in Washington, D.C., "Rapid population growth commenced not because human beings suddenly started breeding like rabbits but rather because they finally stopped dying like flies." For people living in the nations of the South, the postcolonial era of the 1950s brought dramatic improvements in health care, sanitation, and food production and distribution. The arrival of modern medicine helped people take their first steps out of an eighteenth-century-like existence in which many children did not live past infancy and many adults perished before the age of forty. By the mid-1990s, infant and child mortality rates around the world were two-thirds lower than they had been in 1950, and they were still falling.

This transformation surely qualified as spectacular progress, but like many solutions, it created another set of problems. For while death rates in the South plummeted, birth rates did not. Couples were still conceiving as many children as before—eight or nine, in cases of fertile women like Margarita—so the total population naturally increased. And because death rates had fallen so rapidly (in contrast to the industrial world's 150-year-long demographic transition), governments of developing nations suddenly found themselves facing imposing, politically dangerous challenges. How could they possibly build enough schools and hospitals, educate enough teachers and nurses, and create enough jobs and housing to accommodate populations whose annual expansions were now twice (or more) as large as before? And if governments could not supply these things, especially jobs, how could they avoid being forced from power by angry masses?

Though first apparent in the 1950s, this political dilemma grew to awesome proportions as Third World population growth accelerated in subsequent decades. By the 1990s, an estimated one billion people were unemployed or underemployed in the Third World, and continuing population growth was obliging Third World economies to create thirty million new jobs every year—ten million in sub-Saharan Africa alone—simply to avoid deepening unemployment and the political instability that comes with it. Population growth also encouraged the rapid urbanization that was turning so many Third World cities into squalid, ungovernable monstrosities. More than 70 percent of the world's population growth during the 1990s was projected to take place in cities. "No government, no academic expert, has the faintest idea of how to provide adequate food, housing, health care, education and gainful employment to such exploding numbers of people . . . as they crowd into megacities like Mexico City, Cairo, and Calcutta," fretted a report by the usually unflappable Council on Foreign Relations in New York. Even

Iran, whose fundamentalist leaders had urged high birth rates after the 1979 revolution in order to build Islam, has been forced to bow to demographic reality. By 1996, with 45 percent of Iran's population under the age of seventeen, the government was so overwhelmed by the demand for public services that it began requiring anyone seeking a marriage license to pass a family planning course.

Just as population growth makes it harder for a given nation to climb out of poverty, so does poverty make it harder to limit population growth. Poor and hungry people have so many children precisely because they are poor and hungry. The poor tend to lack access both to the contraceptives that could prevent pregnancies and to the decent health care and nutrition that could prevent high infant mortality rates. The high infant mortality rates cause parents to conceive more children than they actually desire, if only to compensate for expected losses. But the extra children then make it harder for the family to escape poverty, and the poverty in turn keeps infant mortality and birth rates high. Thus the cycle perpetuates itself, even as it encourages environmental degradation. "Three-quarters of the people in the South live in ecologically fragile areas," according to the United Nations. "Poverty forces them to exploit their limited [resources] just to survive, leading to overcropping, overgrazing and overcutting. . . . A vicious circle of human need, environmental damage and more poverty ensues."

When I arrived in Brazil in 1992, the global population stood at 5.5 billion, which meant that the human species had nearly doubled in size in a mere forty years. By comparison, it had taken all of human history for the species to reach the one billion mark (in about 1800) and another hundred years for it to reach two billion (in 1900). By the 1990s, a billion new people were being added every eleven or twelve years. The rate of population growth had fallen from its historical high of 2.0 percent in 1970 to 1.7 percent by 1992. But because the base amount—the total number of humans—was already so large, human numbers were growing faster in absolute terms than ever before. In 1992, the population was increasing by approximately eighty-eight million people a year, the equivalent of two new South Koreas. Average fertility rates, at 3.0 births per woman, were still well above the replacement level of 2.1, which guaranteed that absolute growth would continue for decades. And even were the fertility rate magically to fall to 2.1 overnight, world population would still increase to nearly eight billion before stabilizing sometime around 2050. A more realistic projection, according to UN demographers, was stabilization at ten billion.

To make matters worse, the majority of this rapid growth was occurring in the very economically and ecologically impoverished places least able to absorb it—places like Urucará.

On my last day in Urucará, Joao and Margarita, traveling without their kids this time, picked me up at 4:30 in the morning and ferried me upriver to the industrial outpost of Itacoatiara. For ten steamy hours, the boat's belching engine propelled us past the dense, low growth of the rainforest. The Amazon was calm and flat that day and in places it was too wide for me to see the far bank. As befits one of the four great rivers of the world, the Amazon was both a natural marvel and a bustling commercial thoroughfare. I saw many large boats on the river that day, though from too far away to tell what cargo they were carrying.

We finally nudged our way ashore in Itacoatiara at 2:30 in the afternoon. In my rudimentary Portuguese, I invited Joao and Margarita to join me for lunch before they headed home. Smiling warmly, they declined. When I tried to insist, they explained: two of their kids were very sick. Making a cradling motion with her arms, Margarita indicated that one of them was the baby. Joao said that even with the current in their favor on the way back, they still wouldn't reach home until midnight. That would make it a twenty-two-hour round trip for them, almost half of it in the dark. So they pushed off, Joao at the helm, Margarita beside him, waving farewell, her smile of calm, brave faith luminescent in the equatorial sun.

I never learned whether Joao and Margarita's two children survived that day, for I never returned to Urucará. But my travels with Joao and Margarita did convince me of one thing. It is not the planet or some other noble-sounding abstraction that pays the greatest price for excessive population growth; it is the supposedly overpopulated people themselves. Like an estimated 150 million other couples around the world, Joao and Margarita had no real access to the contraceptive technology needed to space and limit pregnancies—technology the world's elite took for granted in their own lives. Because Joao and Margarita's family lived at little more than subsistence level, they exerted much less pressure on the earth's ecological carrying capacity than, say, the shoppers for peach shoes I saw in San Francisco. Yet in most discussions of the population issue, it is Joao and Margarita, not the San Francisco shoppers, who are cast as the source of the problem, as the people who have to change their irresponsible ways.

Joao and Margarita dropped me off in a decidedly malodorous section of Itacoatiara, just yards from a logging mill whose sour, starchy smell had greeted us well before we reached shore. Perhaps because it was lunchtime, there was no visible activity at the mill. Its grounds, however, were covered with huge, jumbled pyramids of recently felled trees, now debarked and delimbed and drying in the sun. From here the wood would be shipped, either ten hours upriver to Manaus if destined for the Brazilian market or five days downriver to the Atlantic port of Belem if headed overseas.

If Joao and Margarita's eleven-member family personified the population issue, this logging mill symbolized the destruction of Amazonian rainforest that had recently become one of the most infamous environmental problems in the world. In the late 1980s, activist pressure and media coverage had made "the rainforest" a household phrase in numerous countries; McDonald's felt enough public relations heat in the United States that it posted notices in its stores professing its deep respect for rainforests—this in response to the "hamburger connection," the charge by activists that much of the burning of rainforests (at least in Central America) was to clear land for cattle ranches whose beef ended up in fast-food restaurants. Pictures of vast swaths of tropical forest going up in flames appeared throughout the media, accompanied by warnings that the burning was occurring at the rate of a football field every second, a pace that would eliminate most of the world's rainforests within fifty years.

Dramatic visuals may have been what caught people's attention, but what kept it were the larger implications of forest loss. The Amazon is the "earth's greatest engine of photosynthesis [and] its hothouse of evolution," in documentary filmmaker Adrian Cowell's words. Through photosynthesis, the vast amount of plant life in tropical forests acts as a "sink," absorbing some of the six billion tons of carbon pumped into the atmosphere every year by fossil fuel burning and other human activities. (Soil and oceans also act as greenhouse sinks, via different mechanisms.) The process works in reverse as well: when trees are burned down, the carbon inside gets released. The destruction of tropical forests thus intensifies the greenhouse effect in two ways: by adding carbon to the atmosphere and by eliminating one of the counterbalances to that build-up.

The Amazon is crucial to the greenhouse dynamic, for it contains 30 percent of the world's remaining tropical forest territory. Since global deforestation accounts for an estimated 25 percent of all carbon dioxide emissions, deforestation in the Amazon alone is responsible for roughly 8 percent of global emissions—not as much as cars in the United States or coal in China, but a sizable amount. Nontropical forests also act as greenhouse sinks, and the world is losing these, too, at a rapid rate. More than two hundred million hectares of forest—an area larger than Mexico—were cut or burned down between 1980 and 1995, according to the World-watch Institute. Between 1990 and 1995, at least 107 countries suffered a net loss of forest cover, leaving the earth with approximately half its prehistoric forest cover. By 1997, the world's forests were, for the first time, losing more carbon than they were absorbing.

What makes tropical forests even more environmentally precious is that they are home to an estimated two-thirds of all living species in the world. Perhaps because tropical forests were untouched by the advancing prehistoric glaciers that pulverized habitats farther north, a fantastic diversity of plant, animal, and microorganism life evolved there. To level tropical forests (and other biologically rich habitats like underwater coral reefs) therefore causes the extinction of large numbers of species. In his renowned study, *The Diversity of Life,* Harvard entomologist Edward O. Wilson, one of the preeminent authorities in the field, offered what he called a conservative calculation of the extinction rate within tropical forests: approximately twenty-seven thousand species a year, which amounts to three every hour. These losses are part of a global trend of extinctions occurring one thousand to ten thousand times faster than the year-in and year-out "background" rate documented in the fossil record going back hundreds of millions of years. "Clearly," wrote Wilson, "we are in the midst of one of the great extinction spasms of geological history."

Many scientists consider species extinction to be the most urgent, far-reaching environmental challenge facing humanity, outdistancing global warming, ozone depletion, soil erosion, population growth, air and water pollution—everything but the civilization-snuffing threat of all-out nuclear war. Well-publicized threats to such high-profile species as the panda bear and Siberian tiger hardly suggest the true dimensions of the threat. "Biodiversity represents the very foundation of human existence," declared the fifteen hundred scientists who contributed to the *Global Biodiversity Assessment,* a report published by the UN Environment Program in 1995 that attempted to quantify the number, status, and distribution of the world's species. Among the UNEP report's disturbing findings was that nearly half of all primates—including humans' closest genetic relatives, the apes—are threatened with extinction. In 1998, the World Conservation Union announced that at least one out of eight plant species is also under threat of extinction, according to a twenty-year global survey conducted by, among others, the Smithsonian Institution's National Museum of Natural History. Although the one-in-eight ratio is a huge proportion, said David Brackett of the World Conservation Union, it is probably an underestimation, since it doesn't reflect rapid habitat loss in certain tropical nations.

Activists and scientists try to convey humanity's deep dependence on other species by appealing to people's self-interest—pointing out, for example, that "nine of the top ten pharmaceuticals in the United States are derived from natural sources." For humans to eliminate other species thus forecloses agricultural, medical, and other breakthroughs that cannot yet be imagined. "It's as though the nations of the world decided to burn their libraries without bothering to see what is in them," lamented Daniel Janzen, a biologist at the University of Pennsylvania who has studied tropical forests in Costa Rica. Yet cures for cancer and the aesthetic pleasure of watching creatures in the wild are perhaps the least of the benefits that humans derive from other species. A basic tenet of ecology is that everything is connected to everything else, and nowhere is this more apparent than in the interplay of the planet's species; without it, there would be no such thing as pollination of plant life, purification of air and water, and decomposition of waste matter. In his elegant essay, "The Little Things That Run the World," E. O. Wilson reminds readers how dependent humans are on lowly invertebrates—the ants, insects, and other tiny creatures whose processing of dead vegetation keeps soil fertile and the wheels of life turning:

> The truth is that we need invertebrates but they don't need us. If human beings were to disappear tomorrow, the world would go on with little change. . . . But if invertebrates were to disappear, it is unlikely that the human species could last more than a few months. . . . The soil would rot. As dead vegetation piled up and dried out, narrowing and closing the channels of the nutrient cycles, other complex forms of vegetation would die off, and with them the last remnants of the vertebrates.

So it was no wonder that the outside world in the 1980s began paying attention to the torching of the Amazon; self-interest dictated nothing less. Yet to people who actually lived in the rainforest, concerns about forest degradation sometimes were secondary. What looked to outsiders like destruction was, to forest dwellers, simple economic calculation. If one's family was to eat, the forest's wealth had to be exploited. For most people, that meant either working for the timber and cattle bosses, or tapping rubber trees, or, most commonly, cultivating one's own patch of land.

Cultivation requires clearing, however, and clearing is what destroys trees and species. In São Sebastião, a tiny village upriver from Urucará where I spent several days, the consequences of past clearing were plain to see. There was literally no grass. The ground was hard-packed dirt; people swept their yards with homemade brooms as if they were sidewalks. A single dirt road led out of São Sebastião, and in the adjacent fields food was grown. Or at least it used to be grown. The fields closest to town were now bare and abandoned, reflecting the Achilles' heel of tropical agriculture: the soil was poor and quickly exhausted. Yields could be high in the initial year of cultivation, but the nutrients were soon spent; within three years at most, the land was good for nothing but cattle grazing (and that, too, was soon doomed). The cultivators had no choice but to press onward, clear yet more forest, and begin the cycle anew. Indeed, the only traffic I saw during my stay in São Sebastião was a truckload of local men who waved as they rumbled down the village's single dirt road into the forest, heading for their distant fields of manioc and beans.

Small cultivators are responsible for an estimated 30–40 percent of the destruction of the Amazon's tropical forests. So, given the high birth rates in places like Urucará, it seems plausible to assign ultimate blame for Amazonian deforestation to rapid population growth. After all, Brazil's population is growing by 2 percent a year. Although average fertility rates had dropped from approximately six births per woman in the 1960s to three births per woman by 1992, the growth rate remains high enough that Brazil's population of 160 million people (already the fifth largest in the world after China, India, the United States, and Indonesia) will double in size in thirty-six years.

But is it that simple? Is the tendency of people like Joao and Margarita to have lots of children the reason the Amazon is disappearing at such a frightful pace?

Most experts who have studied the question do not believe so. "It would be rubbish to say that population growth has no effect on deforestation," said Norman Myers, the Oxford scientist who first sounded the alarm about global rainforest destruction with his 1979 book, *The Sinking Ark*. "But the question is, how much and under what conditions? Population pressures *are* acute in places like Madagascar and the Philippines, where much of the forest has been destroyed, but they aren't so acute in Brazil. You have to differentiate by location."

Myers noted that the most comprehensive data on rainforest loss comes from satellite imagery collected by the U.S. National Aeronautics and Space Administration (NASA). According to images from 1997, about 55 percent of global deforestation is caused by slash-and-burn agriculture, while logging accounts for 20 percent, roads and infrastructure construction for 15 percent, and cattle ranching for 10 percent. But those four processes, said Stephan Schwartzman, an expert with the Environmental Defense Fund who has worked in the Amazon since 1980, are also "intimately related. In the Amazon, for the past ten years logging—and mahogany logging in particular—has been what opens new roads, which then become the means by which colonists come in to farm and ranch.

So the satellite data would indicate that logging has relatively little role in overall deforestation, but if you look at where deforestation happens, it's along roads."

But are farmers and ranchers traveling along those roads in such large numbers because of excess population growth? Not in Brazil, said both Myers and Schwartzman. "The farmers who follow those logging roads are migrating from other parts of Brazil," said Schwartzman. "But that migration is not caused by population pressure. It's about land tenure, and the fact that 1 percent of the landholders occupy 50 percent of the land." Myers agreed that rainforest destruction would be much less if Brazil were not "one of the most unequal societies in the world. There's enough farm land for everyone in Brazil if the land were properly shared out. But instead, 90 percent of the farmers have to make do with 10 percent of the land."

The main impetus for deforestation in Brazil, then, is not too many people but too little equality. Land-hungry peasants from other parts of the country migrate to the Amazon not because the rest of Brazil is physically crowded (Brazil's population density is, outside the big cities, actually quite low) but because land is not available elsewhere. Brazilian landowners maintain their control through open corruption and unapologetic violence—buying police, intimidating judges, hiring gunmen. There were at least 539 assassinations of peasant organizers, activist priests, and other rural leaders in Brazil between 1985 and 1989 alone. The most famous to die was Chico Mendes, the proponent of "socialist ecology" who headed the rural workers union in the state of Acre. Mendes was assassinated in his home on December 22, 1988, after organizing countless actions by rubber tappers and other rural people determined not to be pushed off their land.

Mendes and his fellow tappers were beset by aggressive cattle ranchers who constantly had to expand onto fresh land if their herds were to flourish. I saw the same dynamics at work when I visited land squatters in Goiás, a state in Brazil's central highlands whose lopsided distribution of property has spawned some of the country's worst violence. My guide in Goiás was Sergio Sauer, a young religious worker whose shoulder-length blond hair and Aryan features seemed atypical for a Brazilian but were in fact common among those born to European immigrant parents in the south of the country. Sergio was a field organizer with the Catholic Church's Pastoral Land Commission, a project of the liberation theology wing of the church that actively assisted the dispossessed in their fight for land and dignity.

Sergio and I left the Goiás capital of Goiânia in the early morning, driving north through dry heat. Parched yellow pastures stretched on all sides as far as the eye could see. After a couple hours, we stopped for lunch in a small town, then drove north for another ten minutes before pulling off the paved road onto a dirt path. After ten more minutes, we turned left into a field of high grass, and Sergio nosed the car into a dense grove of trees that obscured it from view. From there, we hiked five minutes along a creek bed, emerging into yet another uncultivated field. At the far side of the field were more woods, and it was there that the squatters made their home.

The group's lookout saw us coming, so by the time Sergio and I ducked under the low-hanging branches bordering the camp a contingent of adolescent boys was showering us with greetings. The boys led us past rough-hewn sleeping huts and lean-tos down to the cooking area, where plastic buckets holding rice, beans, and other supplies sat beneath crude wooden tables. There were about forty squatters, ranging from grandparents to toddlers. With their patchy clothes and unwashed faces, they looked a bit like the Swiss family Robinson, except they were much poorer.

And on the run. According to Francisco, a large, middle-aged man with fierce, tired eyes, the squatters had recently been farmhands on a large plantation south of here. Francisco and some of the older squatters had worked at the plantation all their lives. But one day, the foreman told them they had to leave—no explanation, no grace period, just go. Such evictions were becoming more common, Sergio interjected, as landowners mechanized their operations; some landowners were now converting their holdings into soybean plantations for the export market. Workers discarded in the process were usually trucked to the nearest town and left to fend for themselves. After the group that Francisco led was evicted, they decided to make their way back to a remote corner of the enormous plantation and try to farm there.

"We lived there two years," recalled Francisco. "We pooled our savings to buy tools and seeds. It was very hard. Then came disaster. We were far from the main complex, but somehow the boss learned we had snuck back onto the land and he sent men with guns to make us leave. This time they drove us many hours away, to a town we never heard of. When they pushed us off the truck, they said that if we came back again they would shoot us dead."

Francisco said the squatters had stumbled onto the land they now occupied with divine help—"We prayed to Jesus Christ to deliver us from our suffering, and He led us here"—and that with help from Sergio and the church, they hoped someday to own this land. It was decent land, there was plenty of it, and it was lying idle. Why shouldn't they use it?

"It is a sin for one man to have so much land when others have nothing," Francisco insisted, adding, "We will never leave this land."

"Never?" I asked. "But what if this owner kicks you out like the last one did?"

"This time we will fight," replied Francisco softly.

He clearly meant it, but it was hard to imagine: a motley crew of forty forest dwellers, many of them elderly or children, none of them armed with anything more threatening than a machete, as far as I could see. How could they resist armed men?

"What if the other men bring guns?" I asked.

"We will fight," Francisco repeated. "We cannot keep running. We would rather die fighting than watch our children starve to death."

Such feelings are understandable in rural Brazil, where fifteen million people are landless and one of every two people endures daily hunger. The obvious solution is land reform, and in 1993 the Brazilian government passed a law allowing the expropriation of idle land. But landowners resisted a more equitable sharing of Brazil's natural wealth, defending their territory with armed men. The result was more violence. A gruesome example involved land in the western state of Rondônia. In August 1995, after landowner Helio Pereira de Morais rejected a compromise that would have granted local squatters 250 of the 44,000 acres he owned, military police and hired gunmen from surrounding ranches stormed the squatters' camp with machine guns. As Diana Jean Schemo reported in the *New York Times,* at least ten squatters and two policemen were killed; some squatters were tortured, and hundreds were wounded.

In response, organizations demanding land reform—including trade unions, the National Bishops Conference, and the Landless Workers Movement—pledged that land occupations would increase, and they began mounting nationwide demonstrations under the call "Cry of the Excluded." In 1996, another police massacre left nineteen squatters dead and forty wounded in the state of Pará. Videotape of the attack outraged the nation, but government policies did not change, thus ensuring a continuation of both land conflicts and rainforest destruction. (In fact, nine of the police involved in the 1996 massacre later took part in a second massacre in Pará in March 1998 that left two peasant leaders dead.)

All of which illustrates how easy, and wrong-headed, it can be to blame population growth for environmental problems that often have much more complex roots. Deforestation in Brazil is fueled far more by economic inequality and political favoritism than by demographic pressure, and it cannot and will not be stopped until those underlying causes are addressed.

To be sure, population has been a decisive contributor to deforestation in the Philippines, Haiti, Nigeria, and certain other nations that do not enjoy the low population density of Brazil. But in the southeast Asian countries whose deforestation rates were the highest in the world in the 1990s, overbreeding peasants were but a minor causal factor. In Indonesia, Malaysia, Cambodia, and Thailand, the chief culprits were instead corrupt government and military officials and politically well-connected logging and agricultural companies. Indonesia, for example, had banned forest burning in 1994, then simply failed to enforce its environmental laws. As a result, the nation suffered through several months of uncontrolled forest fires in 1997 that made front-page news the world over. Singapore, Kuala Lumpur, and other distant Asian cities were blanketed by smoke that, in the words of one foreign correspondent, turned the sky "a milky twilight" at noon and sent thousands to hospitals. The fires released more greenhouse gases in a few months than all of Europe's industrial activity did in the entire year.

Brazil, too, erupted in flames in 1997. Although Brazil's fires received less international media coverage, they destroyed an even larger amount of forest than Indonesia's fires did. Deforestation in Brazil had declined briefly in 1990 and 1991, but in 1992 it began increasing again, and by 1995 it had reached an all-time high. The fires of 1997, however, surpassed even the 1995 record, and they were unprecedented in another respect as well. A drought struck the Amazon in 1997—another manifestation of global warming?—and it combined with an increase in logging to create a truly extraordinary situation. Virgin rainforest, which is usually too moist to ignite easily, began catching fire in record amounts. Scientists feared a vicious circle was developing in which the drought encouraged fires, the fires dried out the rainforest even further, which led to still more fires and thus more deforestation and more carbon dioxide emissions. If their hypothesis was correct, it was a classic example of a positive feedback mechanism: global warming was causing fires that in turn accelerated global warming. Smoke from the Amazonian fires of 1997 was so thick that, in Manaus, wrote one observer, "the sun disappeared for days at a time."

Eduardo Martins, president of Brazil's environmental agency, blamed the fires of 1997 on desperate peasants like Francisco and his fellow squatters, who would starve, Martins said, if they could not clear land to grow crops. But the truth is that most forest clearing in the Amazon (as opposed to globally) has been done to create cattle pasture, not cropland. And the cattlemen and commercial loggers have been encouraged in their deforestation by generous government subsidies. To make matters worse, there is virtually no environmental law enforcement in Brazil; the constitution passed in 1988 during the transition from military dictatorship to electoral democracy included no such provision, and the landed and industrial elites have blocked subsequent efforts to pass one. Eighty percent of logging in the Amazon is illegal, according to an internal government report, but only 6.5 percent of the fines imposed are ever collected.

In 1998, the Brazilian government pledged to turn over a new leaf. With funding from the World Bank and political endorsement from the World Wildlife Fund, a private environmental group, President Fernando Henrique Cardoso committed to put sixty-two million acres of Amazonian rainforest under federal protection by 2000. Brazilian NGOs criticized the program, however, calling it an ill-conceived showcase that ignored the wishes of tribes indigenous to the rainforest and thus was bound to make matters worse. In any case, it remained unclear whether federal protection meant very much in practice. Just months before Cardoso's May 1998 announcement, Brazil's Congress voted to give the environmental protection agency the authority to enforce environmental laws, another key potential step forward. But under pressure from large landowners and agribusiness, the Congress watered down the bill by greatly lowering the possibility of jail terms, fines, or other real punishment for those caught breaking environmental laws. In August 1998, President Cardoso went even further when he signed an executive order that, in effect, declared a ten year moratorium on environmental law enforcement in Brazil.

If population growth is not the all-powerful culprit that some people claim, it is nevertheless a major contributor to numerous environmental problems around the world. At a certain point, the laws of mathematics cannot be denied; even at low consumption levels, more people means more demand for food, water, and other material necessities, as well as more pollution. When rising consumption levels are factored in, the environmental burden increases even further. Advocating "zero population growth within the lifetime of our children," a joint statement in 1993 by fifty-eight of the world's national science academies argued that "ultimate success in dealing with global social, economic, and environmental problems cannot be achieved without a stable world population."

India is a powerful example of how a country's ability to feed itself can be stressed by a rapidly expanding population size. In 1960, when India had 442 million people, it had 0.36 hectares of arable land for each person. By 1990, when India's population was 850 million, the ratio had shrunk to 0.20 hectares per capita. By 2025, it will be 0.12 hectares, even if the country manages to hit the lowest of the UN's three population projections for the country—that is, 1.3 billion people. The reason these ratios diminish is simple: a country's supply of land cannot expand (except by conquest).

The same equation applies to fresh water. There is "no more fresh water on the planet today than there was 2,000 years ago when the earth's human population was less than three percent its current size of 5.5 billion people," analysts Robert Engelman and Pamela LeRoy of Population Action International, a research and advocacy group in Washington,

D.C., pointed out in 1993. Humanity's use of water quadrupled between 1940 and 1990, noted Engelman and LeRoy; while half of that increase could be attributed to rising per capita consumption, the other half was purely a function of population growth. Although many regions had plenty of water, 430 million people lived in countries considered water stressed (not to mention 1.2 billion people whose water supply was dangerously polluted). Shortages were especially severe in such countries as Egypt, Syria, Namibia, and China, where rapid population growth had combined with naturally arid conditions to lower per capita availability to pitiful levels; in northern China, some fifty million people had to walk for miles or wait for days to obtain even minimal water supplies.

There is, of course, an obvious objection to the notion of physical limits: human ingenuity. Supplies of land and water might not expand per se, but if humans manage them more efficiently—learning how to grow more food on the same amount of land, for example, or reducing the appalling amount of waste in most irrigation systems—the practical result would be an expansion of supply. This was the essence of the so-called Green Revolution of the 1960s and 1970s, when the introduction of higher-yielding strains of rice and wheat dramatically increased food production in many developing countries.

The Green Revolution was another example of how technology could revolutionize humans' relationship with the natural world. It also undercut the dark warnings of Paul Ehrlich and others who had predicted in the 1960s and 1970s that a rapidly increasing population in the global South would lead to massive starvation. Although global population roughly doubled between 1950 and 1990, food production nearly tripled, even as prices of such staples as rice, wheat, and maize fell. "The world can take significant pride in the fact that we have been raising food production—without raising its real cost—faster than the population has been growing, and that we have done so at the height of the largest population expansion phase in world history," wrote Dennis Avery of the Hudson Institute.

Nevertheless, there are still many hungry people in the world, and the relentless arithmetic of rapid population growth is part of the reason. The proportion of chronically malnourished people in developing countries fell from 36 percent in the early 1960s to 20 percent by 1992. However, pop-

ulation grew so fast in the meantime that the absolute number of hungry
people declined by less than 10 percent. Another reason hunger remains
widespread is that many people are too poor to take advantage of increased
supplies of food, even at lower prices. "Average income and food production per head can go on increasing even as the wretchedly deprived living
conditions of particular sections of the population get worse," Amartya
Sen, an economist at Harvard University and author of the classic study
Poverty and Famines, has noted. This is especially true when land distribution
is grossly unequal, a disadvantage hardly unique to Brazil; colonialism bequeathed severe landownership inequality to virtually all of South America, as well as Kenya, Zimbabwe, Bangladesh, and many other nations.

The Green Revolution actually reinforced such inequality. The rev
olution's success depended on imported seeds, extensive use of irrigation,
fertilizers, and pesticides, and economies of scale that were simply beyond
the means of many smaller, poorer farmers, as the FAO later admitted.
Wealthy farmers who could afford such investments reaped high yields,
but small landholders unable to compete got squeezed off their lands or,
if they were landless laborers, like Francisco and his band of squatters,
thrown out of work altogether.

Population growth and the environmental effects of the Green Revolution also cast shadows across the future of humanity's food situation.
Food production will have to double (again) over the next thirty years
just to keep pace with projected population growth, according to the
FAO. Since this doubling of production must occur on proportionately
less land—0.17 hectares per capita, if the UN's projection of a global population of eight billion people by 2025 proves accurate—agricultural productivity will have to rise substantially. That will be a daunting challenge,
not least because of the extensive damages inflicted on ecosystems by the
first Green Revolution.

If the Green Revolution proved that human ingenuity could overcome technical barriers that seemed insurmountable, it also recalled the
old trade-off: greater production means greater environmental degradation. Expanded irrigation was critical to the Green Revolution's success,
but, as in ancient Sumeria, it encouraged waterlogging, salinization, and
other damaging long-term trends. By the 1990s, only 17 percent of the
earth's arable land was irrigated, so increasing this proportion appeared

to offer an easy path to further production gains. But irrigating more land actually promised to be quite difficult. Twenty percent of the irrigated land in China and Pakistan was already plagued by salinization. In the Punjab region of India, the irrigation stresses imposed by the Green Revolution could begin turning the nation's breadbasket into barren desert by 2015, according to the World Bank.

Meanwhile, the world is losing twenty-five billion metric tons of topsoil a year, and various forms of land degradation affect 3.7 billion acres of land, about 30 percent of the world's vegetated surface. "Soil erosion is arguably our single greatest environmental hazard, because 99 percent of humanity's food needs come from the land," David Pimentel, a professor of ecology and agricultural sciences at Cornell University, has observed. "Yet this hazard is often overlooked, because it's a gradual problem. Like population growth, soil erosion nickel and dimes you to death." According to Pimentel, soil erosion rates around the world are the highest ever, about thirty times faster than the sustainability rate. (It takes about five hundred years for nature to create one inch of topsoil.) By the late 1990s, the world's average per capita cropland was a mere 0.27 hectares—about half the amount of land needed to sustain the diets of the United States and Europe. Along with population growth and rising consumption levels, land degradation led to the need to bring under cultivation about fifteen million hectares of new cropland a year. Most of this cropland was being created by felling the world's forests, hardly an environmentally sound solution.

The problem of land degradation is especially acute in Africa, where a harsh climate and poor soil combine with burgeoning populations and unequal property relations to turn vast areas into barren wastelands. During my travels through Kenya and Sudan, for example, I passed through many villages that must have fed themselves at one time but were now surrounded by sand and dust as far as one could see. The causes of such degradation were plain enough—overgrazing, overcultivation, and inadequate fallow periods—and the peasants in question were aware of the harm they were doing. But their lack of access to decent land and capital left them little choice. Their degrading of the land only encouraged more drought and famine, which in turn increased the pressures for further man-made degradation.

Global grain production per capita peaked in 1985 and has remained essentially flat ever since. If production stays constant while population growth advances, there will be less food available per capita, which in turn should push prices up and increase hunger among the world's poor. In theory, the oceans offer an escape from this predicament. Although the oceans provide only 1 percent of humanity's food supplies, food from the sea is an important source of protein, especially for the poor. The global fish catch more than quadrupled between 1950 and 1995, from 19.2 million to 90.7 million tons. But again, the increased production was undercut by population growth. The per capita fish catch in 1996 was sixteen kilograms, essentially the same as in the mid-1960s. Future production increases are very unlikely; overfishing has already pushed eleven of the world's fifteen major fishing areas into decline, according to FAO data.

Of course, more intelligent fishing practices could restore fisheries to sustainability, just as more careful farming could protect agricultural resources. But increased production, while necessary, is unlikely by itself to solve the problem of world hunger. At the 1996 World Food Summit, the FAO announced that 841 million people around the world were chronically malnourished—but not because there was "too little food" for "too many people," to borrow the titles of two sections in *The Population Bomb*. "Agricultural production has basically kept pace with population growth over the last fifty years, and at the moment there is enough food to go around," said the summit's secretary general, Kay Killingsworth. "It's just inequitably distributed—with the result that food does not reach the needy."

In other words, just as there is enough land in Brazil to support everyone if the land were more evenly shared, so there is enough food produced on the planet to feed everyone if the food were equitably and efficiently distributed. If people go hungry, it is usually because they lack the money to buy the food that is available. More than population excesses or food production shortages, poverty is the reason that one out of every six humans doesn't get enough to eat every day.

The relationship between population and environment cannot be reduced to easy generalizations. On the one hand, "population control" has been invoked to terrible human effect, notably in India, China, and Bangladesh, where in the 1970s and 1980s coercive governments forced sterilizations and late-term abortions on large numbers of women. (See chapter 7.) On the other hand, family planning programs have brought great benefits to hundreds of millions of couples, providing them with the technology needed to space and limit pregnancies. Indeed, the main reason average birth rates in the South have fallen so far and fast—from approximately six children per woman in the late 1960s to three per woman in the late 1990s—is that the proportion of the world's couples using contraceptives has risen from 10 to 56 percent. But that still leaves nearly half the world's women without access to contraceptives, with the result that one out of every four pregnancies is undesired, while 585,000 pregnancy-related deaths occur every year.

Many women—an estimated 150 million—want to use contraceptives but cannot, often because their male partners will not allow it. "When I talk to the Vatican, I tell them that when they call for abstinence, I don't know what kind of world they are living in," Nafis Sadik, executive director of the UN Population Fund, told me. "How many women can tell their husband that, when they have no power? We have to realize that in many parts of the world, women want [birth control] methods they can hide from their husbands and families. Many of them are desperate, saying, 'Can't you give me an injection, or a pill, because I don't want to be pregnant again.'

"Women in developing countries have lives that are so predecided," added Sadik, who began her career as a physician in her native Pakistan. "Their role is to be married off. Girls are pulled out of the school system to look after other children. They start serving others in their own family and they're raised to have no other ambition than being a service provider in one way or another and especially to reproduce, regardless of whether they themselves want to, are ill, or can manage. Anyone who says the reproductive role of women isn't the most important in the emancipation of women doesn't know what really goes on in our countries."

During a long train ride in Kenya, I had a conversation with two Kenyan high school teachers that illustrated Sadik's point. The teachers were intelligent, articulate, good-humored men in their late twenties who spoke at least three languages and had a canny understanding of Kenyan politics and a keen interest in the outside world. Like most Nairobi residents I had met, they were also well informed about the re-

CULTURAL VIEWS of women

cent arrest in the United States of heavyweight boxing champion Mike Tyson for raping a young woman in his hotel room. The men thought the arrest a travesty: either the young woman had fabricated the story to extort money from Tyson or the white police were in the pay of one of Tyson's competitors, or both—how else to explain a night of sex landing a man in jail?

"But if the young woman repeatedly told him no," I said, "doesn't that make it rape?"

"Oh, we rape our women all the time in Kenya," replied the teacher nearest me while his friend chuckled his assent. "They are used to it. That is how men are."

Challenging patriarchal privilege around the world is an ambitious undertaking; the gulf in privilege between men and women makes the one between rich and poor almost look small. Yet bridging that gulf is critical, because more and more evidence indicates that the status of women is the single most important determinant of a nation's population growth rate. Traditional demographic theory holds that economics is the key variable: as families or nations grow richer, fertility declines—hence the slogan made famous at the UN population conference in Bucharest in 1974: "Development is the best contraceptive." But while there is historical evidence to support this theory, there are also important exceptions. Arab states, for example, are among the richest in the world, yet they have some of the highest birth rates, apparently because of the low status and restricted freedom of women in those societies.

Where the status of women is raised, on the other hand, birth rates tend to decline without exception, even in very poor regions like the Indian state of Kerala, where birth rates fell from forty-four births per one thousand women in the 1950s to a mere seventeen births per one thousand women in 1991. Indonesia, Thailand, and Mexico likewise reduced their population growth rates by 25 percent, 53 percent, and 38 percent, respectively, between 1972 and 1992. According to Jyoti Shankar Singh of the Technical and Evaluation Division of the UN Population Fund, each of these states pursued a similar strategy: increase access to child health and family planning services; support literacy and education programs, particularly among girls and women; and implement measures to enhance the role of women.

This women-centered approach to population policy was strongly endorsed at the UN International Conference on Population and Development in Cairo in 1994. The Program of Action signed by 180 governments held that providing universal access to family planning services is a necessary but insufficient step; equal attention has to be paid to fighting poverty, improving child and reproductive healthcare, increasing literacy, and, above all, raising the status of women. Although the Program of Action outlined population goals—7.27 billion people by 2015 and 7.8 billion by 2050—its methods differed sharply from the target-oriented policies that had given family planning a bad name in so many countries. Rather than forcing birth control on unwilling couples—the Vatican's nightmare—the Cairo program reaffirmed the right of all people to decide the number and spacing of their children. The underlying assumption was that no coercion would be necessary if women had the right to control their own fertility. As Nafis Sadik explained, "To say that you should have targets is to say that women do not want family planning, whereas our experience is that women do want it but are impeded [because] they are not the decision makers."

Emancipate women. Educate them. Help them space their pregnancies. Give their children healthcare. Allow them options beyond motherhood. This is a revolutionary prescription in today's world, where women routinely get paid less money, work longer hours, receive less education, and exercise less control over their lives than men. Male privilege is especially entrenched in certain African and Middle Eastern nations, where young women sometimes endure genital mutilation and are forced into unwanted, adolescent marriages to husbands who feel it their right to beat, rape, and treat them as little better than farm animals.

Yet examples from the Indian state of Kerala and elsewhere show that progress can be made, not only in reducing fertility rates but also in improving the lives of women and thus the health of society. For 180 nations to embrace this goal in Cairo was an enormously hopeful development. Like the Soviet and American retreat from nuclear brinksmanship, the Cairo Program of Action showed that environmental news is not always bad, that intelligent choices are sometimes made, that our species is capable of pursuing its self-interest. The great success in Cairo of women's rights organizations also demonstrates that individual people can, through persistent organizing with others, bring commonsense solutions to bear on overarching global problems—an achievement that must be replicated on many fronts in the years ahead if the environmental threats to human civilization are to be overcome.

Nevertheless, there remains a long way to go. The successes of the Cairo conference and family planning in general are impressive, but it is much too soon to exult, as *The New Yorker* did at the time, that humanity had "defused the exponential increases in global population" that were a "seemingly permanent feature of modern history." True, rates of fertility and population growth have fallen. But the absolute increases are still massive—eighty million additional people every year by the late 1990s, 95 percent of them in the already strained Third World—and they will remain so because, as *The New Yorker* itself acknowledged, "the largest cohort of females in history is now of child-bearing age." In short, the past is still catching up with the human species, and sheer momentum will propel global population well past the six billion mark of the year 2000.

This population momentum discredits any complacency about humanity's demographic future. Clearly it is good news that the proportion of couples in developing nations using contraception has risen. But because of population momentum, the absolute number of people who lack access to contraceptives is roughly the same today as it was in the mid-1960s, when human population was exponentially increasing. Likewise, by 1997, 40 percent of the world's population lived in countries where fertility rates were at or below replacement level. But their low birth rates were almost canceled out by the 850 million people in countries where fertility still averaged over five births per woman.

The Cairo Program of Action promises to address these problems—if it gets implemented. But the program is nonbinding, and an even greater danger is that it might be nonfunded. Its central goal of making reproductive healthcare and family planning services universally accessible by 2015 means that family planning programs will have to serve roughly twice as many people as they did in 1994. Poor nations are to provide two-thirds of the estimated $21.7 billion annual budget necessary to reach this goal, with the rest of the money coming from international institutions and rich nations. At the Cairo conference, representatives of both camps expressed doubt that such monies would be forthcoming, and subsequent events have borne out those fears. After an increase in rich-nation contributions in 1994, donations declined in 1995 and 1996. In the United States, Congress voted in 1996 to slash contributions to international family planning programs from $547 million to $356 million.

Would any members of Congress have voted differently had they visited places like Goiás and Urucará beforehand? Would they have recognized that sensible population policies can make a crucial difference in the lives of people like Joao and Margarita, and even more of a difference in the lives of their eight daughters? Great progress has been made in population reduction in recent decades, but the big picture is hardly reassuring. It is hard to find a plausible scenario in which the world's population will stabilize at much less than eight billion people, and an eventual total of ten or twelve billion is quite possible. Global, regional, and local ecosystems are already crumbling beneath the weight of today's six billion people, yet many of those six billion are living in indescribable squalor and billions more people are on the way. Under the circumstances, the future looks troubling indeed.

Seven

The Hurricane of Hell

China's population is so large that if the
environment here is not in good condition,
the world's environment cannot be in good
condition

—HU JIQUAN, government economist,
Chongqing, China

In the seventh month, the woman finally broke down and had the
abortion. She had resisted as long as she could, and the local party
leader was adamant. He had told her all along she had to terminate the
pregnancy. After all, she and her husband already had a child, a young
daughter. But like all Chinese peasants, they wanted a son. Their reasoning was not so much sexist as economic. In China, it was sons who lived
with and supported elderly parents; they were the only old-age insurance
available. Sons, the saying went, were "like heavy cotton quilts in winter.
If you don't have one, you will freeze to death."

The woman and her husband had pleaded, but the party leader insisted. There was a quota to meet and superiors to impress. Had not the
party declared the one-child policy vital to China's future? Had not the

woman been warned well in advance of her responsibility to "think clearly" about such matters?

And so, as recounted in Steven W. Mosher's *The Broken Earth,* a detailed memoir of village life in southern China in the early 1980s, the deed was done. When the woman was cut open at the village clinic, it was discovered she had indeed been carrying a son. In fact, two sons—twins. In the eyes of the peasants, this twist of fate cast the abortion in an even harsher light; it was as if the gods themselves were condemning its wickedness. The news spread through the village as fast as one horrified resident could tell another.

When word reached the father, he exploded into a murderous rage. Screaming about the violence done to his sons, he ran through the village to the house of the party leader. The leader was not at home, but his two sons, aged eight and ten, were. Howling in pain and fury, the father grabbed the two boys and dragged them, terrified, to the courtyard well. There, before anyone could stop him, the man hurled first one boy, then the other, down into the darkness. He then leaped into the well after them, closing the circle of death with his own suicide.

Incidents like this one have made China's one-child policy infamous the world over. Initiated in 1979, when Deng Xiaoping was on the verge of consolidating control of the Communist Party leadership after Mao's death, the policy was a desperate attempt to slow the growth of a population already approaching one billion. But despite the policy, or perhaps because of it, Chinese birth rates did not decline substantially during the 1980s and 1990s. By the end of 1996, the population was officially calculated at 1.22 billion people—nearly one out of every four people on the planet.

Deng's death was still a month away when I left China in January 1997, yet his legacy pervaded every discussion I had about his country's environmental future. Deng had revolutionized China with the marketplace reforms of the early 1980s, igniting one of the most fantastic, if uneven, booms of economic growth in modern history. Putting profits in command did wonders for Chinese paychecks, doubling average real income by 1997, but the environmental costs were high. The purchase of millions of new refrigerators and air conditioners, for example, had made China the world's leading producer of the CFCs that destroyed the ozone layer. Deng also oversaw the massacre of unarmed demonstrators in

Tiananmen Square in 1989 and then used the tragedy to renew his campaign for economic liberalization. These twin attachments—an eagerness for rapid economic growth on one hand and a fear of popular unrest on the other—are likely to remain grounding assumptions of Chinese governance under Deng's chosen successor, Jiang Zemin. Stay in power by raising living standards if you can but by deploying tanks if you must.

China thus brings together two of the most disturbing trends in global environmental affairs: the large, growing population typical of poverty and the high-impact consumption patterns promoted by Western capitalism. This combustible union makes China a sort of environmental superpower in reverse. Like the United States, the other environmental superpower, China wields what amounts to veto power over the rest of the world's environmental progress. China and the United States are each responsible for such a large share of global consumption that any international attempts to reduce greenhouse gas emissions, limit oceanic overfishing, or phase out ozone-destroying chemicals simply cannot succeed without their cooperation. The United States, with 275 million people, casts its long environmental shadow largely through its extravagant consumption patterns: gas-guzzling sport utility vehicles, an advertising-saturated culture of fast food and throwaway plastic packaging, and twenty-four-hour shopping malls crammed with all manner of unnecessary junk. China's environmental heft, on the other hand, still derives mainly from the size of its population; at the moment, less than 10 percent of its people can afford even a pale imitation of American excess. But if incomes keep rising, that percentage will increase, and the environmental effects of China's gigantic population could be fearsome to behold.

I was lucky enough to see a good bit of China during my six weeks of travel there. From Beijing I took the train north to Manchuria, then arced west and south across the Yellow River and through Shaanxi province to Sichuan, where I took a boat down the Yangtze to Wuhan before making my way south to Guangdong province and finally to Hong Kong. With my interpreter Zhenbing's invaluable assistance, I was able to talk with scores of families during my travels, and talk freely. In some cases, especially in the countryside and above all among children, I was the first foreigner these people had ever seen. They acted like it, too, unapologetically staring at me for minutes at a time, remaining expressionless even when I smiled or frowned at them to provoke a reaction. "They think you are like a statue, so they stare," explained Zhenbing. "Then they see: 'Statue move! Statue talk!' "

One of the most striking findings of my travels in China was that the one-child family was virtually unheard of, at least in the countryside, where seven out of ten Chinese live. Because of incidents like the double-murder described above, the party had had to relax the one-child policy in rural areas in 1984. In its place was substituted the so-called one-and-a-half-child policy, which allowed rural couples to have a second child under certain circumstances—most notably, if their first child was a girl. The one-and-a-half-child policy supposedly remained in force in 1997, but I saw little evidence that rural people obeyed it. Each of the scores of families I met in China had at least two children. Many had three, four was not uncommon, and a couple families had five or more. In a village near the Pearl River in the far south of China (by coincidence the same area where the double murder had taken place years before) I shot baskets with a lad of ten who shyly replied that he was the youngest of seven children.

No wonder a top Chinese demographer told me, "The one-child policy is more slogan than reality." Officially, the government claims that Chinese women average 2.0 births apiece. My travels left me skeptical of even that figure. The only Chinese couples who seemed to adhere to the one-child target were urban dwellers. For them, especially those who worked for the government and could be easily monitored and penalized through withholding of salaries, promotions, and the like, the one-child policy was real. As one professional woman in Beijing who had a two-year-old daughter told me, "My husband would like another child, but we would be punished at work." Only three of every ten Chinese live in cities, however, and more and more of them are abandoning government jobs for the private sector. This suggests that urban birth rates could soon rise. More to the point, though, is the discrepancy between urban and rural birth rates. If only 30 percent of the total population are obeying the one-child policy, while more than twice that many—the rural 70 percent—are having at least two children and often three or four, it is impossible for the fertility rate of the total population to be 2.0. The math simply doesn't add up.

This discrepancy casts strong doubt on other official claims about China's population as well: that the country's total population was only 1.22 billion as of year-end 1996; that the population will not reach 1.5 billion until 2030; that it will peak at 1.6 billion in 2046; that the problem is, in short, only somewhat out of control. The actual numbers are almost certainly higher. The truth is, no one knows exactly how big China's population is or how fast it is growing, largely because of the practice—rampant throughout the Chinese system—of massaging statistics in order to tell superiors what they want to hear.

"Ten years ago, China had a reputation for having the best population statistics in the world, because there was no way for its people to hide what they were doing from the government, but today Chinese figures have become very questionable," Gu Baochang, associate director of the official China Population Information & Research Center, told me over lunch one day in Beijing. "The problem is, the local party leaders compete with one another to post the lowest birth rate, just like they compete to have the highest economic growth rate. . . . So at each level of authority, the targets get tightened. If the central government sets a target of eighteen births per one thousand people this year, provincial leaders tell county officials no, they must achieve sixteen, and county leaders tell village officials no, it must be fourteen."

The regrettable results, added Gu, include underreporting of the nation's true birth rates, renewed coercion of pregnant women to have abortions, and increased abortions of female fetuses. Discrimination against female births has produced the so-called missing girls syndrome—many girls are either aborted before birth or mistreated to death after birth. According to William Lavely, a demographer at the University of Washington, Chinese census figures from 1995 show that among four-year-olds there were 115 boys for every 100 girls; among three-year-olds, 119 boys per 100 girls; among two- and one-year-olds, 121 boys per 100 girls; and among children less than a year old, 116 boys per 100 girls. Such prejudice has deep historical roots in China, and in other poor, patriarchal, rural societies such as India, where there were 108 boys per 100 girls in 1991. Recently, the arrival of ultrasound technology has provided peasants a foolproof tool with which to indulge their preference. " 'Everyone has boys now,' said one peasant in a tone of awe," wrote Nicholas Kristof and Sheryl WuDunn in *China Wakes,* describing a village in Fujian province where twenty-seven of the year's twenty-eight births were boys. The difficulty these boys will have in finding mates one day is but one of the many distorting effects the one-child policy has on Chinese society.

Just how inaccurate are China's population statistics? Chinese demographers admitted in 1996 that rural fertility could be undercounted by as much as 37 percent. If true, that would push China's national fertility rate up to between 2.5 and 2.7 births per woman, which, for what it's worth, would square with my own limited firsthand observations. The demographers' admission was buried in a technical report issued by the Food and Agriculture Organization, which asserted, without explanation, that the UN's population statistics on China had been corrected to take account of any undercounting. Western demographers later told me that such correcting would have been done by checking census data. Even if birth rates were being undercounted, they said, the undercounting would be caught when the extra children showed up a few years later during the next census. To be sure, this method assumed that census data were not "adjusted" the way birth rates were, but according to Judith Banister, a professor at the Hong Kong University of Science and Technology and a leading expert in the field, China's full-scale censuses were indeed carried out honestly.

In any case, even scrupulously accurate reporting could not change the basic dilemma of China's population problem, which is that its leaders waited too long to attack the problem. Mao, in the late 1950s, brushed aside warnings about the approaching difficulties, arguing that China could always produce its way out of trouble since "every mouth is born with two hands attached." The warning came from a friendly source: Ma Yinchu, the president of Peking University. Ma realized that the social welfare gains delivered by the Communist government in the early 1950s—land reform and the dispatch of "barefoot doctors" had greatly increased health and nutrition levels—would lower death rates. Since birth rates remained high, Ma feared that the resulting population growth would restrain China's modernization. Although Ma was a staunch Communist, Mao branded him a "rightist" for his demographic prescience and cast him into internal exile. "Professor Ma's argument was unfortunately considered more on political than on demographic grounds," commented Gu Baochang. But Gu pointed out that "it wasn't just Mao who wasn't ready [to contain population growth in the 1950s], it was the Chinese population and the rest of the world, too. Remember, the United Nations World Population Conference didn't take place until 1974."

By the 1970s, even Mao recognized that China had to do something. The country began pursuing birth control in earnest in 1971, when its population already exceeded 850 million. The "later, longer, fewer" program of the 1970s urged later marriages, increased spacing between children, and a two child per family limit. It was both more successful and less coercive than the subsequent one-child policy. "Later, longer, fewer" reduced average births per woman from 5.8 in 1970 to 2.8 by 1977, a remarkable achievement.

But it was not enough. Because the total number of Chinese was already so large, even this lower rate of growth translated into huge absolute increases. The same dynamic has remained in place ever since, which may be why Jiang Zemin floated the idea of reviving the one-child policy in 1995. Even with women supposedly averaging only 2.0 births per lifetime, there are now twenty two million births a year in China. Subtract seven million deaths a year, and the net annual increase is a staggering fifteen million people. "So even though China has reduced its fertility almost as much as possible, to replacement level," Gu told me, "the total population is still growing as much as it was in the early 1970s, when women were having four or more children each."

Thus China mirrors the population dilemma of the world as a whole. Success, even remarkable success, in reducing birth rates is all but canceled out because it was achieved too late; the absolute size of the population is already so large that sheer momentum guarantees continued massive growth. And there is no assurance that birth rates will stay low, especially in China's case. True, China has completed "the demographic transition"—the historic passage from a society of high birth and death rates to one of low rates—and in most societies, this transition is permanent. However, because the transition was achieved artificially in China, any relaxation of official pressure quickly results in higher birth rates, as was seen after the one-child policy was relaxed in 1984. In the 1990s, most Chinese still wanted more children than they were allowed to have, according to surveys cited by Judith Banister. And now that people are richer, local officials are often happy to allow the extra births because they can pocket the hefty fine placed on parents who disobey the policy (as much as $1,800 per child). Of course, the officials are then obliged to lie about such extra births in their year-end reports, but that is the work of a moment.

Population growth is probably China's most fundamental environmental issue, for it underlies all others. Officially, the Chinese account for at least one out of every five humans on the planet. If official statistics are as inaccurate as they appear to be, the true ratio could be closer to one in four. And China's population is certain to keep expanding for decades. Even the government's projections envision a population of 1.5 billion by 2030; that means adding one-fourth again as many people to what is already a very crowded country. China already ranks near the bottom in global comparisons of per capita availability of natural resources. The country's per capita supply of arable land is only 28 percent of the global average; its supply of forests and wilderness areas, 15 percent of the global average; its water, about one-third of the average. Each of these figures will, of course, shrink commensurately as China's population size increases.

The country's water crisis is even more severe than the one-third global average figure suggests, for nearly all China's water is in the south, below the Yangtze River (where too much water has subjected millions of people to devastating floods over the years). By contrast, northern China has only one-fifth the per capita water supply of the south. Approximately three hundred cities are short of water in China, and one hundred of them are extremely short. The water table beneath the North China plain fell an average of five feet a year during the mid-1990s, as cities, factories, and farmers all increased their demand for water. There were so many cities, factories, and farmers drawing water from the Yellow River that it was dry for 133 days in 1996 and for 226 days in 1997. "That means no water at all for more than four consecutive months in a river that is one of the largest on world maps," said Edward Vermeer, a professor of Chinese studies at the University of Leiden in Holland.

Like any first-time visitor to China, I was looking forward to my first glimpse of the Yellow River, but what an underwhelming sight! As our train approached the bridge that leads over the Yellow one gray and windy morning, I could see the river's wide, grassy flood plain down below. Not until we were directly overhead, though, could I see what remained of the river itself: a tiny, strangled stream, perhaps ten yards wide, that trickled through the spacious channel. Granted, this was during the dry season, but in 1997 the dry season lasted most of the year. And from that bridge crossing, the Yellow still had to travel hundreds of miles, past countless thirsty towns and factories, before it would reach the cropland of Shandong province, which, as Lester Brown has observed, is responsible for a larger share of China's grain harvest than Iowa and Kansas together are of America's.

Water is so scarce in Beijing that party leaders have debated whether the city should remain the nation's capital, admitted Yu Yuefeng, staff director of the Environmental Conservation Committee of the National People's Congress. With a nervous chuckle, Yu told me that the problem had eased in the previous two years, thanks to higher than normal rainfall, but he conceded that "this is a roll of the dice. We have to rely on the gods to keep the rains coming." Turning serious, he said that northern China's per capita water resources are but a tenth of the world average, which is "a real headache for the government. For the next ten years, solving the problem is hopeless."

Of course, population growth is not the sole cause of these problems; geography and government policies deserve blame as well. Much of China's land is arid or semi arid, especially in the north and west. And during the Great Leap Forward of the late 1950s, Mao's lunatic pronouncements about maximizing grain and steel production by turning lakes into cropland and having peasants build backyard furnaces brought destruction to millions of trees and many fragile ecosystems. "Of course China has a population problem, but it now functions mainly as an excuse for a badly governed society," argued demographer Lavely. "When people can't get a train ticket or a job, when there's a shortage of housing, the government can say, 'Well, it's because we have too much population.' It becomes the all-purpose alibi."

But it's a credible alibi. There is a limit to what even the best managed government can achieve in the face of endlessly rising demands for service. "We made a great historical error in losing control of population growth, and now we have fifteen million new laborers to employ every year," complained Chen Qi, director of the Environmental Protection Bureau in Liaoning. "The loss of control meant we had to impose the one-child policy [to compensate], but that cannot continue. It's irrational and people won't accept it. Meanwhile, it is a great challenge to feed everyone, because the population is growing and arable land is shrinking. Economic development is difficult, too. China is opening to the world economy, which should mean more jobs for us. But it's not that simple, because many workers are not educated enough for those jobs. So unemployment is also a big problem."

One way to grasp the dimensions of China's population crisis is to envision a map of the United States, which happens to cover almost exactly as much land area as China does. But imagine that almost all of the people living west of the Mississippi River were transplanted to the eastern half of the United States—the reason being that the deserts and mountains that fill the western half of China are so foreboding that less than 10 percent of the total population lives there. (In Ürümqi, an oil-rich province in the far west, there is a prison that has no walls; inmates are free to run away into the waterless desert, if they are foolish enough.) Then, since China's population of (at least) 1.22 billion people is nearly five times larger than the 275 million of the United States, imagine life east of the Mississippi with nine times as many people living there as in real life. That is daily reality in China.

China feels as cramped as that sounds. In all the thousands of miles of scenery I observed during my travels, there was not a single place that did not betray the signs of intense human settlement. Open space was for farming, period, and cultivated within an inch of its life, with furrows stretching right to the edge of any road and curling into hollows as small as a pitcher's mound. Forests? Except in the southwest of the country, Chinese forests seem to have been annihilated; the most one saw were woods that might cover a hillside or two before giving way to more farmland. In the cities, I never saw a deserted street, not even when a bus deposited me in the Yangtze city of Wuhan at 4:30 in the morning. During my two-block walk to a hotel, I passed dozens of people—some sweeping the street, others firing up cooking pots for morning dumplings, still others patronizing a public lavatory and waiting for a local bus. To be alone in China seemed impossible.

In daylight hours, the cities became a churning mass of congestion. Although cars are still scarce on a per capita basis (China has one car per every five hundred inhabitants, while the United States has one for every two), China's population size translates that ratio into a very large number of automobiles. Many of the cars I saw were taxis, usually shabby domestic models like the tiny Xia Li. There were also growing numbers of Lexuses, BMWs, and other luxury imports for the party elite and newly rich businesspeople derisively known as *dahu,* or "big money bugs." Jockeying for space alongside the cars were sky-blue cargo trucks, ancient city buses, an occasional horse-drawn wagon, and of course an endless fleet of bicycles. Traffic jams were the rule. Since no vehicle seemed capable of forward motion without frequent burps of its horn (Chinese drivers said this was necessary to clear the way forward, just as they honestly believed that using headlights at night wasted gas and caused accidents), making one's way across town was invariably a noisy, stressful adventure.

Negotiating sidewalks was no better; bumper cars on foot about describes it. Chinese pedestrians didn't necessarily try to crash into one another, but they certainly didn't shrink from it. Intersections were bedlam. There were no stop signs, very few traffic lights, and no concept of right-of-way, so everyone simply pressed forward at all times. Crossing the street boiled down to a game of chicken, as pedestrian and driver each inched ahead in apparent disregard of the other until someone blinked. I marveled at the nerve of cyclists who forced their way across avenues of nonstop traffic, but the truth was, they had little choice. In the swarming cacophony that is urban China, one pushes ahead or is pushed aside.

Because of the tight quarters, pushing and shoving is nearly as common as breathing and spitting. On trains and buses, one's body is constantly pressed against, usually from two or more sides, by the bodies of other passengers, who seem neither to notice nor care. As they have done with so many discomforts over the years, the Chinese have long since gotten used to such close proximity. After all, conditions were much worse during the Cultural Revolution, when Mao gave China's youth unlimited free train travel. Compartments became so crowded in those years that the aisles were literally impassable; passengers routinely urinated in their seats and exited by climbing out the windows.

Zhenbing and I were once standing in line at a train station, waiting to offer the bribe necessary to gain a sleeper ticket, when suddenly I felt myself grabbed from behind and moved aside as roughly as if I were blocking Patrick Ewing's path to the basket. I'm not the fighting type, but I instinctively whirled around with fists raised to find . . . no one. The culprit, a man in his sixties, had already hurried past me, intent only on his destination. There had been no malice in his gesture, only the natural impatience of an animal who had been confined in too small a space with too many others for too long a time. I tried to take a calming deep breath, but my lungs couldn't reach it. Not for the last time in China, I felt like I had stumbled into some fiendish laboratory experiment that was mushrooming out of control.

I concluded my visit to China in Guangdong province, in the far south of the country. It seemed a fitting spot, for Guangdong has the highest birth rates in China, and it is also where Deng's economic liberalization first took hold. Located just across the border from Hong Kong, Guangdong used to be one of the poorest provinces in China; now it is a much-envied symbol of the new prosperity. Nothing is more important in contemporary China than getting rich, and the example of Guangdong shows that this is no mere dream; it is attainable reality.

For Zhenbing, a poor peasant boy from the extreme north, our visit to Guangdong was an eye-popping experience. "Paradise!" he kept exclaiming, an awed grin on his face, as we toured a region where peasant families owned motorbikes and city dwellers enjoyed air-conditioning and decent clothes. We had arrived in the provincial capital of Guangzhou (also known as Canton) by plane, and from the moment we stepped out of the terminal and into the tropical sunshine, it was clear this was a very different China from the one we had been traversing the past few weeks. We gratefully peeled off our heavy coats and boarded a bus headed downtown. Inside, three of our fellow passengers were shouting into mobile phones—at business associates, Zhenbing said—while an elegantly coifed middle-aged woman in a jade business suit arranged two bulging shopping bags on the seat beside her. Soon we were speeding down the highway, past traffic circles that actually had flowers planted inside them, a splash of color that would have been inconceivable up north.

After finding a hotel, we took a walk along the Pearl River. High above the far side of the river, billboards flashed names like Colgate, Honda, Seiko, and San Miguel in blue, green, red, and white neon. We passed two busy McDonald's, countless kiosks offering Salem cigarettes and Snickers candy bars, and shop after shop selling knockoffs of European watches and clothing. Some buildings even boasted the polished blue mirror glass I recalled from skyscrapers in Hong Kong.

The consumerism of Guangdong rested on a genuine industrial base. One day, I took a bus from Guangzhou to Hong Kong, a ninety-mile trip across the Pearl's flat, steamy delta. At one point, gazing out through the afternoon haze, I counted fifty different smokestacks at once—the exhaust pipes of the factories and power stations that had transformed this recently rural area into a manufacturing dynamo, and an environmental casualty. Electricity production had jumped from a mere 2.5 billion kilowatt hours in 1981 to 15 billion kilowatt hours in 1996; since 89 percent of this electricity is produced by burning coal, 60–90 percent of the rainfalls in Guangdong are acid rain. Most of the investment capital that created the factories was Chinese, but Hong Kong Chinese. For Hong Kong capitalists, the opportunity to slash labor costs by moving their factories across the border had been irresistible. In the 1980s, manufacturing had accounted for 40 percent of Hong Kong's economic output. Now, it was down to 10 percent, as production of toys, textiles, and other forms of light industry migrated to the mainland. The ideological contradiction of turning to Hong Kong capitalists to finance a new beginning for Chinese communism might have stopped another man, but not Deng Xiaoping, the man who once famously asked, "Who cares if a cat is black or white, as long as it catches mice?"

The comrade I met at the Guangzhou Environmental Protection Agency embraced an equally flexible definition of socialism. In the five hours I spent with him, in an interview and a working dinner, the comrade cheerfully bragged over and over again about socialism's clear superiority to capitalism. How lucky he felt to live in China, where everyone shared the fruits of their labor! He had been to Germany some years ago, rich Germany, but its restaurants were small and dingy compared with

Guangzhou's! To clinch the point, he told a joke. It seems a man had hijacked a plane in the city of Fuzhou and demanded to be flown to Hong Kong. The pilot obediently headed south, and after a couple hours told the hijacker they had arrived. The hijacker looked down, saw a cluster of tall office buildings and highways bulging with cars and shouted for joy; at last, his dream was in sight. The pilot landed the plane and taxied to a halt. The hijacker leaped onto the tarmac, eager to begin his new life in Hong Kong. But something was wrong. Approaching him were a dozen soldiers wearing the familiar olive green uniforms of the Chinese military. Spying a sign on the terminal behind them, the hijacker realized, too late, that he had been tricked. The sign read "Guangzhou International Airport."

The comrade couldn't stop laughing at this joke, but it was more revealing than he knew. To people in Guangzhou, Hong Kong was the dream, the model they measured their progress against. But to be likened to Hong Kong was not necessarily a compliment, and the wealth gap between Guangzhou and Hong Kong actually remained profound. Not just at the tycoon level, where Hong Kong boasted some of the richest individuals on the planet, but also on the street; conspicuous consumption was like a religion in Hong Kong. "Shopping here is relentless, and it's the locals more than the visitors," one longtime Hong Kong resident told me. "You'll see people carrying their Louis Vuitton or Armani shopping bags around town—it's what they live for." On the main shopping avenues in the Kowloon district, the neon signs were so numerous and densely stacked together that they blurred into one unreadable mass of glowing, frantic enticement. If the mainland wanted capitalism, Hong Kong had it, in its most insatiable consumeristic variety.

For Guangzhou, one dark side of emulating the Hong Kong model involved the environmental costs, which were all the more burdensome for a city that lacked Hong Kong's wealth. Traffic jams, for example, were at least as frequent in Guangzhou as in Hong Kong, and the air pollution was much worse. The comrade had insisted on coming to fetch Zhenbing and me for our interview, but a jam delayed him by half an hour. Apologizing as we drove back across town, he explained that Guangzhou now had five hundred thousand cars, trucks, and buses and two hundred thousand motorbikes, "so the air pollution is quite serious." At 300 micrograms per cubic meter, particulate readings in Guangzhou far exceeded Hong Kong's 82; indeed, they nearly equaled readings in Beijing, even though Guangzhou needed no coal heating in winter. In fact, motor vehicles were the prime cause of Guangzhou's air pollution, followed by construction dust and power station exhaust.

The comrade was confident that the car problem would soon be solved, however. As we inched along an overpass, he said, "The government plans to tackle the traffic problem by building more roads. With help from the World Bank, we will build an elevated inner-city ring road and an outer ring road." Did mass transit have any place in their plans? "Oh, yes," he replied, "we plan a three-line subway system, with the first line due to open on June 28, 1997, in time for Hong Kong's return to the motherland. We already have special traffic lanes reserved for buses in order to encourage people not to use their cars." At that moment, I saw that the special lane next to us was in fact overrun by cars, taxis, and trucks. When this departure from official policy was pointed out to him, the comrade smiled broadly and said, "Now, we close our eyes."

We would have had to close our noses, too, to miss another similarity between Guangzhou and Hong Kong: the abysmal water quality. In Hong Kong, there was so much debris floating around in Victoria Harbor that the city had to send out scavenging boats to collect it, for fear of losing the tourist trade that was now the city's second leading source of income. In 1995, the boats collected seventeen tons of floating refuse every day. But the harbor still smelled like an outhouse when I visited in January 1997. No wonder. It received 1.7 million tons of human sewage a day, 70 percent of which was untreated.

"Up until fifteen years ago, there was a cross-harbor swim race every year in Victoria Harbor, but you'd have to be crazy to get in that water today," William Barron, a professor of urban planning at Hong Kong University, told me. Hong Kong was rich enough to do better, it had simply chosen not to. "In 1989, when city leaders were deciding whether to add a second airport or install a working sewage system, they chose the airport, which shows you the complacency here about the environment," said Lisa Hopkinson of Friends of the Earth, Hong Kong. In 1997, a modern sewage system was at last on the way, but it would spare only the harbor. The system's outflow pipes would instead contaminate nearby ocean waters, which were already grievously overstressed by the chemical, industrial, livestock, fertilizer, and human wastes pouring down from Guangdong's Pearl River estuary.

"You have to accept that six million people go to the toilet every day in Hong Kong, and it has to end up somewhere," a Hong Kong government marine biologist told me, defending the new sewage system. "But look, Hong Kong can never be environmentally sustainable. You can't have six to eight million people living on such a small land area and have it be sustainable. We have virtually no fish here anymore, partly because of pollution, but mainly because of overfishing. So to feed itself, Hong Kong must import a tremendous amount of fish."

The biologist was unwittingly debunking one of the favorite arguments employed by those who claim that high population growth is nothing to worry about. Look at Hong Kong, they say, or Holland—two of the most densely populated places in the world, and two of the richest. A true statement on the surface, the claim overlooks a deeper truth: Hong Kong and Holland can afford high population density only because they are able to buy the natural resources they need from elsewhere. Of course, external reliance is not in itself objectionable—that is what world trade is all about. But for a densely populated entity like Hong Kong (or New York City, for that matter), such reliance assumes there is always an elsewhere that remains sufficiently untouched to provide the raw materials it needs. As more and more of the planet is claimed for human settlement and economic activity, that assumption is growing increasingly tenuous.

In one crucial respect, Hong Kong's elsewhere is Guangdong, source of 70 percent of its drinking water. Literally translated, Hong Kong means "fragrant harbor." This it putridly is, and during my walks along the Pearl River I observed that Guangdong's main waterway was no less so, and for much the same reason. The people of Guangzhou generated 630 million cubic meters of household sewage a year, and less than 10 percent of it got treated before being dumped in the river.

Industrial wastes also polluted the Pearl. In fact, the runoff from Hong Kong's own factories in Guangdong was the biggest source of the river's pollution, according to Ma Ziaoling of the South China Institute of Environmental Sciences. "The eastern tributary of the Pearl River is reserved for Hong Kong's drinking water, and every year the Guangdong government spends 20 billion yuan closing down heavy polluters along the East River and treating that water before we sell it to Hong Kong," said Ma. The water used by mainland residents was not accorded such care, so water throughout the Pearl delta region was "undrinkable" on health grounds, admitted Dr. Ma. She added, "But people have to drink it, or they have no water."

Even the comrade acknowledged that "the Pearl River isn't so clean." But this was a temporary shortcoming, he assured me. I was un-convinced, not just because of his high-handed manner but because of the odor of stale urine that suffused the conference room where we were talking but that neither he nor his two colleagues seemed to notice. When they took a cigarette break I ducked out into the hall and discov-ered the source of the problem. Ten feet away was the toilet, its door wide open, its commode rank and unflushed. Back I went to the conference room, where the comrade was soon enthusing that the government planned to build new sewage plants by 2000 that would double the city's treatment capacity—in other words, to 20 percent. "You must return to Guangzhou to see for yourself," the comrade said, flashing his coaxing smile. "Our lives will be even better then."

One thing that did trouble the comrade was migrant workers, whom he blamed for much of the city's water problem. Guangzhou's metropolitan area boasted an official population of 6.8 million people, 3.6 million of whom lived in the city itself. Another 1.5 million migrant workers also lived in the city, making the actual city population 5.1 million—on pa-per, at least. But the comrade, perhaps letting his urban disdain of mi-grants get the better of him, confided that the real number of migrants was much higher than 1.5 million—another example of the unreliability of China's official population statistics. "The migrants have helped build our economy," he conceded, "but they are poorly educated and have bad sanitary habits, so they are a big stress on our environment, especially on our water supply."

I saw plenty of migrants in Guangzhou. One night, I spent a couple hours crisscrossing the streets of downtown and found scores of bodies lying in sleep before the roll-down metal security doors of storefronts. Many of the bodies belonged to old men, who were covered with nothing more than old newspapers and plastic bags. In a pedestrian underpass, a boy of nine or ten slept alone, his filthy bare feet curled beneath him in the fetal position. These street people were part of China's so-called floating population, peasants who had come to the big city to seek their fortune, but they seemed to be missing out on the superiority of socialism so evident to the comrade.

Officially, there are 80 million rural migrants in China; unsanitized estimates range as high as 180 million. Whatever the exact number, China's migrants are part of a broader, global phenomenon: a vast stream of peasants descending on Third World cities in search of better lives. Though the trend dates back to the 1950s, it has accelerated tremendously since the 1970s. The population of Mexico City, for example, doubled between 1970 and 1986, from eight million to sixteen million. Lagos, the capital of Nigeria, is swelling from 288,000 inhabitants in 1950 to a projected 13.5 million by 2000. Cairo, Jakarta, Manila, São Paulo—these and dozens of other mega-cities have all experienced the same explosive increases. Simple population growth is responsible for 60 percent of this urban expansion, George D. Moffett reported in his comprehensive book *Critical Masses,* but rural-to-urban migration caused the rest. As a result, it is projected that humanity will cross a historic threshold in the year 2000: for the first time, fewer humans will live in rural areas than in cities. Since most Third World cities are not equipped to handle this influx, rural migrants end up living in shantytowns on the cities' peripheries that are environmental and public health disaster areas.

On my last day in Guangzhou I toured one such shantytown— Kangle, a neighborhood in the south of the city whose name means "a healthy and comfortable place." My guide was professor He Bochuan, the author of *China on the Edge,* the first book written by a Chinese native about the Chinese environmental crisis. The book became a bestseller in China shortly after its publication in 1988, only to be banned after the Tiananmen Square uprisings the following year because students had passed around copies of it during the occupation. When I first met Professor He at his apartment on the campus of Sun Yat-sen University in Guangzhou, he explained that the book had branded him "a dangerous professor" in the eyes of the party but, oddly, had not cost him his job. "They could not identify anything in the book that was incorrect, so they just said I'm unsatisfied with socialism," he told me. A handsome man of about fifty, with a pronounced cowlick that swept the hair back from his forehead, He (pronounced like *hook,* without the *k*) seemed amused that *China on the Edge* had provoked the government to issue a directive declaring it "impermissible to concentrate all the negative aspects of Chinese society in a single book."

Though forbidden to receive foreigners, professor He had welcomed us onto campus and even borrowed some bicycles so Zhenbing and I could accompany him on a tour through Kangle. Before we left campus, He stopped beside a four-story, windowed building that overlooked a playing field. The ground in front of the building was littered with dozens of white Styrofoam food cartons. Even before Zhenbing finished translating his words, I could tell the professor was upset by the trash. I sympathized, but why the fuss? Every place I'd been in China looked like this. Then Zhenbing explained: this building was where government officials on study sabbaticals lived, and it was they who had tossed the food cartons out the windows. "Our media contains many articles about environmental problems like the ozone hole," He said. "But our people, and even our officials, do not even care enough to dispose of trash properly, so they obviously will not care about things like ozone."

The professor climbed back on his bike and led us off campus. We dashed across a very busy avenue, then turned down a winding dirt and stone path, where we bounced past a gauntlet of street vendors selling vegetables, fruits, and other household items. The path narrowed and the foot traffic thickened, obliging us to dismount as we snaked past two- and three-story concrete buildings that housed apartments above and shops below. A dentist was drilling the tooth of a woman in one shop; next door, a barber cut hair while a tape deck blared a Hong Kong pop song. When the path widened briefly, I saw, far in the distance, peasants in conical sun hats working in a field, the sun glinting off their swinging hoes. The professor saw the question forming in my head and said that this whole neighborhood had been farmland just six months ago.

A bit farther on, we passed a painted wall slogan urging residents to "Create a Safe and Civilized Zone," a motto that made the professor laugh, since he knew what was around the corner. Passing through a tunnel, we emerged onto a footbridge that stretched across a long canal connected to the Pearl River. The canal was perhaps one hundred feet across and literally black with effluent, garbage, and feces; it was like an overgrown version of the sewage stream behind the Chongqing Paper Factory. An older woman with tangled hair and a firm stroke was paddling a small boat through the muck; when the professor attempted to take a photograph, she angrily shouted him off.

Onshore, the ground was again thick with litter—plastic bags, food cartons, and discarded strips of synthetic fabric that a young woman in filthy pants was methodically stuffing into a sack to sell as recycling. A foursome of construction workers, hoisting a load of bricks on shoulder slings, looked me up and down as they marched past. Pieces of rotted wood leaned precariously over the water, the remains of the pathetic shacks that residents here used to live in. Now, thanks to men like those brick carriers, residents lived in the concrete houses with pink and white tile fronts that lined both sides of the canal.

"Look at the houses here," gestured professor He. "They're four and five stories tall, they have television, carpeting, air-conditioning, refrigerator; the indoor environment is quite nice. These houses belong to the richest migrants, peasants who became wealthy after selling their farmland. That farmland is now gone forever, and the peasants are living in these filthy surroundings and causing environmental problems for the entire society. So is this progress? Or is it something else?"

We returned to campus for lunch, and professor He amplified his analysis. "The popular belief now in China is that, whatever problems arise, we can find solutions," he said. "For example, the goal of the government is to stabilize the environmental situation by the year 2000 and then begin improving it. But this is mainly a slogan without evidence to support it, just like the official promise to end poverty by the year 2000. In Guangdong province, fifty-one of the seventy counties still live in poverty, and this is one of the richest provinces in China! The official solutions simply transform the underlying problem into new problems. The migrant peasants' original problem was that they were poor. Now they appear to be rich, but as you have seen, society is left with the negative consequences: environmental destruction and social degradation."

Migrating peasants were but symptoms of the larger problem, He argued: "China is following an approach to development that measures progress only through economic growth, while ignoring the disastrous effects on the environment. Economic growth is supposed to solve the problem of poverty, but it causes so much environmental destruction that poverty continues and development is undermined. It is a vicious circle." The trouble was, most Chinese thought the emphasis on growth was just fine. "In the West, modernization had opposition from certain sectors of society—workers, some scholars and journalists, even some politicians," He said. "But in China, everyone, from top to bottom, supports modernization, fiercely, so no one can point out the problems that exist. Many people claim that the twenty-first century will be China's century. But if we cannot get out of this vicious circle, China will always be living on the edge."

After lunch, Zhenbing and I were leaving the campus when, to my surprise, I suddenly heard a female voice speaking forceful English. Up ahead, standing straight and tall beneath a tree, a student was earnestly practicing her pronunciation by reading aloud from a lesson book. "The future of China is bright," she intoned. "Very bright indeed."

Over the weekend, Zhenbing and I went to the countryside. Dr. Ma, the scientist from the South China Institute, had told me that if I wanted to observe the problem of lost farmland, I only had to travel fifty miles in any direction from Guangzhou and I'd see the evidence all around me. So we headed east on a bus that was filled with migrant peasants returning to their home villages. They were easy to spot, not just by their weathered complexions and ill-fitting clothes but by the looks on their faces—at once innocent, befuddled, and determined.

Not all the farmland had been destroyed in this area—we saw patches of fields where vegetables and rice were growing, and one brief flash of pretty forest—but the forces of development clearly had the upper hand. Just past the forest, for example, we came upon a golf course—one of twenty that had sprung up in the province, said Dr. Ma, to cater to the transplanted Hong Kong elite. On each side of the four-lane highway was an unbroken string of shabby, single-story buildings: auto repair shops, restaurants, gas stations, hotels, factories, building supply stores. Only the highway was paved, so the cars and motorbikes pulling in and out of these establishments kicked up endless clouds of dust. Monotonous and ugly, the scenery went on like this for miles, like an interminable commercial strip through American suburbs, but much, much poorer.

After two hours of driving, our bus had covered the forty miles to Zengcheng, a "town" of 750,000 people. There, we took a local bus another seven miles into the countryside, got off, and started walking toward the nearest village. By this time, Zhenbing was shaking his head in what appeared to be a mixture of envy and disbelief at the riches enjoyed by his southern compatriots; although the village road we were walking on passed a stinky, garbage-strewn sewage pond, the road was paved. To our right, two men were shoveling sand and rocks into a cement mixer beside the skeleton of a half-completed building. Behind them, a field stretched into the distance. Peasants and a group of children were hacking the ground with hoes, while behind them three cows and a dozen geese fed on leftover rice stumps.

A group of boys saw us coming and swooped past on bicycles, shouting greetings in pidgin English. Upon reaching the village, we learned that, besides bicycles, nearly every family had a motorbike. Many had also built new houses in the past few years, generally two-story structures of concrete or brick. Again, to Western eyes the houses had little to recommend them—bare cement floors and blockish architecture—but they were replacing single-story dwellings of mud and straw.

How could the peasants suddenly afford such extravagance? Zhenbing and I visited three villages that afternoon, walking from one to another, and had five conversations in which as many as thirty people participated at a time; we drew such crowds partly because of Zhenbing's charm but also because none of these villages had ever had a foreign visitor, and word of our arrival spread fast. Time and again, the peasants told me the same thing: we built our house after we sold some land. (Actually, what they sold was not the land per se, which was owned by the state, but rather the use of the land, a right granted them under Deng's liberalization.) Usually the buyer was the highway administration, though a few people had dealt with Hong Kong investors who planned to build factories. "Why don't *you* invest here?" one smiling gentleman asked me, a question I heard dozens of times during my China travels. Most of the peasants had kept some land where they grew food for themselves, and everyone agreed that life had improved in recent years. Nevertheless, most of these peasants considered themselves poor. For they knew that life was even better in Dongguan, a city twenty miles to the south, where some of them had worked in textile factories.

The next morning Zhenbing and I took a bus to Dongguan, and I saw what those peasants meant. The landscape we passed on the way was a carbon copy of what we saw the day before: a hodgepodge of farmland shrinking amid a proliferation of roadside shops and factories. The shops and factories grew more numerous as we closed in on Dongguan, and then, like the Emerald City rising on the horizon, there it was, a roiling metropolis whose gleaming white apartment buildings towered twenty stories into the sky. These buildings were even taller than the downtown citadels of Guangzhou—which stood to reason, since Dongguan was closer to Hong Kong and had garnered more foreign investment sooner. As we left the bus station, Zhenbing pointed to a woman steering a push cart stacked with crushed cardboard boxes down the sidewalk. The woman's casual trousers and cotton shirt were equivalent to what middle-class office workers wore in Beijing. "Even the scavengers are well dressed here," he murmured approvingly.

Walking through Dongguan's traffic-clogged downtown, we came upon block after block of stores offering paint, lamps, pipes, toilets, kitchen equipment, paneling, electrical wiring, tile, and countless other building materials and interior decorating items. Zengcheng and Guangzhou, and the highway strips between them, had also boasted a preponderance of such stores. If Hong Kong–owned factories were what jump-started the economic boom in Guangdong province, housing construction seemed to be what was propelling it now. My observations were in keeping with what I had been told in Beijing by Li Yining, a grand old man of market economics who taught at Beijing University and was one of the masterminds of China's transition to private enterprise. Li said he was confident that China's economy would continue growing at 9 to 10 percent a year "for a very long time." His faith was rooted in the very rural-to-urban migration and housing boom that was transforming places like Dongguan, Zengcheng, and Guangzhou. Throughout China, the urban population had been growing by at least 4 percent a year since 1980. It was projected that 42 percent of the population would live in cities by 2010; by 2020, it would be 50 percent.

There was enormous pent-up demand for new housing in China because both the quality and quantity of the existing housing was very poor. Despite a burst of rural construction in the 1980s, which peasants financed with their first gains from Deng's reforms, most peasants still inhabited the kind of mud and straw houses that the villagers near Zengcheng had so recently replaced. With an average residential living space of 7.9 square meters per capita (in 1995), Chinese people lived crammed together in ways that no American would tolerate. Married children and their offspring routinely shared small apartments with one of the spouses' parents, making privacy impossible and doubtless causing considerable personal friction.

Of course, the housing boom would gobble up a considerable amount of land and raw materials, as would another increasingly popular emblem of prosperity: the car. "If you talk to Chinese people, many of them will tell you, 'To have a car is my dream,' " said Ma Zhong, the professor of environmental economics at Renmin University in Beijing. "The car represents affluence to the Chinese, and until they have a chance to own one, it will be difficult to convince them not to use a car because of its environmental effects."

Notwithstanding its sometimes belligerent rhetoric about the United States, China is embracing the American model of economic development, and automobiles are a central element of that model. Automobiles are one of the five "pillar industries" that government planners have identified for special support in China's modernization drive; the others are (housing) construction, petrochemicals, machinery, and electronics. The development of these industries is supposed to stimulate growth in related industries and thus spread benefits throughout the economy; housing construction, for example, is calling into existence all those building supply shops I saw in Dongguan.

When automobiles were first named a pillar industry, in 1994, they were projected to account for 5–10 percent of China's gross national product by 2010. Government officials told me they now believed auto use would develop more slowly than that. After all, Li Yining told me, most urban Chinese did not need cars and most rural ones could not yet afford them. (Since when, I mentally replied, do people buy cars only when they need them?) Li added that the first priority was housing; once the suburban housing sector was established, in 2010, demand for cars would follow.

Even if auto development moved more slowly than originally expected, the Chinese market promised to be huge. China's fleet of motor vehicles had grown by more than 10 percent a year since the mid-1980s. The government projected that there would be twenty-two million cars in China by 2010 and an even larger number of trucks and farm vehicles. Beijing had insisted that foreign automakers enter into joint ventures with state-owned firms to access this market, and Volkswagen, General Motors, and others had been happy to comply. The government also favored automobiles through its infrastructure spending, which mimicked the industrial world's bias against mass transit. Subway and light rail projects had been canceled or delayed in Beijing, Qingdao, Shenyang, Nanjing, and Tianjin, while construction of expressways and ring roads continued across the country.

Highways, housing, and factories are only accelerating the loss of farmland that for decades has been undermining China's ability to feed itself. China has lost some forty million hectares of arable land since the late 1950s, which amounts to nearly one-third of all the land currently under cultivation. Most of the losses in the years prior to Deng's reforms were due to erosion, salinization, and other forms of environmental degradation; now the modernization drive is making things worse, not least because the suburban areas being turned into housing developments and highways are often very productive land.

China's gathering agricultural crisis drew international attention in 1994 when Lester Brown of the Worldwatch Institute published an article entitled "Who Will Feed China?" Brown argued that China was on a collision course with environmental limits and that early in the twenty-first century its inability to grow enough food for domestic needs would throw world grain markets into chaos, causing steep price increases and rising death rates. Brown's argument began with the fact that China's population was both increasing and getting richer and therefore eating more meat. This dietary shift would increase China's grain consumption, since it took two pounds of feed grain to produce one pound of chicken, four pounds of grain per pound of pork, and seven pounds of grain per pound of beef. Meanwhile, China's farmland was shrinking, its inadequate water supply was under increasing stress, and productivity gains from fertilizer use had flattened out. Crop yields appeared to be stagnating in the 1990s, Brown argued, and with land, water, and fertilizer already at the limits of their productivity, dramatic future yield increases seemed unlikely. The coming gap between demand and supply could therefore be closed only by importing food. By 2030, Brown projected, China would need to import between 207 million and 369 million tons of grain. Since both figures were in excess of the entire current volume of world grain exports, Brown predicted that the world was on the verge of steeply rising food prices, grain embargoes, food riots, political instability, and widespread hunger.

The Chinese government was displeased, to put it mildly, by Brown's analysis, which it categorically rejected. Food demand would indeed rise, the government conceded, but there would be no crisis, because food production would nearly double by 2025. Per capita grain production had risen throughout the post-1949 era, the government pointed out, and it would continue to do so in the future as China limited the loss of arable land, reclaimed idle land, and increased crop yields with better seeds, irrigation, and other technical improvements. When China's harvests in both 1995 and 1996 surpassed the 1994 level, the government crowed that Brown's vision had been disproved. "China can feed its people in 21st century" declared the *China Daily* headline I woke up to one morning in Beijing. The accompanying story went out of its way to reproach Brown, and government officials I interviewed were no less combative. The one time the comrade in Guangzhou became less than cheerful was when I asked about farmland losses, at which point he brought up Brown as an example of a mendacious foreigner who knew nothing about China.

The government's touchiness was not surprising, given the Communist Party's responsibility for the horrific famine of 1959–1961, when an estimated thirty million Chinese died. As Jasper Becker documented in *Hungry Ghosts*, granaries in many parts of China had adequate supplies during the famine, despite a drought; the famine developed because Mao ordered that the grain be withheld as part of a campaign to root out sup-

posed enemies of the people. Though the Chinese government never acknowledged its central role in the famine, the disaster of the "Three Bad
Years" still loomed large in the national psyche. The government was so
sensitive about the food issue that it now insisted on maintaining 95 percent self-sufficiency in grain production.

The Chinese government was not alone in criticizing Brown's analysis, however. Václav Smil, a professor of geography at the University of
Manitoba and one of the world's top experts on China's environmental
problems, also took strong issue with Brown. Smil, who had analyzed
China's agricultural problems in rigorous detail in many publications, including his seminal book, China's Environmental Crisis, agreed that concerns
about China's long-term food production capacity were valid. But Brown,
he said, was "a professional catastrophist" who had relied on "highly se-
lective evidence." Smil noted, for example, that while Brown had cited
Smil's own descriptions of water scarcity in China, Brown had ignored a
less convenient but fundamental point: that China had much more farmland than official figures indicated. Land was undercounted because peasants hoped to evade taxes and the government wanted crop yields to
appear higher, Smil reported, adding that this suggested that China's agricultural productivity could indeed be increased in coming years.

Both Brown and Smil's arguments were borne out in 1998, when the
U.S. government's National Intelligence Council released an analysis of
satellite photographs of China. The NIC analysis showed that China had
at least 130 million hectares of arable land—nearly 50 percent more than
the 95-million-hectare figure that had been cited by both Brown and the
Chinese government. This was obviously good news in one respect, but it
also implied that Chinese agricultural yields were much lower than previously thought. The challenge for the future would be to increase those
yields. A second piece of bad news was that China's arable land was indeed
disappearing at a very rapid rate. China was losing 500,000 hectares of
farmland a year, according to the satellite photos, not 191,000 hectares, as
the government had claimed. The NIC study concluded that China
would need to import 175 billion tons of grain in 2025, not quite as high
as Brown's low-end estimate of 207 billion tons but still uncomfortably
close to the current level of global grain exports of 200 billion tons.

The key to feeding China in the twenty-first century, Smil argued,
was to improve agricultural efficiency. Just as China could burn less coal

by installing more efficient boilers, so could it grow more food by reducing waste and implementing more realistic prices, especially for water, which was in even shorter supply than land. Since end users were charged but a fraction of water's real cost, waste was endemic. At least 60 percent of irrigated water in China never reaches the roots of crops. Smil argued that efficiency measures like low-cost sensors and better-lined canals could cut such losses significantly. Fertilizer could be more productively utilized as well. To be sure, there are limits to technical fixes: only one-third of China's soils are highly productive, one-tenth of its land annually suffers from drought or flood, modernization makes more land losses inevitable, and the government is vastly overstating how much idle land could realistically be reclaimed. But if China acted intelligently, Smil argued, it had "at least a plausible hope for a well-fed future."

One had to hope Smil was correct, because neither he, Brown, nor the government disputed that China's demand for food would be much higher in 2025 than it was at present. Another specialist who worried about China's agricultural prospects was Joshua Muldavin, a professor of geography at the University of California in Los Angeles. Muldavin, a Chinese-American fluent in Mandarin, had been visiting Heilongjiang, the province on the Russian border that was northern China's breadbasket, since 1982, working with local peasants and studying the effects of market reforms. Muldavin rejected what he called Brown's "alarmist" approach, and he agreed with Smil that better irrigation canals and other infrastructure improvements were vital. The trouble, said Muldavin, was that those were precisely the kinds of long-term, collective investments that had declined in rural China under the new market rules of get rich quick and every peasant for himself.

"Much of the recent rapid economic growth has been achieved through mining ecological capital," Muldavin argued. Chinese agriculture had traditionally relied on organic fertilizer—human and animal waste—and other ecologically sustainable practices to support its highly intensive land use, but the shift to a market system had changed that. "Peasants say they can't afford to use organic fertilizer anymore," Muldavin told me. "Now that everyone is on his own, rather than sharing risks through the collective, they need to intensify production and aim for immediate returns. Since many peasants now work part-time in local factories, they also don't have time to use organic fertilizers, which is a more labor-intensive process. They have to go for the sure thing in the short term, which means chemical fertilizers that result in poisoned water supplies and other environmental degradation that undercuts long-term productivity."

Like fellow professor He Bochuan, Joshua Muldavin argued that China's pell-mell pursuit of economic growth at all costs was leading to environmental disaster. Many of the consequences are already apparent—for example, in the staggering amount of pollution fouling China's rivers and the growing frequency and severity of floods and other natural disasters. (Eighty percent of China's rivers are so polluted that they no longer support fish.) Moreover, the damage is likely to increase over time as production patterns intensify. The danger is that, in twenty years, when they will be needed most, the water and soil will have been poisoned beyond repair.

It is not in the nature of markets to care about underlying ecological conditions today, much less twenty years hence, and it is market forces that are increasingly shaping Chinese agriculture. The government could, of course, intervene to encourage the kind of efficiency reforms Smil advocates. But Muldavin argued that, as market forces widen economic inequality in China, the government is distracted from long-term planning by its need to keep popular discontent in check: "Necessary state investment in long-term production . . . has been redirected primarily to price subsidies in order to 'keep the peace' with peasants and workers." This bodes ill for China's twenty-first-century agriculture. For China to feed itself in the future, everything will have to go right: yields must rise to international levels, which means that efficiency improvements in land, water, and fertilizer use have to be made and made soon; population growth has to be further slowed; and there is no room for unplanned but highly possible disasters, such as the droughts and floods projected to occur under global climate change. The dilemma facing China's leaders, and its farmers, is encapsulated in the words of an old Chinese saying: "If you don't look far ahead, you will have immediate difficulties."

Like government officials the world over, China's leaders in the 1990s learned to at least say the right things about the environment. In 1992,

China was an enthusiastic participant in the UN Earth Summit. In 1996, President Jiang Zemin and Premier Li Peng began to speak out against environmental destruction and urge a shift toward sustainable development. "We can choke on the air and drink filthy water with all our new money, but is this really a better life?" asked Li in one speech.

But the future is shaped less by official rhetoric and regulations than by what actually happens on the ground. And as the opening scene of this book—the chlorine waterfalls pouring out the back of the supposedly closed Chongqing Paper Factory—illustrates, environmental laws are often simply not implemented in China. This is no state secret; most of the dozens of government officials I interviewed acknowledged the pervasiveness of the problem, frequently without prompting. Often the culprit is straightforward corruption; factory owners and government and army officials use *guanxi*—personal connections—or bribery to get local regulators to look the other way. The central government in Beijing either cannot or will not stop them. As the ancient Chinese adage goes, "The mountains are high and the emperor is far away."

"Americans have a hard time understanding this about China, because you think of China as a communist dictatorship," said Liang Conjie, the Friends of Nature founder. "But enforcing the law is a problem in many areas, not just the environment. An article recently in the *Beijing Youth Daily* described a court employee who visited the apartment of a husband who owed money to his [former] wife. The husband just shut the door on the man and refused to see him. We at Friends of Nature have caught loggers cutting down trees in Yunnan province months after they were ordered to stop by both the provincial and central government. Sometimes, a court will order a polluting factory to close down, and the factory will be sealed shut, with boards across the doors and windows. But at night, the factory manager sends his workers in, they unseal the factory, work a midnight shift, and then seal everything up again by morning."

In addition to corruption, there is the so-called soft law syndrome. Under soft law, the government excuses state-owned companies from full compliance with environmental laws and standards; the law is "softened" to spare the companies (and the often insolvent state banks supporting them) from bankruptcy and shield their workers from unemployment. In contrast to corruption, soft law is not something Chinese officials like to talk about.

Right after my visit to the Chongqing Paper Factory, I had lunch downtown with Hu Jiquan, a top government economist. Keen to encourage foreign investment, Hu was pledging that the local environment would improve in years to come, thanks to tougher law enforcement. "We will close factories if we have to," he said. "We've already closed more than two hundred of them." Having just returned from the chlorine waterfalls, I couldn't help but challenge this rosy vision, and Hu was honest enough to concede that short-term economic considerations often overrode environmental goals in China. "The trouble is, if we close that factory, many workers will lose their jobs, and our government would rather support the workers than protect the water," he said with a shrug.

Hu then extended his explanation, though he first took care to tell Zhenbing not to translate this part for the foreigner. The government of Chongqing knows perfectly well, Hu confided to Zhenbing, that the paper plant should be closed immediately. In fact, it had tried to shut the plant months ago, "but the local people and leaders complained a lot, so the government backed off. It was afraid of social unrest."

This is the crux of the Chinese environmental problem. The government knows that the environment needs protecting, but it fears the social consequences. In short, it worries that doing the right thing environmentally could be political suicide.

The government wants to protect the environment for a very simple reason: senior officials have come to realize that environmental degradation costs money; indeed, it threatens to derail the entire economic modernization program. Li Yining told me that "inadequate ecological protection" is one of the few things that could prevent China's economy from growing at 10 percent a year "for a very long time." The floods of 1998, for example, cost at least $4.8 billion. Caused by deforestation along the Yangtze river, the floods left twelve million Chinese homeless and destroyed 3 percent of the nation's crops. The list goes on. The official *China Daily* has estimated the annual cost of China's environmental degradation to be 7 percent of the gross domestic product. The World Bank estimates the cost of air and water pollution alone at $54 billion a year, or roughly 8 percent of GDP. Václav Smil has calculated the cost of environmental damage at 10–15 percent of GDP. In short, the growth of the economy is being canceled out by the associated environmental degradation. The economy is running hard but poisoning its own future.

Recognizing this threat, the Chinese government has adopted strict and comprehensive environmental laws and regulations that, on paper, compare favorably with—indeed, are often modeled on—their Western equivalents. "China began its effort in 1993, with the establishment of an environmental group within the State Council [China's cabinet] and the convening of annual environmental conferences," said Ye Ruqiu, a deputy administrator of the National Environmental Protection Agency (NEPA). "At the 1994 conference, environmental protection was established as an official state policy, equal to family planning policy. We also determined that environmental degradation must be halted by 2000 and improved by 2010." Ye, a tiny, elegant, older man who spoke superb English, said that he and his NEPA colleagues had been promised a budget that eventually would equal 1.5 percent of China's GNP to underwrite the so-called Trans-Century program of environmental regulation. Beginning in 1996, and divided into three five-year plans, the Trans-Century program was based on principles that would be applauded by ecologists the world over: preventing pollution at the source (rather than treating it after the fact); making polluters pay to clean up the messes they made; and strict enforcement. The program would spend 180 billion yuan ($22.5 billion) in its first five-year period and focus primarily on air and water pollution. "This is not just talk," Ye insisted. "This is the national plan of China, based on a very strict survey of the situation, a concrete budget, and a determination to halt our environmental deterioration while building the country through sustainable development. Producers will have to meet standards, or they will be shut down."

The problem is that faithfully implementing environmental laws would require closing hundreds of thousands of factories, throwing tens of millions of people out of work, and reducing the fertilizer use that is keeping the country fed. Now that China is at last awakening from its long nightmare of deprivation, the Communist Party's tattered legitimacy depends on keeping the economic expansion going and extending it to the many regions that still lag behind. Yet the marketplace reforms that have sparked double-digit economic growth have also brought pain to vast portions of the population. In Chongqing, for example, 70 percent

of the state-owned enterprises (SOEs) are in the red, including the embattled paper factory. The factory had eight thousand workers on its pay-
roll, which translated into forty thousand family members who relied on
the factory's wages. If that ratio were applied to all the bankrupt SOEs in
Chongqing, it is clear that hundreds of thousands of people would be in
serious trouble if the government were to shut such factories.

Laid-off workers in Chongqing told me it was very difficult to find a
new job, even a poor-paying one, despite the 12.8 percent local economic
growth rate proudly cited by Mr. Hu during our lunch. One woman,
aged forty-one, became unemployed after a cloth-dyeing plant where she
had worked for seventeen years was shut down. She found that her par
ticular skills did not easily transfer beyond the textile field, and she was
able to secure a new job only by bribing a boss to hire her, a common
practice here, she said.

Whether this woman would take to the streets in protest if she were
laid off again was anyone's guess. But there had been much more social
unrest in China in recent years than most outsiders realized. The mass
occupation of Beijing's Tiananmen Square in 1989 and the army's subsequent massacre of unarmed demonstrators were well known. But similar militant protests had taken place at the same time in cities and towns
throughout China, news that did not reach the outside world because
there were no foreign journalists in those towns to report it. More recently, as the transition from state-organized economy to private market
free-for-all had touched the lives of more and more Chinese, thousands
of wildcat strikes and street protests had taken place throughout China,
especially in Sichuan and in Manchuria, a bastion of heavy industry,
where unemployment rates exceeded 30 percent. "We Don't Want
Democracy, We Want To Survive," declared one protest banner in
Shenyang.

All this has left party leaders decidedly uneasy, not to mention
determined to keep the economy growing no matter what. They apparently believe that Tiananmen Square was not primarily about politics—
about the issues of democracy and human rights that dominated Western
news reports—but about economics. There may well be truth to this
view. Hundreds of thousands of average Chinese followed the students
into the streets not only because they yearned to breathe free but because
they were angry about hyperinflation, party corruption, and their own

uncertain economic prospects. The party saw its life flash before its eyes in 1989, and it got a second warning in 1991 when its erstwhile "big brother," the Communist Party in the Soviet Union, fell from power. The Chinese communists are determined not to suffer the same fate. As Deng reportedly warned his fellow party leaders after Tiananmen Square, if the party could not improve the welfare of the people, the people would take to the streets.

On this point, at least, Deng was in agreement with his longtime nemesis, Wei Jingsheng. Wei, a former electrician, was a leading voice behind the Democracy Wall demonstrations of 1979. Deng had at first encouraged the calls for greater freedom that Wei and other activists were making in order to consolidate his authority in the aftermath of the Cultural Revolution. Then, his position secure, Deng had Wei and his colleages thrown into jail in 1980. Wei spent nearly all of the next seventeen years in prison, where he was tortured, starved, beaten, and denied medical care. After Wei was forcibly exiled to the United States in November 1997, I had the chance to meet him and ask whether China's widespread unemployment would spark further social unrest.

"It is a very dangerous situation," Wei replied. "Actual unemployment in China far exceeds the 30 percent that has been reported in the West. But factories must follow the communist policy of allowing no unemployment, so workers must stay on the job, even though they are paid no wages. For the workers, there are two choices: achieve their rights through peaceful means, without shedding blood, which is what most workers want. But should this fail, workers will have to resort to violent means. To be very frank, this is one reason I accepted the government's conditions for leaving prison and agreed to come to the United States. Unless we take quick action now, there will be great upheaval in China."

But for the government, there is a catch-22. The people will also take to the streets if their local environment becomes intolerably polluted—if, for example, they are deprived of drinking water.

"There were social revolts along the Huai River, so the State Council had to react," said one retired senior government official, recalling the most dramatic government crackdown on pollution to date. The Huai, located approximately two hundred miles northwest of Shanghai, is the most densely populated of China's seven major river basins; 110 million inhabitants share 108,000 square miles of land. The river had been severely polluted for years, but it got drastically worse in July 1994, when a sudden flood of toxins turned the river black and deadly for weeks. Hundreds of thousands of people were left without drinking water; several thousand were treated for dysentery, diarrhea, and vomiting, and twenty-six million pounds of fish were killed.

Popular outrage took many forms, including pelting local officials with eggs when they blocked foreign journalists and cameramen from filming the river. The most extraordinary moment came when a top leader from Beijing, Song Jian, the elderly chairman of the State Council's environment committee, arrived to inspect the site. Somehow, one brave and resourceful peasant managed to get close enough to Mr. Song to hand him a glass of river water and say, "I invite you to drink the water that we must drink." Song took a sip of the putrid brew, then turned to the local and provincial officials flanking him and shrewdly invited them to drain the glass. These officials had ignored earlier pleas to close paper, leather, and dyeing factories whose wastes were polluting the Huai. Song told them they would be fired if the offending factories were not shut promptly. In June 1996, the government said it closed 999 paper mills and untold numbers of other factories.

One reason Beijing shut down so many factories, said the retired senior government official, was that "for years, no boy from [certain villages in] the Huai River area has been healthy enough to pass the physical examination required to enter the army." Even more important, said other observers, these factories were township and village enterprises (TVEs)—small privately owned plants that employed, at most, dozens of workers each. Since these workers were local peasants pursuing a second income, unemployment did not condemn them to destitution—they had never stopped working their fields—but only returned them to the pinched circumstances they had endured for decades; this was a setback but not cause for revolt. The TVEs employed at most tens of thousands of moonlighting peasants. Against that fact, the government had to weigh the anger of the many hundreds of thousands of people who relied on the Huai for their drinking water, people who had already demonstrated a capacity for protest. There was no question which group to placate.

Beijing went national with the environmental campaign against TVEs in August 1996, when the State Council, China's cabinet, ordered the closing of an additional sixty thousand heavily polluting factories.

"That sounds like a big number, but in a country as large as China it amounts to only 1 percent of the total number of enterprises and workers," said Ye Ruqiu, of NEPA. Ye nevertheless argued that the closings "show the seriousness of the government in this area." Unfortunately, TVEs account for only a fraction of China's pollution; estimates range between 5 and 30 percent. To make a real dent in the problem, state-owned enterprises like the Chongqing Paper Factory will have to be closed.

But fear of social unrest makes that problematic. At the fifteenth Congress of the Communist Party, in September 1997, party leaders pledged to end state ownership of ten thousand of China's thirteen thousand largest industrial enterprises by an unspecified future date. The man in charge of this transition was Zhu Rongji, the deputy prime minister who had succeeded in the seemingly impossible task of cooling off China's double-digit inflation during the mid-1990s without halting economic growth. Perhaps Zhu, who was named prime minister in 1998, would prove capable of another miracle, but ending state ownership on such a massive scale carried great risks. By 1997, China's cities already contained some twenty million unemployed workers, and confrontations between unpaid workers and state security forces were growing more frequent and violent, especially in Sichuan, where police quelled numerous sizable riots in the months before the party Congress.

Thus China's leaders found themselves in a box. They could, in the name of economic growth, leave the big factories and other environmental hazards essentially undisturbed, and hope that the resulting pollution and ecological destruction did not become acute enough to trigger either unmanageable popular protest or long-run economic stagnation. Or they could clamp down, clean up, and face the double, short-term risk of a stalled economy and a wrathful proletariat. Not an enviable choice, but for Chinese leaders not a difficult one either. If only because economic hardship tended to have a shorter fuse, the environment was usually relegated to second place. As Chen Qi, the top environment official in Liaoning, told me, "Heavy pollution may kill you in one hundred days, but without enough heat and food you die in three."

"To get rich is glorious," in Deng Xiaoping's oft-repeated phrase. Now that Chinese of average means have glimpsed the good life (if only in the lifestyles of the "big money bugs" and the consumer advertising increasingly common in modern China), they want their share of it. Living standards therefore have to keep improving, for nothing more surely invites social unrest than the dashing of rising expectations. Unrestrained growth can destroy the ecosystems on which all economies ultimately depend, but headlong pursuit of wealth is the cornerstone of modern Chinese life. The crowning irony is that even China's top environment officials accept that economics have to take precedence over the environment for the forseeable future. They know this means the environment will deteriorate even further before it can possibly get better, but they say they have no choice.

"I think the understanding of China is very weak in the United States," commented Ye Ruqiu. "Americans do not know how urgent our need for economic improvement is—how little our people have, especially compared to what Americans take for granted."

Zhang Kunmin, another deputy administrator of NEPA, told me he no longer even called himself an environmentalist. Rather, he was "a believer in sustainable development." By implementing a strategy of sustainable development, he said, China would halt environmental deterioration in 2000, stabilize the situation, and then "move toward improvement in 2010." More economic growth was essential because it would generate the wealth needed to finance the cleanup. "The enterprises and households must pay the true costs of a cleaner environment, so they need more wealth," Zhang explained.

Zhang and Ye took this position because, as costly as economic growth is to the environment, things could become even worse if economic growth should falter. Few Chinese appear to respect the Communist Party any longer. But everyone is desperate to avoid a relapse into the kind of chaos, waste, and stagnation that befell China during the Cultural Revolution.

"This is the terrible dilemma of China's environmental crisis," argued one Chinese environmental expert who must remain nameless. "Rapid economic growth is the most critical prerequisite for improving China's environmental situation. If economic growth stops, people will go back to the old, dirty, cheaper methods of production. Worse, there will be political instability, and that will overshadow everything; in that case, no one will have time to worry about the environment. Of course, this rapid economic growth will cause additional environmental damage; there are some things in the environment that are irreversible. That's why I think China will have to lose something—some species, some wetlands, something. We are working very hard to strengthen our environment. But, as much as I regret it, you cannot save all the things you would like. You cannot stop a billion people."

That is all the more true when many of that billion (and a quarter) people are prepared to tolerate truly staggering levels of pollution in return for economic growth. Pondering the high rates of death caused by air pollution in China, one might almost question whether industrialization has been such a good thing for China. After all, the Chinese are still dying like flies from lung and bronchial diseases, just as they have done for centuries; but now, it is coal rather than the cold that is killing them. In effect, the Chinese have traded the epidemiology of a peasant society for the new and improved industrial model. But it is no joke. The coal that is killing so many Chinese people is also letting them live longer, warmer lives in the meantime. The same point applies to the fertilizers and pesticides accumulating in China's rivers and soil. They might be poisoning people's drinking water, but they are also helping to produce the surplus food that, for the moment, is shielding most Chinese from the famines that claimed so many of their parents and ancestors. It is the old trade-off once again: greater technological prowess but also greater environmental degradation.

Of course, people's tolerance of these hazards would likely decline if they knew how dangerous they are. But the party's control of information prevents that, at least for the time being, and the public's low level of education reinforces its ignorance. One afternoon in Sichuan province, Zhenbing and I passed a lovely couple of hours with a peasant family whose children stole our hearts with their shy gifts of fresh tangerines. The grandpa was proud to tell me about the high yields in the family's fields that year; he said it was the extra fertilizer the government agricultural agent had taught them to apply that did the trick. I asked if he worried about the fertilizer polluting their drinking water. Oh, no, he said, the fertilizer only ended up in the crops, not in the river.

It is hard to be optimistic about the environmental future after visiting China. Virtually all the key environmental trends there are moving in the wrong direction, often rapidly so. True, parts of the ruling system recognize the danger and are trying to respond. In addition, there have recently been reports of greater official tolerance of democratic ideas, though whether this new openness will evolve into genuine political reform, or get crushed as the Tiananmen Square and Democracy Wall movements were, remains to be seen. And there is even a silver lining to China's rapid economic growth, for it means that 90 percent of the nation's factories, buildings, and other infrastructure will have been constructed anew by 2020, according to the World Bank. This turnover offers a window of opportunity in which to install clean, efficient technologies in China. "The situation in China is not that dreadful, as dreadful as it is," Václav Smil told me. "The question is how much will they invest to fix these problems, and where will they invest it. The Chinese are so poor, it is remarkable they have invested so much already."

But some environmental issues have a cumulative quality to them—wait too long to address them and it could become too late. China's experience with population growth is a prime example. Because birth rates went unlimited for so long, the country now faces much steeper problems on many issues, including the potentially explosive challenge of keeping the Chinese masses fed, housed, and employed. Under the circumstances, there is no time for delay, and even haste carries no guarantee of success. As a government scientist in Chongqing told me, "It is never too late to learn, but it is very late."

Eight

Sustainable Development and the Triumph of Capitalism

> I have become very impatient with my own
> tendency to put a finger to the political winds
> and proceed cautiously. . . . [E]very time I pause
> to consider whether I have gone too far out on
> a limb, I look at the new facts that continue to
> pour in from around the world and conclude
> that I have not gone nearly far enough.
> — AL GORE, *Earth in the Balance*

*P*ara ingles ver"—"for the English to see."

Brazilians were using that expression a lot during the two weeks Rio de Janeiro hosted the Earth Summit, in June 1992. It was an inside joke, a laugh at the fact that Rio was a much-changed city during the summit, but only in ways that would fool the *ingles:* the thirty thousand diplomats, journalists, and political activists who had descended on the city from all over the world to talk about saving the planet.

But why change a city that seemed to be the ideal place to hold an earth summit? After all, Rio's natural splendors—Ipanema and the other world-famous beaches, the spectacular seaside mountains of Sugarloaf and Corcovado, and the tropical rainforest on the edge of town, to name just a few—marvelously invoked the beauty and diversity of the planet's many different ecosystems. The Brazilians themselves were among the most physically beautiful people on earth, and part of what made them so attractive was that so many different races were evident in their faces—African, Caucasian, Asian, Indian. And Rio was plagued by the same ecological and social woes that afflicted many nations around the world, including heavy air and water pollution, runaway urbanization, and an extremely unequal distribution of wealth. For better and worse, Rio represented everything the Earth Summit was supposed to be about.

But better and worse did not interest Brazilian authorities; they wanted to make the best possible impression on the visiting international elite. Thus, the newly built highway in from the airport was *para ingles ver.* So were the telephone and travel offices, ludicrously overstaffed with gorgeous young women inside the Earth Summit conference center. The most striking thing *para ingles ver,* however, was the sudden transformation of Rio into a city where one could almost pretend that poverty didn't exist. The city's notorious street crime was all but eliminated during the summit, at least in the neighborhoods where international hotels were located, and the *ingles* were shielded from contact with Rio's sizable population of poor people. The Brazilian press reported that untold numbers of street kids—known to some Brazilians as "schools of fish," for their habit of traveling and attacking in gangs—had simply been rounded up and shipped across the bay to the town of Niterói. In effect, the authorities had treated the street kids the same way industrialized countries sometimes treat their toxic wastes—as a pollutant best disposed of elsewhere.

The secret behind Rio's remarkable transformation was simple enough. Well before the 115 visiting heads of state and government arrived for the summit, Rio was put under military occuption. Tanks were stationed at strategic spots throughout the city, helicopters roared overhead, soldiers carrying automatic rifles were everywhere. It was as if the military dictatorship that had ruled Brazil before 1985 was suddenly back in charge.

On the cliffside road that wound past the Sheraton Hotel, where the U.S. delegation was staying, three tanks patrolled the exit from Rocinha, the largest favela, or slum, in Rio. Accompanied by half a dozen soldiers with fixed bayonets, the tanks sat squat and heavy, their guns pointed up into the favela, like giant metal bugs ready to strike. Inside the Sheraton, President George Bush and his colleagues had at their disposal a splendid ocean view, air-conditioning, tennis courts, freshly cut flowers, and a fleet of vehicles ready to take them wherever they wanted to go. Up the hill, the one hundred thousand residents of Rocinha made do with rather less.

There were no sidewalks in Rocinha, so whenever it rained, as it did the day I visited, walking around was a slippery, muddy experience. After visiting an outdoor market, I stepped gingerly across a causeway of narrow wooden planks that led into a maze of shoulder-width passages that twisted this way and that above open sewage streams whose stench was constant. This was Rocinha's equivalent to an apartment house, only it was stacked diagonally along the hillside. Residents lived crammed into tiny, square, single-family structures of concrete, no bigger than Zhenbing's dilapidated dorm room in Beijing. Because the hillside was so steep, many of the structures rested on skinny wooden stilts, and it looked like one drenching rainfall would suffice to send them sliding all the way down the hill to the Sheraton. Yet Rocinha was a far better address than the burgeoning favelas north of Rio; in Rocinha, some residents at least had electricity and running water.

None of the people I met in Rocinha were very interested in the Earth Summit, and they deeply resented the tanks and armed soldiers at the bottom of the hill. As one man told me, the constant military presence in their neighborhood confirmed that the "Eco," as Brazilians dubbed the Earth Summit, was meant "only for class-A people."

In a world growing more polarized by the day, the tanks of Rocinha symbolized an unpleasant truth: extreme inequality can be maintained only through armed force. And as much as the tanks outside the Sheraton kept the people of Rocinha in their place, they also carried a message for the *ingles,* who had to pass by those tanks every day—after all, the road past the Sheraton was the only way their taxis and limousines could get to RioCentro, the complex of plush amphitheaters and pavillions forty minutes from downtown where the Earth Summit actually took place. Because of the tanks, the *ingles* were able to go about their plenary sessions and press conferences without being reminded, up close and dangerously, of the poverty that punished so many people around the world, and their eventual decisions reflected this privilege.

The Rio gathering's very name—the UN Conference on Environment and Development, or UNCED—was meant to signal to poor countries that the delegates would not discuss cleaning up the global environment without also addressing the deplorable conditions in which hundreds of millions of the poor countries' citizens lived. The rationale was as much practical as moral: it was hard for people to care about preserving tomorrow's environment when they were desperate for something to eat today. In his opening speech, Maurice Strong, the secretary-general of the summit, emphasized that reducing poverty was a prerequisite to environmental progress. "No place on the planet can remain an island of affluence in a sea of misery," he warned. "We're either going to save the whole world or no one will be saved."

Strong, a Canadian business executive who had organized the first UN conference on the environment in Stockholm in 1972, put the onus for taking action on the wealthy industrialized nations that were responsible for most of the world's environmental degradation: "The rich must take the lead in bringing their development under control, reducing substantially their impacts on the environment, leaving environmental 'space' for developing countries to grow." Strong further emphasized that time was short: "If the agreements reached here do not serve the common interests of the entire human family, if they are devoid of the means and commitments required to implement them, if the world lapses back to 'business as usual,' we will have missed a historic opportunity, one which may not recur in our times, if ever."

The assembled delegates listened to these words somberly, applauded them enthusiastically, and proceeded to ignore them almost entirely. In the end, the Earth Summit's accomplishments were pitifully few: two critically wounded treaties—the Framework Convention on Climate Change and the Framework Convention on Biodiversity Preservation; three nonbinding statements of principles—the Rio Declaration, Agenda 21, and a set of Forests Principles; and one almost empty treasury for translating these noble intentions into reality.

This gloomy outcome was predictable, given the diplomatic maneuvering that had preceeded the summit, especially on the part of the United States. "The American way of life is not negotiable," declared George Bush. The United States took the lead in demanding deletion of virtually all references to Northern styles of production and consumption from documents negotiated at UNCED. It refused to sign the global warming convention unless the goal of reducing greenhouse gas emissions was made purely voluntary, and it refused to sign the biodiversity convention at all, arguing that it placed undue restrictions on American companies. Coming from the world's richest nation and its biggest polluter, this unwillingness to accept limits on its own behavior sent a message that undermined progress across the board in Rio. Malaysia, for example, was guilty of rampant deforestation at home. Nevertheless, it was able to stave off pressure for a restrictive, full-scale forests convention by exploiting international annoyance at the American double standard. Why should poor countries limit their logging, Malaysia's delegates demanded, if the United States would not limit its destructive activities?

Because the North was unwilling to accept restrictions on its consumption, the Earth Summit delegates found themselves in a box, for the South, with its widespread poverty, was equally unwilling to limit its economic growth. The buzz phrase that arose to paper over this stalemate was "sustainable development." The concept had first gained worldwide attention in 1987, when the Brundtland Commission of the United Nations published a report on global poverty and environment, *Our Common Future*. There, "sustainable development" was defined as behavior that "meets the needs of the present without compromising the ability of future generations to meet their own needs." It was a fine-sounding summary of what needed to be done, and vague enough in its practical implications that everyone from General Motors executives to Chinese Communist Party bureaucrats to Greenpeace activists could endorse it in principle.

What became clear at the Earth Summit, however, was that most Northern governments regarded sustainable development as something that the South, but not they themselves, needed to practice. The proof was in Agenda 21, the summit's blueprint for pursuing sustainable development in the twenty-first century. At the end of each of its forty chapters, a passage listed how much the chapter's recommendations would cost to implement—but in the South only; apparently, the North had no such obligations.

Nevertheless, within its eight hundred pages, Agenda 21 contained many principles that environmentalists applauded. It pointed out that overwhelming levels of debt, along with volatile raw commodity prices, made development of any sort extremely difficult for poor countries. It advocated a halt to international trade of toxic waste, the extension of credit to poor people so they could create their own means of employ ment (the so-called micro-credit strategy pioneered by the Grameen Bank in Bangladesh), and the transfer from North to South of more effi cient, less polluting technologies. But like most of the Earth Summit's achievements, Agenda 21 was nonbinding. It did nothing, for example, to actually relieve the South's debt burden or stabilize its raw materials prices. Indeed, UNCED assured its irrelevance on such issues by declining even to mention global corporations, who were, after all, the most pow erful actors in modern economic life, not to mention some of the planet's biggest polluters.

"Perhaps our last chance" to avoid unparalleled catastrophe was how Maurice Strong billed the Earth Summit. Unfortunately, that chance was largely missed. The fact that more than one hundred heads of state and government came to Rio was significant. So was their agreement on cer tain basic points: that the environmental crisis was not just the fevered hyperbole of tree-hugging troublemakers but a real and present danger; that business as usual had to stop, and soon; and that this would entail fundamental shifts in human behavior. But while the world leaders talked as though they realized the prevailing course promised disaster, they acted like they didn't really believe it.

Optimists consoled themselves with the fact that the Earth Summit had generated extensive media coverage, thereby raising the environ mental consciousness of millions of people around the world. And there would be future conferences where UNCED's shortcomings could be rec tified. But those shortcomings—the North–South stalemate; the vague, self-serving definition of "sustainable development"; and the failure to confront corporate power—would endure in environmental politics through the rest of the decade, notably at the meeting in Kyoto, Japan, in 1997 where governments were supposed to finalize an agreement on re ducing greenhouse gas emissions. Moreover, the assumption of an open-ended time frame for resolving such problems ignored a central aspect of the environmental crisis: the enormous lag time between cause and effect. As Maurice Strong has explained, "The fate of the earth is likely to be decided in our generation. It doesn't mean the earth is going to die in our time, but the earth does have a cancerous condition. By the time the symptoms are so acute that they are giving us real pain, it will be too late."

In effect, the world leaders who gathered at Rio were betting that the cancer didn't have to be operated on just yet. And so, fervently declaring their continuing dedication to the cause, they bid farewell to Rio, their limos and taxis clogging the new road to the airport one last time. I stayed in town for an extra week and watched Rio return to normal life. The military retreated, pedestrians resumed their watchfulness, street crime returned. As for the *ingles,* they came and they left, and most of them never saw a thing.

Five years after the Earth Summit, at a conference held in New York to evaluate progress since Rio, no one could even pretend the news was good. "By most measures, the world seems to have moved in reverse," noted the *Washington Post.* Despite the Convention on Biodiversity signed in Rio and later ratified by 161 countries (not including the United States), species were becoming extinct at an "unprecedented" rate of fifty thousand a year, the UN estimated. With the exception of population, where the Cairo conference was an unexpected bright spot, most other environmental trends were also still heading in the wrong direction, and some were accelerating. For example, more than 60 percent of the world's fisheries were being harvested at or beyond sustainable limits. The most positive thing Maurice Strong could find to say about the so-called Earth Summit + 5 conference was that sixty heads of state or government had attended. But those leaders proved no more capable of rising above perceived self-interest than their predecessors had at Rio, so the outcome in New York was much the same: lots of rhetoric, few concrete changes.

The glacial pace of progress contrasted sharply with what most environmental experts believed was required. Even as mainstream a figure as William Ruckelshaus, a top corporate executive who had served as the Environmental Protection Agency administrator under Republican presidents Nixon and Reagan, felt there was no time to lose. The shift from business as usual to an environmentally sustainable civilization would be as basic a transformation as the Agricultural and Industrial Revolutions, Ruckelshaus had said. Tinkering around the edges would not suffice; agriculture, industry, transportation, energy, housing and many other spheres of human activity had to be reconfigured from the ground up. And while the Agricultural and Industrial revolutions had taken place over a span of centuries, the transition to environmental sustainability had to be completed within a matter of decades.

The failure of conferences like the Earth Summit led some environmentalists to conclude that they could no longer focus all their efforts on changing government policies. The hour was so late that more direct tactics were necessary. Some Greenpeace activists, led by Jeremy Leggett, the group's climate change expert, began exploring market-based strategies against global warming. These activists did not necessarily have great faith in markets, much less in the corporations that dominated them. Leggett, an ocean geologist who early in his career had helped oil firms search for black gold beneath the seas, had represented Greenpeace at climate change negotiations for years and watched as oil and coal company lobbyists teamed up with OPEC governments to obstruct diplomatic progress. Leggett and his fellow activists had no doubt that, if Exxon were left in charge of the response to global warming, the planet was in deep trouble.

But fossil fuel companies were only part of the corporate sector. Not long after the Earth Summit, Leggett began meeting with top executives in two other rich and powerful industries—banking and insurance—to educate them about how global warming threatened their profits. The idea was to enlist these two industries in a sort of corporate jujitsu, using the power of the marketplace against itself to produce an environmentally healthy outcome. Specifically, Leggett wanted the banks and insurance companies, which happened to control much of the world's investment capital, to initiate a shift of international investment away from fossil fuels and toward solar energy.

The activists helped the financiers come to realize that they had at risk literally trillions of dollars worth of property and long-term investments, and the time of reckoning was fast approaching. Scientists were warning that most of the beaches on the East Coast of the United States could be gone by 2020, victims of rising sea levels. With more than $2 trillion of insured assets along U.S. coastlines alone, the threat to the insurance industry was clear. As Swiss Re, the world's second largest reinsurance company, explained in a full-page advertisement in the *Financial Times,* "Giant storms are triggered by global warming; this is caused by the greenhouse effect; which is, in turn, accelerated by man." Swiss Re said it had paid four hundred million Swiss francs to Florida insurance companies after Hurricane Andrew in 1992; Andrew cost the industry as a whole $17 billion.

"They know that a few major disasters caused by extreme climate events . . . could literally bankrupt the industry in the next decade," said Hans Alders, director of the UN Environment Program (UNEP), in explaining the industry's about-face.

Banks, too, were concerned. Like insurance companies, they were obliged to look beyond the quarterly earnings that obsessed most service and manufacturing companies and focus on long-term prospects. Peter Blackman, assistant director of the British Bankers Association, prepared a memo in September 1995 for his industry's chief executive officers warning that over half of current bank lending was "affected by environmental factors," and that within the 20–40-year "lifetime of loans granted today, climate change is forecast to have a dramatic impact."

In April 1995, Leggett organized insurance executives to lobby at the climate change treaty negotiations in Berlin. It was the first time governments learned that Big Oil did not speak for all business executives on the climate change issue. Next, Leggett hoped to convince the banking and insurance sectors to move beyond such risk-abasement measures and use their leverage over world capital flows to kick-start the solar revolution. He had a formidable ally in this campaign: a German billionaire named Rolf Gerling, who headed Gerling-Konzern Globale, another of the world's largest insurers. Gerling helped bankroll Leggett's activism and had the Gerling publishing house issue a book Leggett edited making the case for diverting investment capital from carbon to solar.

"The insurance industry collects some $1.4 trillion in premiums every year," Leggett told me in a 1995 interview. "Much of that $1.4 trillion is reinvested. A lot of it goes to fossil fuels, which only make things worse, and almost none to solar and other renewables. We'd like to reverse that."

In Leggett's scenario, the solar transition would be market driven, but governments would have a crucial role to play. "Governments have to help break solar out of the price trap," he argued. "Until economies of scale can be realized, solar will still cost more than most fossil fuel alternatives. It's like photocopiers. In the early days, they cost thousands of dollars. But now that a market has developed, they cost a couple of hundred." Leggett believed that government procurement policies were an effective way to prime the pump. He pointed out that in the United States in the 1960s, the military's large-scale purchases of computers had been instrumental to lowering unit costs and sparking that industry's subsequent take off as a consumer giant. By twiddling a knob here and there, by spending not more but more wisely, governments could do the same for solar energy, Leggett contended. The banking and insurance industries could further the transition by giving clients incentives to avoid fossil fuels and other ecologically damaging practices—for example, by writing into a loan contract for a new office building the requirement that energy-efficient design, lighting, and building materials be employed.

Although Leggett's campaign focused on promoting solar power in the OECD countries of Western Europe, North America, and Japan, his ultimate concern was the Third World. "The battle against climate change will be won or lost in the developing countries," Leggett said. "There are literally billions of people in those nations primed for growth, and if they achieve that growth with fossil fuels, as is presently the plan, severe global warming is all but certain. Therefore, over the next decade, the OECD countries must begin generating significant amounts of solar power, both to reduce their own fossil fuel consumption and to show countries like China and India that there is a tested, cost-competitive alternative to coal-fired power plants. When I talk to Third World officials about the solar transition, they are usually interested, but they always tell me, 'You guys have to lead the way. We're not going to be the solar guinea pigs.' "

It was this set of concerns that led Leggett to emphasize the photovoltaic form of solar energy. Mounted on rooftops, photovoltaic solar panels could generate a house's entire supply of electricity or turn factories into stand-alone power plants. "I've gotten flak from some renewable energy advocates for focusing on photovoltaic, because it is farther away market competitiveness than solar alternatives like wind power," said Leggett. "But wind power can never generate the enormous volume of power needed to fuel today's industrial and industrializing economies. If we want to win the endgame on climate change, we have to get photovoltaics cost competitive as soon as possible."

The Greenpeace solar initiative would test whether market forces could achieve that. In theory, market-oriented policies have great potential because capitalists are vigorously pursuing their self-interests rather than reluctantly obeying (or evading) government mandates. But a capitalist's self-interest is not a simple thing, and the imperative of maximizing profits in the short run is often irresistible. As Lenin once remarked, "A capitalist will sell you on Tuesday the noose you will hang him with on Saturday." About a year after my interview with Leggett, I learned that similar reasoning was keeping the insurance industry (including the Gerling company—Leggett's ally!) from initiating the solar investment shift he advocated.

The tip-off came at a corporate-sponsored environmental conference in Paris in February 1997. It was a high-level crowd: participants included the chairman of the Nissan corporation, the environment minister of France, and the corporate cochairman of President Clinton's Commission on Sustainable Development. An executive of Gerling-Konzern, Dr. Walter Jakobi, happened to be speaking at the conference, and one morning I approached him during the coffee break to ask about the progress of Gerling-Konzern's collaboration with Leggett (who had now amicably left Greenpeace to pursue the solar initiative with his own organization, Solar Century). Jakobi said all was well. But the more he talked, the more it sounded like little was happening. There was no chance the insurance industry would carry out the investment shift needed to kick-start the solar revolution, Jakobi told me; market competition simply would not allow it. To be sure, his colleagues at Gerling believed in protecting the environment. But when it came to investing Gerling's premiums, he said, they could not discriminate against fossil fuel companies; they had to seek the highest levels of return possible. Nor could they afford to discourage fossil fuel use by charging customers who used it higher premiums. "If we don't write that policy, someone else will," said Jakobi.

Jakobi's answer unwittingly illustrated some of the real limits to market-based environmental solutions. To speak out about the dangers of climate change, as insurers had done at the Berlin negotiations and in their advertising, was one thing; it cost pennies and made them look like good corporate citizens. But change their daily investment practices? That would put them at a competitive disadvantage in the marketplace, lose them business, and cost them money. No company could make such a move unilaterally. It could do so only if the whole industry moved in that direction at once, and that was unlikely without firm prodding from governments.

But government mandates are anathema to most corporations, as the chairman of Nissan made clear in his remarks to the Paris conference. Yoshifumi Tsuji was not only the top boss at Nissan, he was the chairman of the Environment and Safety Committee of Keidanren, an association of top Japanese corporations similar to the Business Roundtable in the United States. In his conference address, Tsuji described the "Appeal to Industry" that he and his fellow Keidanren members had recently issued, urging Japanese firms to improve their environmental behavior. Tsuji's rhetoric was impressive: "Environmental issues will be at the very heart of the world's concerns in the next millennium. . . . It is essential that we change from a throwaway to a recycling society. . . . Japanese companies operating abroad must follow the same standards that they would at home." Tsuji repeatedly stressed that a voluntary approach to pursuing these goals was crucial. Companies behaved best, he insisted, when governments left them alone.

That evening, a gala dinner was thrown for all conference participants at the Hôtel de Ville, the City Hall of Paris. An imposing, overly ornate building of light gray stone located just across the river from Notre Dame, the Hôtel de Ville was completed in 1882 as part of the Third Republic's assertion of France's imperial glory. All the extravagance, refinement, and self-regard of that era were on display in the hotel's dazzling second-floor banquet hall. Literally every inch of wall space was covered with art, all of it finely illuminated by the soft glow emanating from crystal chandeliers. More impressive than the canvases on the walls were the images painted directly onto the walls; depictions of the founding skills of human civilization—agriculture, astronomy, hydrology, jurisprudence, music, and dozens more—stretched the length of the room.

Mr. Tsuji entered the banquet hall shortly before eight o'clock, but the other bigwigs sharing his table were late, so I took the opportunity to walk over and introduce myself and request further details about his philosophy of corporate self-regulation. With the aid of Tsuji's interpreter, a short Frenchman with a correct, nervous smile, I asked whether self-regulation was adequate in the case of, say, Mitsubishi's logging operations in Indonesia and Brazil, where the Japanese multinational was clearcutting vast tracts of rainforest. Tsuji seemed unaware of Mitsubishi's actions and asked whether the logging was being done with a government permit or not. Either way, I replied, it was bad environmental behavior, as local and international activists had told Mitsubishi on many occasions. Tsuji parried by repeating a point from his speech: Japanese firms would apply the same high environmental standards abroad that they did at home.

Speaking of home, I asked, was Tsuji worried about all the acid rain China was sending to Japan?

Yes, it was a serious problem, he replied.

So, did he believe that coal-fired power plant managers in China should also practice self-regulation?

The interpreter gave me a double take before translating my impertinence, but when I nodded for him to repeat my question, he did his job. Upon hearing the question, Tsuji emitted a sort of growling bark, then turned and stared me hard in the face. He seemed on the verge of a reply when a conference hostess, oblivious to our exchange, interrupted to introduce him to his table mate, the director of environment for the OECD.

I drifted back to my own table, where the wine was already flowing freely. I had enjoyed but my first sip when I felt a tap on the shoulder. It was the interpreter. First the correct, nervous smile. Then the question: "Excuse me, sir. We did not get your name clearly."

I told him.

"And are you a journalist?"

The reigning vision of sustainable development in the 1990s seemed to be that slapping green paint on virtually any government policy or corporate activity made it by definition sustainable. It's no surprise corporations want to regulate themselves. Who wouldn't? But the laws of the

marketplace itself make that a dubious notion. The organized gre
fierce competition that are the essence of capitalism leave little ro
corporations to be nice guys. As the Gerling company's refusal to change
investment practices illustrates, what determines corporations' behavior
above all is not the good of society but the requirements of the bottom
line.

The profit motive is what makes capitalism go, but it is so basic to the
working of the system that it tends to override other social goals. It leads
the factory owner to care more about minimizing operating costs than
minimizing pollutant outflow; it induces the logging company to see
more value in a clearcut forest than in an intact one; it causes corporations to produce products that make money but harm the environment.
When such products generate complaints, the profit motive inclines the
corporation to deny that the products are hazardous and perhaps even to
mount a propaganda campaign to that effect, as the fossil fuel companies
(like their tobacco brethren before them) have done.

In theory, governments are supposed to police corporate greed,
channeling it toward the marketplace competition that yields lower
prices and away from the corner-cutting that threatens public health and
safety. But regulation is an iffy thing. Corporations are constantly pressuring governments to relax environmental regulations if not eliminate
them altogether. This pressure is often supplemented by bribery—most
commonly, the legal bribery known as campaign contributions, which
has turned so many politicians in the United States into spineless corporate supplicants unwilling to bite the hands that feed them. Communism
dispensed environmentally perverse subsidies for political reasons—to
protect the bureaucracy's power base, enrich corrupt officials, undercut
popular dissent—but many of capitalism's environmental failures are
rooted in similar types of political favoritism.

Just as Soviet bureaucrats disregarded environmental limits in the
name of fulfilling production quotas, so do Western capitalists put their
companies' profit margins above society's need for healthy ecosystems.
When John Smith, the chairman of General Motors, announced in 1997
that GM would start building hybrid and fuel cell cars, he explained that
"no car company will be able to thrive in the twenty-first century if it relies solely on internal combustion engines." Environmentalists happily

cited this statement as evidence of "a fundamental change in approach" by the auto industry. But when the statement is read carefully, paying special attention to the adverb, its full meaning becomes plain. Forced by competitive pressure from Japan and Europe, GM would indeed start producing more fuel-efficient cars. Why not? Sales of hybrid and fuel cell cars would bring GM profits. But Smith and his colleagues had no intention of stopping production of environmentally destructive internal combustion engine cars or even, it seemed, of improving their efficiency very much. From society's perspective, getting hybrid and fuel cell cars on the road is a step forward. But getting gas-guzzling internal combustion cars off the road, not to mention reducing the total number of cars in use, is no less important. Needless to say, those goals were not part of GM's business plan.

In addition to the economic laws of capitalism that shape individual firms' behavior, capitalism's need for growth coincides with a broader political imperative: it keeps the social peace. Without growth, the only way to satisfy the demands of the poor majority for a better life would be to share existing resources more equitably. Privileged classes throughout history have been less than keen on that idea; the hard line that Northern governments took at the Earth Summit is but one recent example. Those same governments embraced the rhetoric of sustainable development instead because it seemed to satisfy both sides of the dilemma. On the one hand, it promised the development that both capitalist economics and the politics of poverty required; on the other, it championed the sustainability that environmental survival demanded.

Sustainable development is, in effect, capitalism's answer to the environmental crisis. But can it work? The question is especially important to ask now that the Cold War is over and the triumph of capitalism is being applauded by governments from Berlin to Beijing. If the world is destined to live within a capitalist framework for the forseeable future—during the very years when the environmental transition has to be launched—then it is crucial to be honest about capitalism's environmental pluses and minuses, rather than wallow in the hagiography that characterizes so much public discourse about capitalism. The issue is not whether communism is a superior system, for clearly it is not. (I say communism for brevity's sake; the system that prevailed in the Soviet Union and its empire during the Cold War might more accurately be described as totalitarian state planning.) Anyone with firsthand knowledge of the ravaged ecosystems in the former Soviet empire is quite aware that the environmental crimes of the Soviet system were as horrific as Western journalists and politicians of the early 1990s were forever pointing out. But communism's many environmental sins do not erase the environmental shortcomings of capitalism.

The late French environmentalist Jacques Cousteau, among others, has argued that contemporary capitalism is an environmental dead end. "Sustainable development is wishful thinking," Cousteau told me a few months after the Earth Summit. "If development means growth, then you cannot have sustainable development on a limited planet. It is a principle that is impossible. I think people should change the wording to a 'sustainable future.' That could be achieved, *if* we would use our efforts to grow in wisdom, culture, education, and those things which are unlimited. But if our development is expressed in electricity and motor cars, then it is impossible." In a subsequent interview with the *New York Times,* Cousteau expanded on his critique of capitalism: "We are prisoners of a system that uses more resources per capita every year. . . . Our Western model is not valid for a world of limited resources. We cannot deliver it to the ten billion people or more we will be in thirty-five years."

Yet capitalism's inherent drive for expansion is spreading the American approach to consumption across the planet. Young Thai villagers like my friend Leno are now wearing the same clothes, buying the same cars, and watching the same television shows their New York and Los Angeles counterparts are. As corporate strategists salivate over the prospect of billions of new Lenos joining the global market, fewer and fewer places remain untouched by the consumerist invasion. In Istanbul, I remember, I stayed in a budget hotel within shouting distance of the famed Blue Mosque. Each dawn, I was awakened by the ancient, sonorous drone of the five o'clock call to prayer. Down in the streets, there remained other signs of old Istanbul—the outdoor bazaar, the Turkish baths, the countless minarets punctuating the skyline—but they were being rapidly eclipsed by a clutter of advertisements for Marlboro, Levis, Renault. The same phenomenon is unfolding in Beijing, Moscow, Bangkok, Berlin. The new consumers know perfectly well—because ads and TV programs show them—what pleasures affluent Americans enjoy, and they naturally want the same for themselves.

But the American lifestyle consumes resources and produces waste at thirteen to thirty-five times the rate that Southern lifestyles do. If that multiplier is applied to the South's far larger population, the increase in humanity's environmental impact will be immense. Can the earth's ecosystems accommodate such an increase?

There are those who believe it can. In an article in the *Atlantic Monthly* titled "Do We Consume Too Much?" professor Mark Sagoff of the University of Maryland cited declining raw materials prices as proof that there is plenty of oil, timber, and food to supply humans for many generations to come. But to rely on marketplace prices for environmental judgments is problematic, because the marketplace generally attributes no value to environmental goods and services (like species diversity) that are, in fact, vital to human survival. A second flaw in Sagoff's argument was his unstated assumption that the environmental damages associated with humanity's *current* resource consumption are acceptable. It is true that humans are not "running out" of petroleum—but only if one is prepared to tolerate additional accidents like the Exxon *Valdez* crash and the near extinction of indigenous peoples like the Ogongi of Nigeria and the Huaorani of Ecuador, whose cultures and ecosystems have been brutalized by the modern world's thirst for the oil beneath their lands. Nor is there much cause to worry about a timber shortage, as long as one does not mind continued felling of old-growth trees in the Pacific Northwest and an annual global loss of forest area equal to the size of Washington State. The list goes on, and though the price system usually does not register these losses, the planet does.

In any case, it is not the front but the back end of the production cycle that is the main worry; the most immediate limits on human activity are not issues of resource supply but of waste disposal. For example, humans have the technological capability to produce virtually endless amounts of CFCs, if they so choose. But past production levels have already depleted the ozone layer sufficiently to cause tens of thousands of cancers and other hazards in the decades ahead, so in reality that option is closed. (Or at least it should be closed. Although satellite photos of the ozone holes over the North and South Poles are readily available, there are still lawmakers in Washington, including Representative Tom DeLay, the number three man in the Republican leadership, who maintain that the ozone threat is a hoax.) Likewise, humans may have enough oil and coal in the ground to burn for centuries, but releasing that much carbon dioxide into the atmosphere would be suicidal, considering that global warming has already begun. Nevertheless, greenhouse gas emissions keep climbing worldwide, even as the forests whose photosynthesis could help offset the greenhouse trend continue to disappear, taking with them untold numbers of plant and other species needed to maintain the planet's ecological equilibrium.

Capitalism needs and promotes ceaseless expansion, yet the evidence that human activity is already overwhelming the earth's ecosystems is all around us, and it seems certain to increase as human numbers and consumption levels rise in the coming decades. Environmentally benign new technologies like fuel-cell-powered cars could counter this trend, but they can not save earth's ecosystems from overload if consumption keeps rising indefinitely. Recycling of paper, glass, and other waste rose 20 percent in the United States from 1985 to 1992, but overall consumption of natural resources still increased by 30 percent, because per capita consumption continued its constant upward spiral.

"We have to distinguish between growth and development," the economist Herman Daly told me in a 1998 interview. "We have to shift from pursuing growth, which is quantitative, to pursuing development, which is qualitative." From 1988 to 1994, Daly, the author of *Beyond Growth*, had worked at the World Bank, where he struggled unsuccessfully to persuade his colleagues that the global economy had to be seen as a subsystem of the global environment and thus that there were inevitable limits to economic growth. (One of Daly's detractors was Lawrence Summers, a senior bank executive who distinguished himself with the remark that poor countries should want more pollution, not less, because it meant more economic growth. Summers later became the number two official in President Clinton's Treasury Department.)

Daly argued that the environmental limits to economic growth were fast approaching, not just in regards to the ozone layer and global warming but also the food chain. Biologists have estimated that humans are consuming 25–50 percent of all the solar energy captured by photosynthesis on earth; that is, 25–50 percent of the plant matter that, directly or indirectly, is consumed by humans and all other species as food. Depending on whether the true number is closer to 25 or 50 percent (the role of oceans is critical), such estimates make clear that the environmental scale of humanity cannot keep expanding for long. A doubling of the human scale may seem possible arithmetically (50 percent times two equals 100 percent) but not ecologically, for that would leave no energy for other species. Daly argued that humans therefore have to contain their expansiveness by lowering consumption levels, improving technologies, and limiting population growth.

If development must replace growth, the question becomes unavoidable: can such a shift be achieved under capitalism, an economic system whose very essence is growth for growth's sake? In an era of capitalist triumphalism, to pose such a question may seem akin to tilting at windmills. But unfashionable or not, the question must be faced. Besides, to say that capitalism is flawed by environmental contradictions does not necessarily mean capitalism must be eliminated. If capitalism can be reformed thoroughly enough, it may not need to be replaced. There is a difference, after all, between capitalism *qua* capitalism and the market mechanisms that govern its daily operations. Herman Daly would be the first to agree that market mechanisms are crucial to a properly functioning, and environmentally respectful, economy. But while the market has a genius for efficient allocation of economic resources, said Daly, it cannot be allowed to decide all economic matters, least of all the matter of appropriate scale. Left to its own devices, the market knows only one answer to questions about how big an economy should be or how much its members should produce and consume: more, always more.

"If by capitalism one means the current thrust toward a hyperconsumptive, financially driven globalization, then I'd say we're fried," Daly told me. "That kind of capitalism accelerates resource depletion and waste production and undercuts nations' efforts to raise environmental and social standards." The trend to globalization can be resisted, however, said Daly, through controls over both capital mobility and so-called free trade. And market mechanisms, when properly guided, can help develop the new technologies and practices needed to navigate the transition to genuine sustainability. For example, market forces have helped reduce sulfur dioxide emissions in the United States significantly, and at much less cost than initially anticipated. When the Clean Air Act was revised in 1990, utility industry lobbyists tried to derail new acid rain provisions by claiming that sulfur dioxide reductions would cost $1,500 a ton. The Environmental Protection Agency estimated the costs at $450–600 a ton; environmentalists said $300 a ton. The standards went into effect, though not until 1995. By then, the emissions trading system the act had established had brought the price down to $132 a ton. In 1996, the price was a mere $70 a ton, and even the business press was wondering whether utilities should have been forced to make deeper emissions cuts than the 1990 act had mandated.

This success was realized, however, only because the EPA stood firm in the face of industry scare tactics. Market forces can accomplish a great deal, but only when backed by relentless pressure on corporations to do the right thing; the carrot works best when combined with the stick. Of course, market pressure comes in various forms. Consumers, for example, can mobilize their buying power to encourage environmental advances like solar photovoltaic panels or to discourage abominations like sport utility vehicles. But in the end, there is no substitute for government. When fifty-one of the one hundred biggest economies on earth are corporations, not countries, only governments come close to having enough strength to enforce real limits on the behavior of the marketplace and the giant corporations who are its most powerful players.

Many American environmentalists thought that the 1992 election had brought them a government committed to just that kind of activism. Democrat Bill Clinton had defeated Republican George Bush, who had been nearly as hostile to the environment as his predecessor, Ronald Reagan. True, Clinton had compiled a poor environmental record as governor of Arkansas. But as his vice president he had chosen Senator Al Gore of Tennessee. Gore was among the most ardent environmental advocates in Congress, and in January 1992 he had published a book, *Earth in the Balance,* that sounded the alarm in urgent, eloquent terms. The rescue of the environment, he wrote, had to become "the central organizing principle for civilization."

Gore's book was not the usual politician's hodgepodge of cliches but a thoughtful treatise that showed he understood some of the reforms needed to make the market economy less environmentally harmful. Gore supported rewriting the economic accounting system so that prices, GNP, and other measurements reflected the actual social costs of air pollution, habitat destruction, and other damaging activities. He urged eliminating government subsidies that kept markets from telling the environmental truth. He suggested that economic processes be changed at the point of production so no pollution was produced in the first place—preventing rather than controlling pollution. All this was music to the ears of environmentalists like Lester Brown, who told me a few weeks after the 1992 election, "It's not just Gore being there, it's that Clinton consciously chose Mr. Environment as his running mate." The Environmental Revolution may finally have found its Gorbachev, Brown added.

Gore had certainly fought the good fight at the Earth Summit, emerging as one of the most outspoken critics of Bush's foot-dragging on carbon dioxide reductions. Late in the summit, I had the opportunity to question Gore one afternoon after he addressed a roomful of American environmentalists at the Sheraton Hotel. Gore had not yet been chosen as Clinton's running mate, so he was still unencumbered by the hordes of reporters and Secret Service agents who in a few weeks would be surrounding him whenever he appeared in public. As Gore slipped out of the briefing room and headed across the Sheraton's enormous ground-floor hallway, he walked like the politician he had been raised since childhood to become: purposeful stride, eyes surveying everyone in sight, ready to greet any face that recognized his own. Market-based solutions to environmental problems were all the rage at the summit, so I asked Gore if he supported "emissions trading," a mechanism that allowed countries or companies that reduced pollution below mandated levels to "sell" leftover emission rights to those who did not meet the standard.

"Well, I wrote about that in my book, didn't I?" said Gore, as if that settled the matter. Some environmentalists criticized emissions trading, I replied, charging that it amounted to buying and selling the right to pollute. Did Gore worry about that? He shot me a wary look out of the corner of his eye and uncoiled a very guarded, meandering response: market-based solutions had much to recommend them, although the advisability of any particular solution could be determined only by examining the specifics of a given situation, and of course one had to consider all options before making a final decision, and so on. Gore's hemming and hawing continued while we rode the escalator up to the Sheraton's street level. By the time we reached the top, he seemed to have said he supported emissions trading, but in such a vague and cautious way that I suggested that he sounded, well, vague and cautious. Apparently taken aback, Gore replied, "Well, I *am* cautious." He then rushed out of the hotel and into a limousine that whisked him away, up the cliffside road, and past the tanks of Rocinha.

"I *am* cautious." It was a curious but telling admission. Curious, because the book Gore wrote about the environment was anything but cautious. *Earth in the Balance* warned that humanity was bringing on "an environmental holocaust without precedent" and that "nothing less than the current logic of world civilization" had to be overturned if catastrophe was to be avoided. Gore was searing in his denunciations of those unwilling to change course; apathy and inaction he condemned as the moral equivalents of Nazi appeasement. Indeed, Gore twice implicitly likened himself to Churchill, who had repeatedly called for taking a firm stand against Hitler in the 1930s, when many Britons preferred to believe that war with Germany could still be avoided.

But once they entered the White House, Gore and his boss Clinton resembled not so much Churchill as the accommodationists he despised. The environmental proposals the Clinton administration advanced in its first months in office included only a pale version of Gore's original vision of reform, and even that was abandoned at the first sign of opposition from corporate interests.

During the transition between the November 1992 election and Clinton's January inauguration, one of the ideas floated by the incoming administration was a carbon tax, so named because it would increase taxes on the burning of fossil fuels. Representatives of the energy industry quickly attacked the idea, and Clinton just as quickly dropped it. When Clinton gave his first State of the Union address in January, he instead proposed taxing only the heat content of fossil fuels. This so-called BTU

(British thermal units) tax would amount to fifty cents per gallon of gasoline and was intended to lower air pollution and greenhouse gas emissions. However, this weaker substitute was no more acceptable to the fossil fuel industry than a carbon tax was, and the industry and its friends made it clear they would play hardball to defeat it. Well fortified by industry campaign contributions, Democrat David Boren of Oklahoma and other senators from western oil and coal states publicly informed the White House it had to cancel the BTU tax or they would vote against Clinton's entire federal budget.

Similar resistance greeted the administration's proposal to cut certain fiscally egregious and environmentally harmful subsidies. One subsidy program dating back to 1872 gave away the nation's mineral rights for as little as $2.50 an acre. Another program actually paid logging companies to cut down the national forests. A third subsidized ranchers who overgrazed their herds on public land. On the subsidy front, it was Democrat Max Baucus of Montana who led a contingent of western senators, again from the president's own party, who promised to sink Clinton's overall budget if he pursued the cuts.

It was make-or-break time for the new president, and he broke. Clinton caved, the subsidies stayed, and the energy tax was scaled back to a meaningless 4.3 cents a gallon. Caution was only part of the problem, though; the White House also blundered tactically. The BTU tax would have been difficult to pass in any case, given the antipathy it was bound to excite among energy companies. But the administration further hurt its chances by ignoring a crucial political angle of environmental tax reform, even though Gore had noted it in his book: environmental tax increases must be offset by simultaneous reductions in income or other taxes. Otherwise, the government has no hope of building a constituency in favor of the tax shift (especially with a tax like the BTU, which was incomprehensible to the average citizen); instead, it gets accused of "raising taxes." This is exactly what happened to Clinton, as Republicans seized the opportunity to portray him as a tax-and-spend liberal.

Gore might defend the administration's tactical choices of 1993 (I say "might" because the vice president's office declined my repeated requests for a formal interview with Gore) by arguing that Clinton's main goal upon taking office was reducing the deficit. Therefore, the BTU tax could not be revenue neutral; it had to generate funds to help lower the deficit.

Leaving aside the debatable assumption that deficit reduction was an urgent national priority, there was an obvious solution to the deficit problem: the peace dividend. After all, the Cold War was over. The Soviet Union had collapsed. A ground war in Europe between the former superpowers was now inconceivable. Since the forces earmarked to deter or fight such a war accounted for more than half the U.S. military budget, logic said that some $150 billion was now available for alternative purposes. But Clinton was cautious on this front as well. He maintained Cold War military spending levels throughout his first term. When he ran for reelection in 1996, the Pentagon budget of $266 billion was barely less than what Bush had projected spending in that year.

The 1993 budget debacle was but the first in a series of environmental retreats, defeats, half-measures, and worse on the Clinton administration's part. On Earth Day 1993, Clinton pledged a vigorous effort against global warming, but the plan his administration later released relied almost exclusively on voluntary measures that Gore himself conceded were inadequate. Perhaps the lowest point came in 1995, when Clinton signed into law a bill that, in the name of "salvage" logging, allowed unlimited clearcutting of national forests, including ancient, old-growth trees. Not until 1997 did the administration deliver a major environmental victory, when Clinton endorsed the tougher air pollution regulations insisted upon by EPA administrator Carol Browner, the one Clinton administration environmental official who seemed willing to stare down corporate opposition.

In *Earth in the Balance*, Al Gore had called global warming "perhaps the greatest danger this country has ever faced." But as the Clinton administration prepared for the global warming conference in Kyoto in December 1997, Bill Clinton typically tried to be all things to all people. Rhetorically, he agreed with environmentalists that global warming was serious and had to be stopped. Concretely, he put forward a proposal that was arguably worse than the Bush administration's had been. Clinton proposed that so-called Annex 1 countries—the industrialized nations of Europe, North America, Japan, and the former Soviet bloc—reduce greenhouse gas emissions to 1990 levels by 2012. While Clinton's proposal envisioned a binding agreement—an improvement over Bush's insistence on voluntary goals—it delayed the emissions reductions an extra twelve years, until such time that both Clinton and Gore (if Gore succeeded Clinton as president) would have left the White House.

Clinton also endorsed a key demand of the fossil fuel lobby, one apparently intended to scuttle any agreement at Kyoto. The fossil fuel lobby's claim that global warming was but an unproven theory had usefully delayed progress for five years now, but its effectiveness was waning in the face of accumulating scientific evidence and journalistic exposés of the propaganda campaign. Thus a new message took its place, backed by $13 million worth of print and broadcast advertising. The ads warned that the Kyoto treaty would single out the United States for special punishment and ruin its economy: American energy costs would rise, but countries like Mexico, India, and China would be exempt from any cutbacks. The same argument was made in the Senate, which voted 95–0 for a resolution advising the White House to accept no treaty in Kyoto that exempted developing nations. Clinton quickly agreed. All this was high cynicism on the fossil fuel lobby's part and revealed as such when Exxon chairman Lee Raymond visited China a few weeks before the Kyoto conference. At the very time when the industry's advertisements were demanding that developing nations like China be forced to limit their greenhouse gas emissions, Raymond gave a speech urging his Chinese hosts not to let unfounded fears of climate change reduce China's fossil fuel consumption.

The one aspect of global warming policy where Clinton did show boldness was in public relations, where he tried to redefine failure as success. At a White House briefing two months before the Kyoto conference, the smooth-talking president explained that his climate change policy would limit the concentration of carbon dioxide in the earth's atmosphere to 550 parts per million, a mere doubling of preindustrial levels and thus, according to Clinton, an admirable goal. Writer Bill McKibben pointed out two days later in the *New York Times* that Clinton's policy "move[d] the goalposts of the climate game in dangerous ways." Humans had increased the carbon dioxide concentration from 275 parts per million before the Industrial Revolution to 365 parts per million by the 1990s, noted McKibben, who added, "When scientists warned about the really scary probabilities [of global warming]—coastal flooding, widespread drought and so on—they always said these would happen if we didn't keep carbon dioxide from *doubling*, to 550 parts per million." Yet Clinton had now made a concentration of 550 parts per million the target of U.S. policy.

Clinton's climate change position disgusted even former supporters like Lester Brown, who labeled it "pathetic." Some of Clinton's aides seemed equally dismayed. Eileen Clausen, the assistant secretary of state in charge of climate change policy, resigned in September 1997. Her superior, Tim Wirth, the undersecretary of state, also resigned just days before the Kyoto conference. Meanwhile, the European Union charged that Clinton's proposal promised "environmental disaster." The one party to the debate that seemed pleased was the fossil fuel lobby. A coal industry newsletter applauded the White House with the headline "Clinton Saves Place in History, Does Nothing."

The Clinton administration's position was so at odds with the rest of the world that Kyoto seemed destined to fail, which would have been bad news for both the planet and the presidential hopes of Al Gore. Climate change was Gore's signature issue, environmentalists were key members of his political base, and the United States was sure to be blamed for any failure. After weeks of noncommittal silence, Gore finally announced he would attend the Kyoto talks. He ended up spending even less time in Kyoto than Bush did in Rio de Janeiro—a mere nineteen hours. But in a widely covered speech, Gore instructed the U.S. delegation to show "increased negotiating flexibility," even as he repeated the fossil fuel lobby's demand that an acceptable accord had to include "meaningful participation of key developing countries." Gore was rewarded with media coverage that credited him with salvaging an agreement.

But when the dust settled, the Kyoto talks had produced little change in actual policy. Industrial nations pledged to reduce greenhouse gas emissions 5.2 percent below 1990 levels by 2012, while the contentious issue of Third World participation was delayed to a follow-up meeting in Buenos Aires in November 1998. Of course, a 5.2 percent reduction was far from the 50 to 70 percent reduction the IPCC had said was needed to maintain the atmosphere's current carbon dioxide concentrations (concentrations that might or might not deter severe global warming). And there was less to the 5.2 percent figure than met the eye, thanks to a bevy of loopholes, including Al Gore's old friend, emissions trading.

Under the aptly named "hot air" provision, the United States and other industrial nations would be able to maintain their high levels of greenhouse gas emissions by buying emissions rights from countries that had them to spare. The leading candidates were Russia and Ukraine, whose economies had collapsed after the fall of the Soviet Union, lowering their industrial output and carbon emissions. Russia and Ukraine would now be able to sell their extra emissions rights to Western nations, thereby allowing the latter to avoid making significant reforms at home. The hot air loophole alone threatened to cancel out the entire 5.2 percent reduction mandated by the Kyoto treaty. Annex 1 countries would instead be able to satisfy their treaty obligations with a mere 0.7 percent decrease in overall emissions by 2012. This target was virtually indistinguishable from Bill Clinton's original goal of returning to 1990 levels by 2012, as White House economics adviser Gene Sperling later all but admitted to reporters. And even this paltry target was less than airtight. Notwithstanding official claims about a legally binding treaty, "the compliance section [of the Kyoto Protocol] has no binding consequences at all," complained Ozone Action, an American environmental group.

Some observers nevertheless regarded the Kyoto Protocol as a historic achievement. For the first time, binding limits had been set on greenhouse gas emissions. Ross Gelbspan, author of *The Heat Is On,* called the Kyoto Protocol "a puny beginning" but a useful "dress rehearsal" for the larger reductions needed to meet the IPCC's 50 to 70 percent target. Gelbspan drew inspiration from the Montreal Protocol of 1987, which was repeatedly tightened in subsequent years as evidence mounted that ozone loss was more advanced than originally thought.

But the Kyoto Protocol recalled even more the agreements signed at the Earth Summit. Like their predecessors in Rio, the Kyoto negotiators did make progress, but of an incremental scope and at a snail's pace. In Kyoto as in Rio, progress foundered on the refusal of Northern governments to reduce consumption levels. In Kyoto, this refusal took the form of the American demand that China and other developing nations be required to limit carbon dioxide emissions just like the industrialized nations. The developing nations rejected this demand, offering the same objection they voiced in Rio: the North's historical patterns of production and consumption were responsible for most of the world's carbon dioxide concentrations, so the North had to take the lead in fixing the

NO oNe waNTS To Be firsT-

problem. Besides, the pervasive poverty in Southern countries made economic advancement their top priority. As one Chinese delegate charged, "In the developed world, only two people ride in a car, and yet you want us to give up riding on a bus."

Another Rio–Kyoto parallel was the negotiators' timidity about challenging the political and economic status quo and pressuring corporations to change their behavior. Invoking the example of the Montreal Protocol was all very well, but corporate resistance to phasing out fossil fuels was much stronger than it had been to abandoning CFCs. After all, fossil fuel was not one product among many for corporations like Exxon, as CFCs were for DuPont—it was the heart of the business. Besides, DuPont and its allies had not truly ceased their resistance to abandoning CFCs until they cornered the market on such substitutes as HCFCs. True, the Kyoto talks led some fossil fuel companies to begin a similar repositioning. Royal Dutch Shell and British Petroleum, for example, both pledged large investments in renewable energy. But these corporations had in mind a slow and gradual transition away from fossil fuels, and during the intervening decades they did not intend for their traditional products to be taken off the market the way CFCs were.

The Kyoto negotiators' failure to establish a much brisker schedule for the phaseout of fossil fuels reflected a third similarity with Rio: a conception of sustainable development that envisioned few fundamental changes in the practices and power relationships of modern civilization. Somehow, marginal reforms and appealing rhetoric would magically add up to a revolutionary transformation and reverse our environmental course without undue pain, conflict, or stock market losses. Al Gore implicitly endorsed this same message a month after the Kyoto talks when he proudly announced a new product-labeling program that would enable consumers to choose televisions and video recorders that used 50 percent less power, *while turned off*—an initiative that, at best, could reduce annual American carbon dioxide emissions by less than one tenth of 1 percent. This from the man who complained in *Earth in the Balance* that, instead of fundamental changes, politicians were prescribing "minor shifts in policy, marginal adjustments in ongoing programs, [and] moderate improvements in laws and regulations, rhetoric offered in lieu of genuine change."

If a government boasting Vice President Gore, one of the most environmentally aware senior politicians in the world, could not make progress, what did that say about the larger question of humanity's chances of avoiding planetary catastrophe? Beldrich Moldan, who served as the environment minister of the Czech Republic after the Velvet Revolution of 1989, told me during one of my visits to Prague that he was "surprised at how little Clinton and Gore have accomplished environmentally," and he added that their failure "has a very negative effect on efforts to push for environmental reforms in other countries. The United States is watched much more than Americans realize. As a European, you may like the United States or not like it, but you know it's the future. So when the United States refuses to reform, other countries will refuse as well."

Al Gore was too well informed not to recognize that the environmental accomplishments of the Clinton administration were indefensible. His excuse was to blame the opposition party and the inertia of official Washington: "The maximum that is politically feasible . . . falls short of the minimum that is scientifically and ecologically necessary," he said in 1993. Of course, it is hard to know what is politically feasible unless one puts up a good fight. Which raises a second possible explanation for the Clinton administration's woeful environmental record: Bill Clinton wasn't that interested. Gore himself would never breathe such slander about his good friend. But at the very least, Gore failed to persuade Clinton to fight what Gore had called "a mortal threat . . . more deadly than that of any military enemy" America would ever again confront."

Perhaps Gore will have the chance to do better as president. In *Earth in the Balance,* he likened himself to Churchill, a politician prepared to brave all risks to awaken his countrymen to the unprecedented crisis gathering around them. But translating the noble vision of Gore's book into reality would have required standing up to some of the most powerful entities on earth, including the two largest and richest industries of the century—oil and automobiles. In shrinking from that task, Al Gore and Bill Clinton were no different from the vast majority of lawmakers. But then the vast majority of lawmakers are forgotten by history. It is the Churchills we remember, precisely because they did not just speak, they acted, and their actions changed the world.

Nine

Living in Hope

I can hardly imagine living without hope. As for
the future of the world, there is a colorful
spectrum of possibilities, from the worst to the
best. What will happen, I do not know. Hope
forces me to believe that those better
alternatives will prevail, and above all it forces
me to do something to make them happen.

—VÁCLAV HAVEL

Can the human species bring its behavior into balance with the systems of nature that make our lives on earth possible? The answer to that question may well decide humanity's fate in the century to come, and one need hardly add that there is ample reason for pessimism. After all, our economic systems are predicated on continual growth, and traditionally growth has meant ecological destruction and decline. Humans seem addicted to forever wanting more of everything, yet there are billions who still lack basic necessities. Political leaders talk about changing course, but they rarely do so, partly for fear of retribution from voters, corporations, and others who might be discomforted in the process. Meanwhile, the clock ticks.

But if humanity's environmental progress has been inadequate in the

1990s, it is not because people don't care. During my travels, I encountered a high degree of concern about the state of the environment among average people the world over; it was one of the most consistent and encouraging findings of my entire journey. In hundreds of conversations with individuals from all walks of life, I found that the vast majority of people had not only heard about the gathering ecological crisis, they cared and were happy to talk about it (which is more than could be said about other issues of global import, such as the vicious slaughters taking place in Bosnia and Rwanda). People tended to be better informed about their own area's environmental ills than about such global questions as biodiversity and climate change, and often they did not grasp the scientific details of either, but the overall importance of environmental hazards was accepted without question, as was the need to do something. That such awareness existed in virtually every country I visited (the exception being China) and at all levels of society is all the more remarkable considering the limited information available to many people.

I remember a woman in Uganda who ran a small grocery store near the botanical gardens in Entebbe. Tall, big-boned, she looked about thirty-five and wore a light-blue bandanna with red polka dots that stretched across her forehead and knotted behind her neck. Lack of paint gave the outside of her roadside shop a dingy, run-down look, but inside, the boxes of soap powder, bottles of soda, and other household items were displayed in orderly rows, and the floor was swept clean. Obviously hard-working—"It is not good to be idle," she said with a smile—she had no idea I was a journalist.

We had been chatting for a few minutes about the gardens when she said she was sorry the weather had not been nicer for my visit to Uganda. For the past week, there had been a queer gray tint to the sky, a kind of distant haze that constant sunshine had not been able to burn off. I asked what caused it.

"This happens sometimes," she explained with a shrug. "Some people think that it is because of pollution, some say it is the desert coming. I am afraid it is the desert."

"The desert?"

"Yes. When there is no rain, the winds blow the sand into the air and the desert grows and starts to cover the land. This is why we are told to keep our forests green and not to cut them down. The rule is, 'If you cut a tree, you must also plant a tree.' "

"Do people follow this rule?"

"Most people do, except the stupid ones who care about nothing."

Desertification posed a clear and immediate threat to this woman and her neighbors, but it turned out that she was also familiar with the more distant hazard of ozone depletion, which she had read about in a local magazine and referred to as "the hole in the sky." She did not know what caused ozone depletion, and she had not heard about the green house effect at all (in this she was no different from some urban Europeans I had met), but she seemed genuinely interested in my brief descriptions of these problems. For someone living in a country where the leading newspaper was often a mere four pages long and devoted largely to government pronouncements, she was remarkably well informed.

People in rural areas tend to have even less access to media reports, but this disadvantage can be overcome by their more intimate relationship with the natural world. A journalist who had been born and raised in Leningrad told me about how working seven years for a newspaper in an agricultural district had made him realize how divorced he was from the biological underpinnings of life. "It's a pretty obvious point that without the food raised in the countryside by the peasants we urban sophisticates would starve," said Grigori Churov with a wry grin, "but it had never really occurred to me before. Like most city people, I had lost any connection with the earth. The peasants had a very different attitude toward nature, which was revealed by how fiercely they fought against the pollution of their rivers by a chemical factory in the provincial capital. I must admit, I approached that story thinking that the peasants were just silly, backwards people who were obstructing progress. But I came to respect their views. To them, it was inconceivable to dump chemicals into the river, because they understood that the environment was what sustained them."

Urbanites can become environmentally conscious very quickly, however, said Simona Bouzkova, an economist who helped reform the Czech Environment Ministry after the Velvet Revolution in 1989. "At that time it was easy to convince people about the need to clean up the environment because they could see the problems with their own eyes," she told me. "Our situation was similar to the conditions in the United States in the 1960s, when the pollution was so extreme you could actually see the black smoke pouring out of factories. Our towns and industrial areas always registered higher environmental consciousness than rural areas, because they were the people who had to breathe that air."

Lest these and countless other testimonies I collected be dismissed as random anecdotes, let me note that professional pollsters have reported similar findings. One poll conducted by the George H. Gallup International Institute, based on personal interviews with people of all economic classes in twenty-four countries (including Germany, Nigeria, Japan, Chile, Canada, India, South Korea, Mexico, Hungary, and the United States) claimed to reflect the attitudes of two-thirds of the earth's people. Summarizing the results at the Earth Summit, George H. Gallup Jr. reported that "concern for the environment has become a worldwide phenomenon. Not only do people place a greater priority on environmental protection than on economic health, they also indicate a willingness to pay for environmental protection. . . . These results clearly challenge the view that being concerned about the environment is a 'luxury' that only those in rich nations can afford to pursue."

One of the most striking findings of the Gallup poll concerned the health effects of environmental hazards. Only in the Soviet Union, Poland, Germany, and the Philippines had majorities of people claimed to suffer significant health effects in the early 1980s. But by the early 1990s, majorities in sixteen countries (including well-to-do Canada, Great Britain, and the United States) reported that they had experienced environmentally related ailments. Majorities in all twenty-four countries expected their children and grandchildren's health to suffer over the next twenty-five years. According to the Gallup pollsters, "This belief helps explain the surprisingly high levels of concern about environmental quality among residents of all types of nations. Environmental problems are no longer viewed as just a threat to [the aesthetic] quality of life . . . but are considered a fundamental threat to human welfare."

All this sounds like good news for the environmental cause. The problem is, what humans say is not necessarily what they do. For various reasons, the high level of public concern about the environment has not translated into an equivalent amount of concrete change. One problem, reflected both in the Gallup poll and my interviews, is that environmental consciousness, though heartfelt, tends to be imperfectly informed. For example, the majority of those surveyed by Gallup exhibited "limited—if any—understanding of global warming," a problem they often confused with ozone layer depletion. I encountered the same confusion, not just among people like the shopkeeper in Entebbe but also among well-educated professionals in the media-drenched countries of Europe and North America.

Reliance on the media is itself problematic, however, in an era of rampant press manipulation and trivialization. In the United States, the propaganda campaign the fossil fuel lobby launched in 1991 had considerable effect. When *Newsweek* polled Americans in 1991 about global warming, 35 percent of those surveyed said that they considered it a serious problem. In 1996, *Newsweek* repeated the poll. During the intervening five years, the scientific consensus about global warming had tightened considerably, and the IPCC had issued its landmark report concluding there was "a discernable human influence" on the planet's rising temperatures. But industry's propagandists had also been hard at work, placing their spokesmen on radio and television and posting their articles in newspapers. *Newsweek's* 1996 poll found that only 22 percent of those surveyed considered global warming a serious problem.

Among the loudest voices claiming that the danger of global warming and other environmental hazards was overstated was journalist Gregg Easterbrook, a self-styled "eco-realist" who attracted a lot of attention with a book he published in 1995, *A Moment on the Earth.* Easterbrook's message was that the global environment was in pretty good shape and getting better, thanks to nature's abundant healing powers and the innovative genius of democratic capitalism. This belief was congenial to the status quo on both a social and an individual level, but it was wildly at odds with scientific reality. Scientists who reviewed *A Moment on the Earth* in *Scientific American, Natural History,* and other leading journals were shocked by the frequency and magnitude of Easterbrook's errors of fact and interpretation. The misstatements were so egregious (and yet passing unremarked in such influential mass media organs as *The New Yorker* and the *New York Times*) that the Environmental Defense Fund published a two-volume rebuttal of them, correcting Easterbrook on everything from the effects of the Exxon *Valdez* oil spill (the sea otter population in Prince William Sound was devastated, not untouched) to rainforest fires in Brazil (common, not rare, in the 1980s) to the rise in sea level already experienced in this century (not one but rather four to eight inches, enough to have eroded forty feet of a typical beach on the East Coast of the United States). Its scientific shortcomings made *A Moment on the Earth* unlikely to exert long-term influence within the environmental debate. But in the meantime, it helped to lull many readers and fellow journalists into a happy complacency.

In other countries, ignorance was imposed by force. In China, for example, people knew that their air and water were polluted, they weren't blind. But they didn't know just how dangerous the situation was, because the government took pains to withhold the most alarming information. "As an ugly bride dare not show her face to her parents-in-law, so the facts should be covered by a veil," one official reportedly said in arguing that a report on Beijing's pollution should be suppressed for fear of driving away foreign investment and inviting popular revolt. "If the masses knew about it, there would be an outcry," said another official. "It could even cause social instability." It could indeed. A 1997 opinion survey found that 89 percent of Beijing residents were "unhappy" with the city's environmental situation, while 31 percent expressed "great disapproval." Imagine what might happen if the true pollution levels and their corresponding health effects were known! Even under China's political system, the pressure on the government to clean up could be intense. The environmental revolts along the Huai River in 1996 demonstrated that popular resistance was not impossible under a police state; nevertheless, environmentalists in China noted that the specter of repression squelched most protest before it began. As one observer commented, "In the end, no environmental policy will work without an educated public who can speak out and raise hell about problems in a free way." Which helps explain why, in China and elsewhere, the struggle for human rights is indivisible from the struggle for a healthy environment.

Besides insufficient information, a second obstacle to progress is complacency; many people who care about environmental problems are not animated by a sense of urgency. Even people who, when asked, fervently agreed that the environment was important did not usually pur-

sue the topic on their own. I remember sharing a bus ride in Kenya w
an immensely likable fellow from Kisumu named John who was about as
environmentally informed as the Entebbe shopkeeper. That is, John was
aware of "the need to keep forests thick," had heard a little about the
ozone problem, and knew not the first thing about global warming but
was grateful to learn. "We rely on you and those who are learned to inform us of these things," he said with touching sincerity before merrily
changing the topic to the treacheries of Kenyan president Daniel Arap
Moi, a topic on which he expounded in hilarious detail for the next hour.
The afternoon I spent with John was great fun, but we never did get back
to talking about the environment, despite a number of conversational
gambits on my part.

Likewise in Czechoslovakia, environmental reform had ranked as
the number one public demand immediately after the Velvet Revolution
of November 1989. But by September 1991 the political momentum behind reform had fizzled out, Simona Bouzkova told me, because politicians could talk of nothing but the growing possibility that the nation
would split in two. Former Czech environment minister Bedrich Moldan
told me in 1994 that the environment had dropped to between fifth and
tenth on the list of people's concerns, and he added that this fall was inevitable, even rational. "Now people care more about immediate things,"
he said, "like who will own the shop on the corner? Will the house they
live in be sold? Will they be able to pay the new rent? These are things they
can influence. What can they do to influence emission standards?"

Substitute the names of virtually any other country on earth for
Czechoslovakia and Kenya and the priorities are the same. After all, it's
only human nature to pay more attention to an immediate problem
than to a distant one, and environmental problems are still perceived as
distant by many people, especially those affluent or lucky enough not to
live in directly threatened areas. Like the nuclear executives who assured
me that nothing could go wrong with their waste disposal sites for hundreds of years, many people tell themselves that dangers like global
warming and ozone depletion are so far off in the future that they don't
really exist. On some level, these people may know better than that, but
the possibility that we humans are dooming ourselves is simply too terrible a thought to absorb. It is much easier to pretend the danger doesn't
exist, or adopt a childlike faith that everything will turn out all right in

the end—surely the experts will think of something!—and burrow back into the daily routine of paying the bills, getting the kids off to school, and waiting for the weekend.

Drowsy and content, sheltered in their denial, such people resemble nothing so much as the frog in the famous laboratory experiment. Dropped in a pan of boiling water, the frog jumped out instantly, no harm done. But when placed in a pan of cool water over a low flame, the frog was in no hurry to flee. As the water temperature gradually rose, the frog perhaps even relished the gathering warmth. Soon the water reached the boiling point, but by then it was too late for the frog to jump to safety. He was already dead.

What is needed to shake people out of their lethargy, Lester Brown told me in 1990, is an environmental Pearl Harbor, an event that demonstrates to one and all that the danger is upon us and immediate mobilization is necessary. The Chernobyl disaster of 1986 had certainly gotten people's attention. So had the discovery of the ozone hole in 1985. The 1990s brought additional warnings. By 1998, global temperatures were setting new records virtually every year; nine of the previous eleven years had been the warmest since modern record-keeping began. There had also been unprecedented forest fires in Brazil, Indonesia, Mexico, and Florida. And scientists had discovered a so-called dead zone in the Gulf of Mexico, caused by vast amounts of human sewage and fertilizers carried into the gulf from the Mississippi River watershed. The problem is that by the time such hazards become plain to the person on the street, they have often accumulated too much momentum to be stopped anytime soon. Indeed, this momentum was part of what had led Brown and others to warn in 1990 that humans had ten years to act or it would be too late.

"I don't think it's a question of the human species going extinct," Brown told me. "There will be at least some of us around for a long time. But the social costs of our inadequate response are going to be much higher than I would like to have seen. People ask, 'Is time running out?' The question is, for whom? The two million Ethiopians who've starved since 1970—time did run out for them. Time also ran out for the thousands of kids in Los Angeles who have permanently impaired respiratory systems because of air pollution, and for the 11 percent of Russian infants born with birth defects because half of the drinking water and a tenth of the food supply are contaminated."

But is scaring people with warnings about an imminent point of no return a good way to provoke activism? Not according to Bernard Lown, cochair of International Physicians for the Prevention of Nuclear War (IPPNW), which won the Nobel Peace Prize in 1985 for its leading role in organizing opposition to the nuclear arms race. "I think that's terrible talk," Lown said of the ten-year deadline when I interviewed him in Stockholm in 1991. "You can't be Jeremiah with a calendar because any logical person says we can't turn it around that fast and retreats into passivity and self-indulgence. But the moment you become active, you start reversing the clock."

The historical resilience of the human species is, for some environmentalists, a source of optimism amid the gathering storm. "I have my days when I think, 'This isn't going to work out,' " Paul McCartney, the former Beatle and longtime supporter of Friends of the Earth, told me. "Then the day after, I always change my mind. . . . What makes me believe it is history, and all the terrible things we've gone through, but we're still here. So I do think [humans] have a propensity for sorting things out, or we'd have disappeared already."

At the end of our conversation in Paris about how modern man was playing the sorcerer's apprentice with the planet, I asked Hubert Reeves if he was hopeful about humanity's environmental future. The French cosmologist replied that he was "involuntarily optimistic." He added, "We have no right to be pessimistic, because that will only make things worse. We have to drive between the two extremes of defeatism and alarmism. It's like David and Goliath. On the one hand there is Goliath, the bulldozer of environmental destruction, which began ten thousand years ago with the start of agriculture but wasn't really threatening until this century and really only in the last thirty years. On the other hand there is David, the embodiment of popular consciousness and action, which is small but growing. Over the next ten to one hundred years, David and Goliath will be fighting for the future of the planet. We can't know who will win this battle. But we must act as if we can win, or we will surely lose."

Environmentalism has been one of the ascendant social forces of the twentieth century, but it will not succeed in the twenty-first century if it does not deliver economic well-being as well as ecosystem salvation. Virtually everyone I met during my travels said that, as much as they wanted a clean environment, their first priority had to be keeping themselves above water economically. No one wanted to drink dirty water or see acid rain destroy forests, but for most people these were remote problems compared with the daily struggle for food, clothing, and shelter.

In Istanbul, for example, an able young journalist named Esra Yalazan gave a bitter chuckle when I asked whether Turkish politicians talked very much about pollution. "This is not America or Europe," she scoffed. "The problems here are so much more basic—finding a job, food to eat, and heat for the winter—that any politician who talked about pollution would be laughed at." This in a city whose streets, sidewalks, walls, and windows were covered with a film of black, sooty grime when I visited in November 1991, thanks to the coal burning that had long been the city's main source of heat and power. It was not that the locals were uninterested in environmental problems. While riding a packed public bus through a rain-spattered rush hour I fell into conversation with a young office worker who told me in workable German that she did worry about the air in Istanbul—whenever she blew her nose, the discharge was all black. But she added that most people had simply gotten used to the bad air, and the difficulty of buying food and paying rent left them no time to think about ecology.

The assumption that environmental protection must cost jobs and lower profits is shared by most average people and corporate and government officials around the world. Especially in the United States, that belief has been repeated for so many years by so many people—politicians, journalists, business executives, labor leaders, economists—that it has come to be regarded as fact. Thus, in analyzing the economic effects of carbon dioxide reductions proposed during the Kyoto climate change negotiations, the *New York Times* and *Washington Post* time and again employed such words as "pain" and "sacrifice." *Newsweek*'s economics columnist asserted, "[N]o one knows how to lower emissions adequately without crushing the world economy."

This is simply not true. Not only do some companies and nations know how to lower carbon dioxide emissions without economic pain, they are already doing so—at a profit. The key is efficiency: not doing without but doing more with less. Germany and Japan use half as much energy per dollar of GNP as the United States, not because their citizens suffer lower standards of living but because their economies use energy more efficiently. Even in the wasteful United States, GNP grew by an average of 2.5 percent a year from 1973 to 1986 while energy use grew not at all, thanks to increased efficiency. In Sweden, the State Power Board in 1989 found that it could obey the voters' command to phase out nuclear power and still maintain ample economic growth, even though nuclear accounted for half the nation's electricity. Cogeneration—reusing the heat produced during electricity generation instead of expelling it as waste—was a big part of the solution. Not only did Swedish utilities' costs decline by $1 billion a year, their carbon emissions fell by one-third. And Sweden already ranked among the most energy-efficient nations in the world. Efficiency has even greater potential in developing countries, where technology tends to be outdated and relatively small investments can yield enormous benefits. Efficiency is the key to reducing environmental and economic problems in China's energy and agricultural sectors, for example, and there is no reason other poor nations should not achieve similar gains.

Companies, too, can benefit from "smart" production. Dow Chemical has enjoyed an extra $110 million profit every year thanks to energy-saving initiatives undertaken by its Louisiana division, where investments in superefficient lighting and other commonsense innovations have yielded rates of return over 200 percent. In Singapore, engineer Lee Eng Lock of Supersymmetry Services has designed air conditioners that cost less than the competition and use 65 percent less electricity; the energy savings are captured through such prosaic innovations as redesigning pumps and fans to reduce friction. In Amsterdam, the headquarters building of the Netherlands' second largest bank, ING Bank, uses one-fifth as much energy per square foot as a bank across the street, even though the two buildings cost the same to construct. But the ING headquarters boasts efficient windows and insulation, as well as a so-called passive design that enables solar energy to provide much of the building's needs, even in cloudy northern Europe.

A crucial additional advantage of raising efficiency is that the economic activity it generates is labor-intensive. Drilling for oil, for example, takes lots of heavy machinery but relatively few workers; insulating houses, the reverse. Indeed, investments in energy efficiency and renewable energy yield two to ten times as many jobs as investments in fossil fuel and nuclear power do. Incinerating a million tons of solid waste requires eighty workers, and putting it in a landfill takes six hundred workers, but recycling it takes sixteen hundred workers. Building railroad tracks generates 50 percent more jobs per dollar invested than building highways. Employing all these additional workers would have other benefits as well, including higher tax revenues for governments (and lower welfare costs), greater consumer demand for businesses, and more stable communities. Thus the efficiency solution to the environmental crisis offers sluggish economies in the North and South alike the opportunity to address the other, largely unremarked crisis of our time: jobs. As long as an estimated one billion people around the world lack gainful employment, poverty and human suffering will grow, economies will remain mired in stagnation, and political systems will slide further toward instability and intolerance. Needless to say, the popular will needed to confront the environmental crisis will be difficult to mobilize under such conditions.

"The idea that reducing global warming will harm the United States economy is flatly contradicted by business experience. Climate change is actually a lucrative business opportunity disguised as an environmental problem," said Amory Lovins, the energy specialist who, as noted in chapter 3, foresaw (and helped to develop) the spectacular promise of hybrid cars years before Detroit and Japan put such cars on the market. A physicist by training, a polymath by intellect, and an eccentric by temperament, Lovins speaks fondly of his "taxidermically challenged" (in other words, stuffed) pet orangutan and freely volunteers back massages to total strangers during environmental conferences. But his corny jokes and omnipresent pocket calculator have not kept him from briefing nine heads of state and advising scores of industries and governments over the years. The *Wall Street Journal* named Lovins one of the twenty-eight people in the world most likely to change the course of business in the 1990s; much of his blueprint for change is contained in *Factor Four: Doubling Wealth, Halving Resource Use,* a book he wrote with his wife, L. Hunter Lovins, and Ernst von Weizsacker of Germany's Wuppertal Institute and which was published in 1997.

Cheerful, practical, and visionary, *Factor Four* is one of the most important contributions to environmental thinking and policymaking to appear in years. On the basis of some fifty real-world examples, including some of the ones cited above, the book demolishes the idea that economic prosperity and environmental health must be enemies. Humanity can live twice as well while consuming half as much, argues *Factor Four,* if it quadruples what the authors call "resource efficiency"—the amount of wealth extracted from one unit of natural resources. Super-refrigerators, for example, use 86 percent less electricity while costing the same or less to build and delivering superior performance. *Factor Four* outlines equally impressive gains for construction, office equipment, lighting, irrigation, railways, agriculture, and a host of other products and activities. Improving efficiency may not sound flashy, but the savings produced by small design changes turn out to be dramatic. Take electric motors, which use more than half the world's electricity. Design changes described in *Factor Four* can double the efficiency of those motors, which therefore could keep one-quarter of the world's electricity from having to be produced in the first place.

Most of the alternatives *Factor Four* promotes are available today, under current prices, which puts a radically different spin on such contentious issues as how to respond to global warming. The authors argue that it doesn't matter whether one believes that global warming is scientifically proven or not; the economic savings available are so large that countries, corporations, and individuals ought to be increasing energy efficiency anyway, simply to save money.

This optimistic message repudiates not only mainstream fear mongering about the cost of environmental reform but also the gloom and doom often heard from environmentalists. The authors do not deny the long-term necessity of structural and behavioral changes in building a sustainable economy; indeed, they argue that the *Factor Four* efficiency revolution will buy the time needed to navigate the transition. But rather than lecturing people about sacrifice, *Factor Four* shows them how to live better with less. (The Lovinses speak from personal experience. One of the book's most striking case studies concerns the Rocky Mountain Institute, which the husband-and-wife team cofounded in 1982. Perched seven thousand feet above sea level in the snow-packed mountains of Colorado, the institute's headquarters nevertheless has no furnace; its passive solar design delivers 99 percent of its energy, enough to support an indoor tropical garden, complete with banana trees—guarded by the stuffed orangutan—and work space for a forty-five-person staff. Electricity bills average a mere $50 a month.)

If the advantages of fourfold efficiency improvements are so obvious, why isn't everyone pursuing them? What's holding up the efficiency revolution? I asked Amory Lovins about that at a conference on sustainability consciousness held at the Esalen Institute in Big Sur, California, in 1995.

"The main obstacle is that pesky euphemism known as 'market failures,' " Lovins said, eyes twinkling above his black walrus moustache. Although Lovins said he admired "the genius of private enterprise," he had few illusions about how markets actually operate. The image portrayed in textbooks—of producers and consumers blessed with perfect information and facing perfect competition—simply doesn't match the real world. "I can tell you where to buy a refrigerator that is cheaper than an identical model that uses six times as much electricity," said Lovins. "But if you don't know that such a refrigerator exists, you probably won't shop for it." The same holds true for corporate managers. Lovins recalled a chief executive officer of a Fortune 500 firm who said he couldn't get very excited about saving energy because energy amounted to only a small fraction of his total overhead costs. But the CEO changed his mind after learning that a manager at one of his facilities was adding $3.5 million a year to the company's bottom line through energy savings and that if similar results were achieved throughout the firm overall profits would increase by 56 percent.

Perverse incentive systems are another market failure, said Lovins. The profits of most utilities around the world are based on how many units of energy they sell, rather than how efficiently they operate; thus, most utilities push to build more power plants that burn more fuel, because that is good for profits, even though it is bad for the environment. Likewise, the fees of most architects and engineers are calculated as a percentage of a building's total cost, which discourages them from pursuing cheaper, more efficient designs. Split incentives are another ubiquitous problem—why should a renter pay for energy-saving lightbulbs if the landlord pays the electric bill?

Above all, markets are rarely free. Prices can be distorted by the monopolistic power of corporations and by government subsidies that reflect brute political power more than sound public policy. "When it comes to subsidies," Lovins told me, "the United States is almost as bad as the Soviet Union used to be, especially in the energy and transportation sectors. The United States spends more than $30 billion a year subsidizing the expansion of energy supply. The rule is corporate socialism for political favorites like nuclear and fossil fuels and free market discipline for efficiency, solar, and renewable energy. The same with transportation. Roads are paid for by taxes instead of user fees, so drivers think they're free, which is just one of our many hidden subsidies for automobiles."

The United States is hardly the only government guilty of unwise subsidies. All over the world, governments subsidize deforestation, overfishing, excessive irrigation, air pollution, and other environmentally destructive activities, in amounts estimated between $500 billion and $900 billion a year. In the five years following the Earth Summit, the World Bank funded so much fossil fuel development in China, Russia, and elsewhere that the projects in question will produce more CO_2 emissions over their lifetimes than the entire world was producing in 1997, according to a report by the Institute for Policy Studies in Washington. The bank's $8.75 billion of fossil fuel loans, though earmarked in many cases for development and poverty relief, actually ended up subsidizing such corporate giants as British Petroleum, Exxon, and Royal Dutch Shell. Meanwhile, the bank was spending only 1 percent of energy-related funds on efficiency projects, even as it proclaimed that its notoriously wrong-headed environmental policies had been reformed.

Besides subsidies, another reason prices do not always tell the ecological truth is that markets systematically undervalue the services that ecosystems provide and disregard the interests of future generations. Clean air, fresh water, fertile soil—these and other foundations of human society are often considered free goods, and since people do not pay to consume them, they have no incentive to conserve them. This attitude is rooted in classic economic theory, which considers nature to have a value of zero; production is assumed to be a function solely of capital and labor. The real world is a different matter, of course. A farmer (labor) and his tractor (capital) could no more produce wheat without soil and water (nature) than a cook could bake cakes without flour and eggs. Yet markets treat soil and water as worthless, just as they ignore the rights of people yet to be born. The citizens who will populate the world in 2050 have a keen interest in today's societies not consuming all the available petroleum or pumping too many greenhouse gases into the atmosphere. But these people cannot express their interest in today's marketplace (or in today's political systems), so the market's price signals continue to reward short-term thinking, as do the measurements of GNP that economists, politicians, and journalists habitually use to gauge an economy's health. The depletion of natural resources—the logging of forests and the mining of coal—is counted as an addition to income that makes the GNP go up, rather than a depletion of capital that reduces the economy's ability to produce income in the future.

Just how profoundly the economy undervalues its ecosystems was suggested in a landmark study published in the scientific journal *Nature* in 1997. An international team of researchers coordinated by Robert Costanza of the University of Maryland's Institute for Ecological Economics attempted to put a price on seventeen critical ecosystem services, ranging from soil formation and climate regulation to pollination, pest control, and water supply. The team concluded that the economic value of the world's ecosystem services and capital was at least $33 trillion a year—nearly twice as large as the world's GNP of $18 trillion. The implication was clear: for markets to reflect the environment's true value, prices had to increase.

Doesn't that prove that environmental protection means economic pain after all? Yes, but no. The true cost of fossil fuel use is already being paid, just not in the marketplace. Rather, society as a whole pays—in the form of dirty air and higher health costs—and future generations pay in the form of higher greenhouse gas concentrations. In economic jargon, these costs are "externalized"—kept separate from the market. To reflect the cost of the environmental damage that fossil fuels cause, their market price should therefore indeed go up. That sounds like an economic burden. But in fact, rising fossil fuel prices need not cause economic distress for society as a whole if the prices of other goods and services go down by an equivalent amount.

The mechanism the authors of *Factor Four* and many others advocate for achieving this is environmental tax reform. The basic idea, as economist Herman Daly has put it, is to "tax bads, not goods." That is, raise taxes on the things society wants to discourage—natural resource use, environmental degradation, and pollution—while lowering taxes on things it wants to encourage—employment and business investment. Combined with the elimination of harmful subsidies, such tax reform would enable markets to tell more of the ecological truth, thus making investments in efficiency, renewable energy, and other environmentally superior activities more attractive. Since taxes on labor and investment would be reduced, the reform should also lower prices throughout the economy while sparking job creation and business expansion.

Proponents regard environmental tax reform as a key tool for promoting the transition to sustainable economies. Ingeniously simple in concept, such reform would institutionalize the basic environmental principle of "the polluter pays." Already, partial environmental tax reform has been implemented in Germany, the Netherlands, and the Scandinavian countries. Analysts recommend making any environmental tax shift revenue neutral—cutting labor and investment taxes as much as environmental taxes are raised—to disarm the charge that the shift is actually a tax increase. In addition, the tax shift is best phased in gradually but consistently so that companies and individuals can plan ahead. *Factor Four* suggests raising energy and resource prices by 5 percent a year for twenty years. Finally, the poor need protection. Since basic needs like food and shelter tend to be more resource intensive, a tax that raises resource prices will hurt the poor more than the rich. The Netherlands has addressed this problem by exempting the nation's poorest households from environmental taxes.

Politically, environmental tax reform would seem to be a win–win solution: citizens benefit both from the shift away from environmentally destructive activity and from the promotion of job creation and investment opportunities. Those benefits flow to society as a whole, however. Costs would increase for sectors of the economy that consume the most natural resources and produce the most pollution. Oil, coal, chemicals, logging, mining, iron, steel, and aluminum are the main industries likely to suffer from environmental tax reform. Their pain could be ameliorated by reductions in labor and investment taxes, but only partially. In the long run, companies engaged in environmentally harmful activities—as well as the workers, communities, and investors who rely on those companies—will have to move on to other businesses. Governments can assist in this transition, not least through phasing in the tax shift gradually so everyone has time to adjust and retraining the workers for new, environmentally benign jobs. Nevertheless, the industries in question, most of whom have considerable political clout, are likely to resist tax reform, just as they have resisted efforts to reduce environmentally harmful subsidies. In the end, achieving environmental tax reform requires political victories over some of the most powerful, interests in modern society. Like Hubert Reeves said, David versus Goliath.

Yet it is heartening to realize that practical solutions exist and have even been implemented in some nations. Germany has begun holding manufacturers both legally and financially responsible for the final disposal of their products, causing firms to design products much more efficiently in the first place and, later, to recycle every possible component. Sweden is adopting genuinely sustainable forestry practices. Chlorine-bleached paper is virtually a thing of the past in most of Western Europe. The European Commission has approved a White Paper urging a doubling of the amount of Europe's energy that comes from renewable sources—to 12 percent by 2010. In Japan, 10 percent of new homes will have solar panels on their roofs by 2000. In his book, *A New Name for Peace,* Philip Shabecoff, former environmental reporter for the *New York Times,* describes numerous projects in developing nations that have successfully married environmental restoration with poverty alleviation, such as the tree-planting campaign of Kenya's Green Belt movement (see below). Details still need to be elaborated, and missteps along the way are inevitable, but the basic point seems clear: neither technical knowledge nor economic feasibility stands in the way of transforming our civilization into an environmentally sustainable enterprise. As Shabecoff puts it, "[B]y now we know what needs to be done."

But making it happen, that's the rub.

The history of modern environmentalism is largely a story of ordinary people pushing for change while governments reluctantly follow behind.

The thousands of environmental and development activists who attended the Earth Summit in 1992, for example, were certainly imbued with a greater sense of urgency and more direct approach than their government counterparts. Government negotiators actually spent days arguing about whether the North should (not would) increase developmental assistance to the South to 0.7 percent of Northern GNP "by the year 2000" or merely "as soon as possible." Meanwhile, the activists were criticizing the entire aid discussion as narrow-minded because it focused on boosting Southern GNPs rather than on meeting basic human needs.

Addressing a crowded audience one morning in the so-called women's tent in Rio, activist Wangari Maathai of Kenya described the successes of the Green Belt movement, a project that encouraged poor women to plant trees. For every donated seedling that survived, a Kenyan woman was paid the equivalent of four U.S. cents, an approach that combated land degradation, global warming, and rural poverty all at once. In an interview later, Maathai told me, "If you look at what Earth Summit documents say about sustainable development, it's as if we in the Green Belt movement are carrying out word for word what they suggest. Yet we have never gotten any money from our government. If we are serious about dealing with poverty, the UNCED Green Fund should include a stipulation that some of the money goes to poor women's organizations around the world."

"UNCED should identify affluence, not poverty, as the major destroyer of the environment," Martin Khor, a Malaysian scholar-activist with the Third World Network, a group of two hundred nongovernmental organizations (NGOs) from various developing nations, told me in Rio. "Sustainable development does not mean just conserving present resources for future generations but also reducing the excessive consumption of a minority so that resources are 'freed' to meet the basic needs of the rest of humanity within this generation." The amounts of aid being fought over at RioCentro were meaningless, Khor added, within an international economic system that annually transferred some $200 billion from Southern to Northern economies through debt payments and deteriorating raw materials prices.

The activists' arguments did not carry the day at the Earth Summit, to put it mildly, but thousands of NGOs in North and South alike now routinely collaborate on common projects. One of the issues uniting them is the fight against economic globalization—the trade and currency deregulation embodied in such agreements as the North American Free Trade Agreement (NAFTA), the Maastricht Treaty, the General Agreement on Tariffs and Trades (GATT), and the World Trade Organization (WTO). The corporate and financial institutions that favor globalization argue that greater capital mobility will spark economic growth and reward rich and poor alike. Activists counter that globalization will in fact produce environmental degradation and increased poverty as governments lose the power to enforce national environmental and labor standards and workers see overseas competition drive their wages down to pennies per hour. Eliminating investment and currency barriers will also accelerate logging, drilling, and other forms of resource extraction, while spreading the American model of hyperconsumptionism across the world. Since today's levels of production and consumption appear unsustainable, and current government regulation fails to protect either workers or the environment adequately, the idea of deliberately intensifying these trends with more globalization strikes environmentalists as a form of planetary suicide.

I should say, it strikes some environmentalists that way. In the United States, leaders of such mainstream environmental organizations as the Natural Resources Defense Council, the National Wildlife Federation, and the Environmental Defense Fund broke with grassroots activists and joined the Clinton administration in supporting NAFTA in 1993. "We broke the back of the environmental opposition to NAFTA," bragged John Adams of the NRDC, adding, "We did [Clinton] a big favor." Adams and other NAFTA supporters claimed it was the environmental side agreements that Clinton added to NAFTA that convinced them to support the treaty. Critics charged that the about-face had more to do with the cozy relationships the NRDC and other large environmental groups had with major corporations. Exxon, Chevron, DuPont, and Waste Management are but a few of the notorious polluters that began funding and joining the boards of directors of mainstream environmental organizations in the 1980s. For the corporations, environmental philanthropy was an inexpensive way to burnish their public image. Was it mere coincidence that the recipients of their largesse soon began adopting more corporate-friendly approaches to environmental advocacy?

In 1995, journalist Mark Dowie charged in his book, *Losing Ground: American Environmentalism at the Close of the Twentieth Century,* that the mainstream groups were "courting irrelevancy." By turning their backs on the grassroots in favor of partnerships with corporate America, wrote Dowie, they were diluting the militancy needed to force fundamental change. In an interview in 1998, Dowie stood by his analysis but said he was heartened by recent developments, both in the United States and abroad.

"The international listing of environmental NGOs is thicker than the Manhattan phone book," said Dowie. "So the infrastructure is there, and the popular consciousness exists, it's just a matter of putting them together. In the United States, the radical edge of environmentalism didn't disappear during the 1990s. In fact, thousands of small local groups have formed to oppose clearcut logging, toxic waste dumps, and other single-issues. Now these groups are joining into networks that magnify their impact enormously. And the grassroots memberships of groups like Sierra Club and the National Wildlife Federation are pulling their leaderships away from compromising postures and insisting on things like a zero-cut logging in the national forests. And it's very encouraging to see how the Republican attempt to gut the nation's environmental laws after the Gingrich revolution of 1994 went down in flames. It failed partly because more aggressive activists came to the fore inside the Beltway, like Phil Clapp at the National Environmental Trust, but also because Republican voters themselves were opposed to what Gingrich and his crowd were trying to do."

In Europe, direct grassroots action also scored important victories in the 1990s. In 1995, European environmentalists orchestrated an international campaign of street protests and consumer boycotts against Royal Dutch Shell, the world's biggest oil company. Shell was intent on burying an obsolete oil-storage rig at the bottom of the North Sea. Greenpeace and other environmentalists complained that the *Brent Spar* rig contained toxic sludge that would pollute the marine underworld. Greenpeace activists twice managed to board and occupy the *Brent Spar,* generating extensive media coverage, which in turn aroused public concern. Other activists in Germany began encouraging boycotts of Shell gasoline stations. Shell proceeded with its plan to bury the rig off the western coast of Scotland, but this only fanned the flames of opposition. Soon, motorists in Denmark and Holland were also shunning Shell, and public officials of all stripes were rushing to claim spots at the front of the gathering parade of outrage. German chancellor Helmut Kohl, never much of an environmentalist, was maneuvered into a political corner and publicly urged British prime minister John Major to block Shell's burial plan. But it was the boycotts that hurt Shell most. In Germany alone, service station income fell 30 percent. Losses estimated in the millions of dollars, along with the blackening of the company's name, at last led Shell to cancel the *Brent Spar* sinking. The lesson of the affair, concluded the *Economist,* was that "companies that choose to defy their consumers' political demands are placing their businesses in jeopardy."

The *Brent Spar* affair unveiled a new model for environmental politics, one in which direct action was practiced not only by countercultural monkey-wrenchers but also bourgeois consumers, all united in a militant, multinational mass movement. At a time when millions of people the world over felt alienated from formal political structures and victimized by forces seemingly too remote to challenge, the direct action model offered an effective way to put personal beliefs into political practice. Some 60 percent of British consumers said they were prepared to boycott products on ethical grounds, and they were not alone. When France announced in 1995 that it would resume nuclear weapons testing in the South Pacific, angry consumers and businessmen from Hong Kong to Sydney to Tokyo began spontaneously boycotting French goods; French wine sales fell by one-third in Australia and New Zealand.

Just because people look passive doesn't mean that they're apathetic—they may simply need a concrete means of expressing themselves. It is the job of political activists and leaders to provide such occasions, and these days the consumer marketplace is as valid a battleground as the ballot box. Greenpeace, for example, has forced refrigerator manufacturers to produce ozone-friendly refrigerators by combining scientific expertise with consumer power to divide the industry against itself. In the early 1990s, after years of being told by manufacturers that refrigerators that did not use ozone-destroying HFCs and HCFCs would be unsafe, Greenpeace found independent scientists who designed a refrigerator that safely relied on butane and propane. (HFCs and HCFCs are the chemicals that replaced CFCs after the Montreal Protocol, and while they are far less injurious than CFCs, they still destroy the ozone layer and intensify global warming.) Greenpeace located a state-owned refrigerator company in eastern Germany that was going bankrupt and resuscitated the firm with orders to produce thirty thousand of the new models, which would be sold to Greenpeace members who guaranteed their purchases in advance. These refrigerators made a big enough dent in the German market that leading manufacturers Bosch-Siemens and Miele decided they had to provide ozone-friendly refrigerators themselves to remain competitive. Soon the technology had spread throughout Europe, accounting for 50 percent of all refrigerators sold in 1994, and was heading to Asia.

Jeremy Leggett has pursued a similar strategy with his Solar Century organization. Undeterred by the lack of support from the insurance industry and governments, Leggett is now trying to jump-start the solar power market by organizing environmental, church, and other groups into a sort of buyers' club. The groups commit to putting solar photovoltaic cells on their buildings, and their combined purchasing power calls forth economy-of-scale cost reductions. The first breakthrough has come in California, where a purchasing program involving the Sacramento Municipal Utility District is expected to make solar panels fully cost competitive by 2002.

Dramatic breakthroughs like these are vital to building a political movement because they show that victories are possible. But in the end there is no substitute for the slow, patient work of local education and organizing. The activists cited here were able to achieve things big business didn't like (indeed, had said were impossible) not just because a few leaders had a creative battle plan but because they had the active support of many thousands of average citizens. The refrigerator industry was reformed not only because of the technical ingenuity of a few outside scientists but because thirty thousand consumers put their money where their mouths were and bought ozone-friendly refrigerators. Helmut Kohl publicly castigated John Major over *Brent Spar* not because Kohl suddenly got in touch with his inner planetary child but because the Green Party of Germany controlled the swing votes in Parliament and Kohl needed their cooperation on other issues. And the Greens found themselves in that happy position because Petra Kelly and other activists had been educating and organizing their fellow Germans behind a radical ecological vision since founding the Greens in 1979. Likewise, the success of the Cairo Population Conference stemmed largely from preparatory work by thousands of women's groups all over the world; they championed the pro-female focus the conference endorsed, they continued to pursue their agenda at the Beijing women's conference in 1995, and they are now working to hold governments to their promises.

"There is definitely a race between rapidly growing public consciousness and rapidly growing environmental problems," said John Passacantando, the executive director of Ozone Action, one of the few militant Washington environmental groups. "Kids now recycle automatically. Politicians can't be anti-environmental without opening themselves to attack. There is no constituency for building new CFC factories in this country. These are changes that will not be reversed. Now, I don't know how much longer it will be before people insist on buying energy-efficient cars and houses, living near the Metro, and taking the other steps needed to create a truly sustainable society. So I don't know if we as a civilization will make it. But change can come fast once the times are right. In 1957, Lyndon Johnson was voting against the Civil Rights Act as the Senate majority leader, but by 1964 he was signing that act as president—not because he had changed, but because the world around him had changed. Now our world is changing. So I do have hope. And with hope, you can have magic."

The truth is, magic has struck fairly often in recent human history. In South Africa, apartheid was overthrown and a new government headed by African National Congress leader Nelson Mandela was democratically elected. In Russia, Mikhail Gorbachev introduced reforms that led to the peaceful end of the Cold War, a retreat from the nuclear precipice, and freedom for the millions of people formerly oppressed by the Soviet Union and its client states. The first time I passed through Berlin on my global travels, I met a German man who was returning to Sweden, his home for the past twenty years, after a visit to his elderly parents in Berlin. The wall had fallen eighteen months earlier, and still this man (and many other Germans I met) could hardly believe how the world had turned right side up. "We never thought the system would change," he told me. "Not in our lifetime."

In Prague, the same fatalism had pervaded even the dissident movement. Ales Sulc, a member of the Charter 77 human rights group, told me that even after the Berlin Wall fell, most Czech dissidents believed it would be another ten years before their own nation was liberated. But in fact the opening episode of what dissident playwright Václav Havel christened the Velvet Revolution took place a mere five weeks later, when riot police attacked a peaceful student demonstration, provoking mass outrage and drawing an estimated half a million people into the streets. By the end of December 1989, the communist regime had been forced from power. Havel, whose outspoken criticism of the regime since the 1960s had led to countless terms in prison, was elected the interim president of the country.

Havel always insisted that he was no hero; he was simply driven by an unshirkable sense of personal responsibility. His compatriot, the novelist Milan Kundera, once refused to sign a petition Havel was circulating on behalf of political prisoners in the late 1960s; Kundera questioned whether such symbolic acts of opposition had much practical effect against a totalitarian government. Havel, however, insisted that calculations about what is practical must never stand in the way of doing what is right, not least because such calculations often lead a person not to act at all. Havel called his philosophy "antipolitical politics" and explained it as follows: "When a person tries to act in accordance with his conscience, when he tries to speak the truth, when he tries to behave like a citizen even in conditions where citizenship is degraded, it won't necessarily lead anywhere, but it might. There's one thing, however, that will never lead anywhere, and that is speculating [about whether] such behavior will lead somewhere."

The Czech people's admiration for Havel was abundantly clear when I arrived in Prague to interview him in September 1991. A few weeks before, when Havel attended a concert by Paul Simon at Letna stadium, it was Havel who got the night's biggest round of applause. The Czechs trusted Havel because he stood up to the communists, but they liked him because he was like them: he drank, he smoked, he liked to hang out at his local pub with other regulars. Presidential duties now left him much less time for socializing, but he was still a serious believer in having fun. One night I attended a private party organized by cronies from his theater days. Trailed by bodyguards, Havel arrived at ten o'clock, looking like a schoolboy who'd been given the rest of the week off. Hands in pockets, eyes gleaming with anticipation, he was instantly surrounded and maneuvered onto a crowded couch in the corner. Cigarette and wineglass in hand, he spent the next two hours grinning and chatting with his old friends who were clearly glad to see him but who treated him with no special deference. For his part, Havel listened far more than he talked, though when he did speak, he was engaged and animated, gesturing freely and laughing often.

"Right here, right now, there is no place I'd rather be," went a Western pop song I heard on Wenceslas Square the next day. "Right here, right now, watching the world wake up from history." It was hard to imagine a more beautiful city in which to observe the awakening. Combine the flair of Florence, the fairy-tale charm of Vienna, the elegance of Stockholm, and the grandeur of Paris, and you begin to approach the magnificence of Prague. From Old Town Square, with its rainbow-hued astronomical wall clock and monument to the fifteenth-century free thinker Jan Hus, to the massive Charles Bridge with its baroque statues of thirty-nine different saints, the architecture of Prague summoned all the glory of old Europe even as its politics heralded the coming of the new. Throughout Old Town, crooked streets of cobblestones curled away beneath red-tiled roofs and golden-spired churches, while statues of lions, angels, and departed city fathers stared down from every possible angle. Looming high above the city like some central European Mount Olympus was Hradčany Castle, an awesome complex of palaces, museums, and cathedrals that dated back to the ninth century.

It was inside that castle, in the luxurious presidential office with its twenty-five-foot ceilings and richly patterned oriental rugs, that I interviewed President Havel about the future of his country and the prospects of global environmental survival. At the time, Czechoslovakia's environment was the most polluted in all Europe. Life spans were six to eight years shorter than the European average, largely because of excessive coal burning. In a previous interview, Josef Vavrousek, Havel's environment minister, had informed me that, in Slovakia, every ton of coal burned released a kilogram of arsenic into the air. In northern Bohemia, coal burning left the air so filthy that in winter children were often sent to school wearing gas masks. I visited northern Bohemia one weekend myself and saw an appalling variety of environmental scars, including a huge open-pit mine of gray and brown gravel, easily three times the size of downtown Prague, that had been gouged out of the earth to supply a hulking petrochemical complex nearby. As an example of industrial man's assault on the natural world, that moonscape matched anything I later saw in Russia or China.

Havel arrived for our interview through a back door in the presidential office, tugging his way into a banker's black jacket as aides fluttered around him. Sitting down beside a large black coffee table, he pulled out a gleaming gold pocket lighter and a box of Camels, which he smoked more or less continually during our forty-five-minute interview. Havel was quite aware of the catastrophic state of his country's environment, and he lay much of the blame for it on the totalitarian system. In a speech to environment ministers from across Europe in 1991, he had charged that communism had favored "incessant growth in production at all costs, regardless of . . . the consequences for the environment" while pursuing "unbridled exploitation of the past and the future in favor of the present." Of course, much the same could be said of capitalism, as Havel well knew. Indeed, in his political essays Havel had argued that the essential problem of our time was the ideology that pervaded modern civilization in East and West alike: namely, the arrogant belief that technological man was separate from and superior to nature, rather than a part of it— the conviction that man could play God. "The system of impersonal power" that ruled modern technological civilization had achieved its most complete expression in totalitarian states, said Havel, but parliamentary democracies were by no means exempt from it. Havel maintained that multinational corporations were no more humane than totalitarian bureaucracies, and that citizens of democracies were "manipulated in ways that are infinitely more subtle and refined than the brutal methods used in the post-totalitarian societies." The main source of manipulation was "the omnipresent dictatorship of consumption, production, advertising, commerce [and] consumer culture," a hydra that Havel criticized as a "well-spring of totalitarian thought."

In our interview, Havel said Czechoslovakia was experiencing the dictatorship of consumption and advertising "in an especially absurd way, because in our country the remnants of socialism are mixing with reconstructing capitalism. So we now see on our televisions more commercials than [exist] in America almost, commercials that advertise things no one has money to buy and no one needs. When I watch TV news here," he added with a wry grin, "I always face a dilemma: whether to wash my laundry with Persil or Ariel, because they are apparently both the best in the world."

Havel respected the market economy, but he did not worship it, and this perspective shaped his vision of how to restore his country's environmental health. "I am convinced that competition which has rules that are binding and respected by all is a real motor of economics," he told me. However, "a safety net for those who for whatever reasons are incapable of making their own living" was also essential. "As far as ecology is concerned," he added, "we must create a legislative system and economic policies that will force the directors and managers of firms to choose, for their own economic self-interest, the most ecological technologies." In other words, as Minister Vavrousek had already explained to me, Czechoslovakia needed "to establish a framework for environmental protection within the market economy" by ending wasteful subsidies, enforcing strict standards, and boosting resource efficiency—the very ideas that Amory Lovins and others were advocating in the West.

Alas, Václav Klaus, the nation's powerful finance minister, was strongly opposed to such notions, which he ridiculed as carryovers from the nation's communist past. A conservative economist with a reputation for arrogance, Klaus was imposing shock therapy to accelerate the transition to a market economy: squeeze the money supply, let prices rise, and encourage foreign investment. Prices of food, housing, and other basic expenses had duly risen by 100 percent since the Velvet Revolution, while wages and pensions stagnated. Meanwhile, Klaus dismissed the environmental reforms Vavrousek and Havel favored as unaffordable luxuries, a view that did not prevent his election as prime minister in 1992.

Thus, the struggle to build a humane and environmentally healthy nation continued for Václav Havel, even after the miracle of the Velvet Revolution had vanquished his old foes and put him in the president's office. This twist of fate could hardly have surprised a confirmed absurdist like Havel. As his friend and hero Samuel Beckett had written in the novel *Worstward Ho,* "Try again. Fail again. Fail better." Havel found the presidency an unpleasant job in many respects, but he was not the type to give up. As he wrote in the 1980s in one of his prison letters to his wife, Olga, it is crucial to resist "resignation, indifference, the hardening of the heart and laziness of the spirit"; otherwise, one turns into "a parasite" who "is entirely absorbed in the problem of his own metabolism and essentially nothing beyond that interests him."

Havel once conceded that his philosophy of antipolitical politics—acting on one's moral convictions, no matter their apparent practicality—struck people who saw society "from above" as silly because it yielded results that were "indirect, long-term, and hard to measure." In my interview with him, I noted that one of the most devilish characteristics of the global environmental crisis was its inherent time limits; any remedial actions had to be undertaken soon or they would be irrelevant. Could Havel's antipolitical politics succeed quickly enough to save humanity from environmental self-destruction?

"When I wrote about antipolitical politics, which means politics that understands itself as a service to others rather than simply fighting for power, I didn't fully realize how relevant it was and how necessary it is," Havel replied. "Only my time here in political office showed me that it's the most important thing to strive for, because now I see from the inside how the power ambitions of individuals can badly influence the general situation. Of course I think that to save humankind it is indispensable that people come to their senses. That was my opinion in the past, and I believe it nowadays as well. It is quite clear it will be an eternal fight, full of problems and obstacles. Still, it is necessary to go about it in this way."

I turn to Havel near the end of this book not just because of his critique of the environmental crisis but because of what his life and presidency symbolize: victory over despair. If anyone had an excuse for giving up in the face of long odds, it was Havel during his four-year prison term in the early 1980s, when the prospects of change in Czechoslovakia seemed remote at best. The same point goes double for Nelson Mandela, who served twenty-seven years behind bars under apartheid. Yet in the end, the causes these men championed triumphed, and they themselves became the leaders of free peoples. To be sure, their nations continue to face difficulties; in human affairs, progress is always provisional. But that doesn't make it any less real.

Like Havel said, there will be many problems and setbacks in the struggle to avoid environmental catastrophe. Whether the effort will succeed is impossible to know. But the answer begins with you. Not everyone else—you. As Havel's example shows, waiting passively for the world to right itself is a losing proposition. To borrow one more nugget of ancient Chinese folk wisdom, "You can stand on a mountaintop with your mouth open for a very long time before a roast duck flies into it."

Ambassadors from Another Time

> Climb the mountains and get their glad tidings.
> Nature's peace will flow into you as sunshine
> flows into trees. The winds will blow their own
> freshness into you, and the storms their energy,
> while cares drop off like autumn leaves.
>
> —JOHN MUIR
> American naturalist

Three hours drive east of San Francisco, in the western foothills of California's Sierra Nevada mountains, live some of the largest trees on earth. Local people call them, bluntly, the Big Trees. More properly, they are named giant sequoia redwoods, and in truth they are the largest living organisms of any kind on this planet, and very nearly the oldest as well.

The oldest known giant sequoia is 3,300 years old—it was already ancient when Jesus, Buddha, and Mohammed walked the earth—and 2,000 year lifespans are common. Sequoias live in such exquisite balance with their environment that they do not age. The largest of them stand over three hundred feet tall, and they measure as much as thirty-two feet in diameter near the ground: It takes twenty adults with their arms extended to encircle one. Their limbs resemble the trunks of ordinary trees, their root systems encompass an entire acre of soil. Viewed from a distance through afternoon shadows, giant sequoias appear supernaturally oversized, as if they belong to a separate and mightier reality, which in a sense they do.

The author John Steinbeck called giant sequoias "ambassadors from another time" because, as a species, they have existed since the age of the dinosaurs, some 100 million years ago. Redwoods and dinosaurs, perhaps the largest plants and animals this planet has ever known, were the dominant life forms throughout the northern hemisphere for tens of millions of years. But during the last major climatic change, 65 million years ago, when the earth's temperature cooled dramatically, the dinosaurs failed to adapt and became extinct. Giant sequoias survived. Over time, their geographical reach has shrunk, however, and today they grow naturally only in some seventy-five groves scattered along the western slopes of the Sierra Nevada mountain chain.

It was my good fortune while working on this book to spend a year living near one of those seventy-five groves. This was in Calaveras County, an area made famous by Mark Twain's early writings about jumping frogs and gold mining. My friends, the Smiths, had rented me a log cabin that was nestled so deep in the forest it was impossible to find if you didn't know exactly where you were going. At the end of the paved road that led back to civilization, a dirt path led up to an old logging trail that was inaccessible in winter but sprinkled with pale golden pine needles the rest of the year. There was no telephone in the cabin, nor television, but I did have access to my friends' truck for once-a-week grocery runs.

My nearest human neighbors lived over a mile away, so I saw bears on my porches as often as I saw people. One morning, I was sitting on the cabin floor, writing away, when a mother bear and two young cubs started scratching at the kitchen door. Luckily, the mother couldn't see inside the cabin well enough to feel secure about pushing the door down to get in. Instead, the trio trooped around to the clearing in front of the cabin, while I stepped out onto the porch above them. The bears spent the next hour in that clearing, fully aware of my presence five yards away but paying me no real attention. The cubs scratched for bugs inside fallen tree trunks, the mother drowsed in the sunshine, the insects droned. Just another day in paradise. And then, too soon, the bears padded slowly into the forest, momma in the lead, the youngsters behind, leaving me a memory I shall treasure the rest of my life.

The forest was full of such delights. Besides bears, I regularly saw deer, coyotes, hawks, and owls, even the occasional mountain lion. Best of all, though, it took only forty minutes to hike over the ridge and visit the giant sequoias in the North Grove of Big Trees State Park.

Even at the height of springtime's exuberance, when Douglas squirrels squawked a counterpoint to the insistent tapping of whiteheaded woodpeckers and dogwood blossoms splashed color among the awakening undergrowth, the grove was pervaded by a deep sense of calm. The giant sequoias towered skyward, massive yet graceful, dominating their environment without overwhelming it. Besides handsome black oaks and ponderosa pines, the sequoias were joined by incense cedars with furrowed chocolate trunks, sugar pines whose limbs drooped from the weight of cones over a foot long, and white firs whose branches ascended in a symmetry as perfectly triangular as an old church steeple. From the forest floor rose the rich, fragrant aroma of pine needles and duff warmed by the morning sun, while the creek gurgled with new life imparted by the melting snows of the High Sierra. Dappled by shifting sunbeams, the sequoias' bark seemed constantly to change its hue, from cinnamon brown to red to shades of violet. The air felt still, but far overhead the foliage tossed and swayed, dancing joyfully in the high breeze, singing a song infinitely soothing yet powerful, like the murmured whispers of countless generations past and future. The sequoias, ageless and vibrant, seemed at once inseparable from yet indifferent to the swirl of nature's elements; they had held this ground forever. As moment followed moment, time dissolved into the eternal Now and the splendor of life on this earth stood revealed as a never-ending miracle.

Native Americans had considered the giant sequoias sacred for thousands of years before the white man arrived in the mid-1900s. Members of the Mono tribe would not even touch the mammoth trees, believing a guardian spirit in the form of an owl would punish any who dared abuse them. Some of the first whites who encountered giant sequoias harbored a similarly respectful attitude, including a hunter by the name of Augustus T. Dowd who literally stumbled upon the trees one day in 1852. This was during the gold rush era, when thousands of people were hurrying to

California with dreams of striking it rich. It was Dowd's job to furnish meat for construction crews working in the gold mining operations near the town of Murphys. One day, a grizzly bear he had shot and wounded fled across a meadow and disappeared into the forest. Dowd gave chase. Entering the forest, he suddenly encountered a tree so impossibly large that for a moment he thought his eyes were deceiving him. And the giant was not alone; nearby stood more of the incredible creatures. Forgetting about the grizzly, Dowd spent the rest of the day exploring the grove, the same grove I came to know and love over a century later.

When Dowd returned to Murphys and described what he had seen, the miners dismissed his account as the tallest of tales. But Dowd persuaded them to come have a look, and soon the story was out. Reporters from Stockton and San Francisco arrived to investigate. The stories they wrote were not universally believed—exaggeration was a staple of newspaper writing in those days, and besides, the plain facts about the trees did sound too bizarre to be true—but within weeks visitors were making their way up to the grove to satisfy their curiosity. Where most saw beauty and wonder, however, a reckless few saw the chance for personal gain.

The danger arose in January 1853, less than a year after Dowd's accidental discovery. Three local capitalists, including Dowd's boss at the construction company, Captain W. H. Hanford, decided they would fell the largest of the big trees, cut a huge slab out of it, and display the slab in a traveling exhibition, like a freak in a circus show. The plan was by no means certain of success; many people objected to the idea of mutilating such ancient, majestic creatures. An official history of the giant sequoias published by the University of California speaks of general "expressions of outrage, dismay, and disgust" toward the plan and notes that Dowd in particular refused to have anything to do with it. But appeals to the businessmen's higher selves fell on deaf ears, and since there was not yet any real law enforcement on the frontier, there was no stopping them.

Hanford and his fellow investors had their minds set on the so-called Discovery Tree, the first tree Dowd had seen. It was by far the largest tree in the grove, measuring thirty-two feet wide at ground level and twenty-four feet wide at eight feet above the ground. There was not a saw in the world long and strong enough to topple such a tree. Instead, Hanford and his cronies hired five miners from the Murphys gold camps who decided to fell the tree by drilling holes in it with long metal augers, as wide as a man's fist. They would direct their attack in a line eight feet above the ground, hoisting the augers on specially built sawhorses. If the operation went as planned, the drilling would weaken the base of the tree enough to cause it to fall of its own accord.

The operation began on June 5, 1853. The first step—removing the bark from the areas to be drilled—proved simple enough: after a few incisions, great chunks of bark were pulled off like the peel from an orange. Having breached the tree's skin, however, the men found its underlying bone and muscle less accommodating. Their augers were simple devices, to propel them into the tree, two men turned a wheel at the far end of the auger round and round, as if closing a valve. It was difficult work.

The spectacle was sufficiently arousing that a group of well-to-do ladies and gentlemen journeyed up from Murphys to witness the final stages of the assault. After twenty days of fierce labor, the men had succeeded in punching an uninterrupted line of shafts around the entire circumference of the great tree. This was the moment the drillers and their backers had been longing for, but it turned out to hold a maddening surprise: the tree refused to fall. The sequoia "was so symmetrical, so well balanced, that it continued to stand upright and gave no sign of weakening," says the University of California's history. As if in revenge, the tree resisted the villains another two and a half days before finally giving up the ghost. Adding insult to injury, it waited until the entire entourage was away at lunch before crashing to the forest floor.

Undeterred, the investors pressed on with their plans, which included developing the site into a luxury resort, complete with a saloon and a bowling alley that were carved into the tree's fallen trunk. The stump itself was converted into a dance floor big enough to accommodate a dozen waltzing couples, an act of blasphemy that inspired the title of John Muir's tract *And the Vandals Danced Upon the Stump.* Another grand old tree, known as the Mother of the Forest, was also killed when its bark was stripped off and shipped to New York and London to entice tourists to visit. An accompanying handbill encouraged the curious to take the daily stagecoach from Sacramento City to see the extraordinary "Vegetable Monsters."

Captain Hanford and his fellow investors soon went bankrupt, but by then the larger damage had been done. "Any fool can destroy trees. They cannot defend themselves or run away," Muir later wrote, adding,

"Through all the eventful centuries since Christ's time, and long before that, God has cared for these trees, saved them from drought, disease, avalanches, and a thousand storms; but he cannot save them from sawmills and fools; this is left to the American people."

More than once during my years of working on this book I retreated to my cabin in the forest. The sequoias were good company as I pondered the implications of all the places I had visited and people I had interviewed. Had the Discovery Tree not been cut down in 1853, it would now be the largest living thing on earth. Its destruction, I believe, ranks as one of the great crimes modern man has committed against nature. Yet the beauty of the remaining sequoias, and of the forest surrounding them, has never failed to lift my spirits, and their ancient history encourages me to take the long view of the human enterprise. As poet Edwin Markham wrote of the Big Trees in 1914, "Many of them have seen more than a hundred of our human generations rise, and give out their little clamors and perish. They chide our pettiness, they rebuke our impiety."

Now, as I complete this book, after one last visit to the Big Trees, I return once more to the question that sent me off around the world: will the human species survive the many environmental pressures crowding in on it at the end of the twentieth century?

Without reiterating all that I have reported, it seems plain to me that most of the key global environmental trends are moving in the wrong direction, and many are picking up speed. If, as Mostafa Tolba of the UN Environment Program put it, the 1990s were the decisive decade when humanity's environmental decline had to be reversed or it would accelerate beyond our control, it seems clear that we failed the test. Yes, progress has been made, but it has been too incremental, too grudging and slow. This book has focused above all on the threat of climate change, where the IPCC's call for immediate 50 to 70 percent reductions in greenhouse gas emissions dwarfs what the world's governments (provisionally) agreed to in Kyoto. But similar discrepancies exist in regard to many environmental hazards. The gap between what science demands and what our political structures deliver remains vast, and it is vigilantly patrolled by powerful interests that profit from the existing order. The one important exception is population growth, though even there the good news is not all that encouraging: our numbers may eventually stabilize at a mere eight billion, rather than ten or twelve billion.

The so-called environmental skeptics and eco-realists say we should relax, most environmental fears are much ado about little. Gregg Easterbrook's favorite straw man in *A Moment on the Earth* was to cast environmentalists as psychologically and financially adverse to good news. If the public knew how inflated the doom-and-gloom scenarios are, he argued, environmentalists would be out of work. This conceit allowed Easterbrook to pose as bravely open-minded, but intellectually it is bankrupt, even silly. As a journalist who has written about many different subjects over the years, I for one would be delighted to declare the environmental battle over and move on to other issues; there's no shortage of problems needing attention in this world. The reason I don't share the self-satisfied outlook of the eco-realists is simple: it does not jibe with what I have seen while exploring this planet, nor is it supported by the preponderance of other people's reporting and scientific research. For example, it's true that air and water quality have significantly improved in the United States over the past thirty years (though largely because of laws and regulations won by the very environmentalists the eco-realists love to deride). But improved is not the same as satisfactory. According to the Environmental Protection Agency, 40 percent of America's rivers, lakes, and coastal waters remained unsafe for fishing, swimming, or basic recreation in 1996, in part because industry had dumped more than a billion pounds of toxic chemicals into waterways between 1990 and 1994. This, in a country whose environmental laws and technology far surpass those of most nations.

The self-styled environmental skeptics and realists are in fact raging optimists, for at bottom they believe humans can easily adapt to any hazards. We can build dikes against the rising sea levels of global warming, wear sunscreen to block the ultraviolet radiation streaming through the ozone hole, develop genetically based supercrops to compensate for lost farmlands and fresh water. It's true humans are incredibly adaptable; adaptability has been our great evolutionary advantage, perhaps the biggest reason we have flourished while so many other plant and animal species have died out. But humans are so adaptable, so used to escaping adverse natural conditions with technologies like air-conditioning and prescription drugs that we can easily forget that we are still Nature's subjects, not its masters, and there are limits to our adaptability, especially in the face of Nature's unbridled power. El Niños, for example, are natural phenomena that are expected to become both more frequent and more volatile in a globally warmed world. Californians may have survived the 1998 El Niño tolerably well, but what if future El Niños are twice as powerful and start coming every *other* year rather than every seven?

No one can be sure what the future will bring, but in a way that is precisely the point. Although our scientists have done an admirable job of investigating the environmental side effects of our behavior, we do not and perhaps cannot know for certain what will happen to us if we maintain the present course. Of the perils facing us, only nuclear war seems capable of extinguishing all human life on the planet (and then probably only in the event of a global war). Global warming, deforestation, species loss, land degradation, and other dangers threaten dire but more limited consequences: food scarcity, flooding, disease, population displacements, wars over fresh water and arable land. While such disasters could, as discrete events, spell misery and death for millions of people, the suffering would probably be confined, as it is today, to certain geographic areas or social groups; it almost certainly would not mean the end of the species. If, on the other hand, more than one of these perils reaches a crisis point at the same time, resulting in, say, a major fall in food supplies simultaneous with widespread flooding and ecosystem-destroying temperature increases, what then? Would the disasters so compound one another that everything would begin to unravel, ushering in an era of chaos and decline? The truth is, no one knows; these are uncharted waters. But at some point, the difference between the extinction of the species and the mere disintegration of social order might seem trivial indeed.

By not taking decisive remedial action today, we are in effect gambling that the warnings of our most capable experts will turn out to be false alarms. "The climate system is an angry beast and we are poking it with sticks," Wallace Broecker, a geochemist at Columbia University who was one of the earliest investigators of global warming, has said. If provoked, said Broecker, the climate system could undergo *within a decade* a drastic warming or, worse, a return to Ice Age temperatures; such a shift could so devastate global agriculture that citizens of the wealthy North might suddenly discover they have more in common with their Dinka brethren than they ever imagined.

"The inhabitants of planet earth are quietly conducting a gigantic environmental experiment," warned Broecker. "So vast and sweeping will the consequences [of this experiment] be that, were it brought before any responsible council for approval, it would be firmly rejected." Broecker was referring to the buildup of greenhouse gases, but the point applies to other environmental hazards as well, including the vast array of man-made chemicals that pervade our plastics-saturated societies. In their book *Our Stolen Future*, zoologist Theo Colborn and coauthors Dianne Dumanoski and John Peterson Myers argue that DDT, dioxin, PCBs, and countless other bedrock chemicals of industrial civilization are severely undermining the reproductive health of wildlife and humans the world over by disrupting the hormones that regulate pre- and postnatal development. The most dramatic example is the 50 percent drop in human sperm counts reported in various places around the world during the last two generations. In addition, dozens of species—from Florida alligators to Lake Michigan mink to Baltic Sea fish—have suffered documented cases of shriveled penises, female sterility, and other fertility malfunctions.

The authors of *Our Stolen Future* grant that man-made chemicals have delivered benefits, but they caution that our ignorance of these chemicals is as great as our knowledge. They note that because scientists earlier in this century did not know what they did not know, the world celebrated CFCs and DDT for decades before their drawbacks were recognized. Whether our current reliance on synthetic chemicals is undermining our species' ability to reproduce is still a hypothesis, not a fact, and an airtight case may be impossible to make. In today's chemical-saturated world, it is difficult to prove a cause-and-effect link between a given chemical and a corresponding malady; individual substances cannot be sorted out from the overall mix. Our present course thus amounts to another case of experimenting, with our offspring once again serving as the guinea pigs.

So, is there any hope? During my travels, I often encountered despair about our chances of reversing course, even among people who were not particularly involved in environmental affairs. When people learned I was writing a book about whether the human species was going to survive its environmental problems, the question they invariably asked me was, "Well, will we?" And often, before I could reply, my interrogators would ruefully add words to the effect of, "It doesn't look good, does it?"

No, it doesn't, and there's no sense denying it. But there's no sense being paralyzed by it, either. To be sure, the task at hand is enormous, complex, and difficult, but the answers to it are not obscure; such august bodies as the UN Brundtland Commission and Earth Summit have been outlining the way forward for years now, and dedicated NGOs and enlightened government and business leaders have fleshed out many of the concrete details. In the short term, we must accelerate changes already underway in our technologies to make them more efficient and environmentally friendly. Furthermore, these technologies must be diffused throughout the planet, which means in concrete terms that the North must help transfer them to the South. In the medium term, population size must be stabilized both in the South and the North, and the hyperconsumption that is now common in the North and among elites in the South must be cut back. In the medium to long term, capitalism will probably have to be transformed so that the constant expansion in material terms of production, consumption, and waste is no longer a central feature of the system. Development, not growth, must become our motto.

Hope is the foundation of action, and as I review what I saw and heard during my travels around the world, I see two principal reasons for hope. First, most people want to do right by the environment and, if given the chance, they will—as long as they are not penalized economically for it. Second, far from being enemies, economic and environmental health can reinforce one another. In fact, if humans are smart, repairing the environment could become one of the biggest businesses of the coming century, a huge source of profits, jobs, and general economic well-being. "The potential profit . . . is limitless," one Japanese official has said, because the market is all but limitless. Fuel cell cars and solar power panels could be just the beginning. In the words of an executive from AT&T, one of the few U.S. corporations that recognizes what is possible, "We are talking about restructuring the technological basis of our entire economy . . . integrating environmental considerations into all technology and economic behavior."

The idea is to renovate human civilization from top to bottom in environmentally sustainable ways. Humans would redesign and renovate everything from our farms to our factories, our garages to our garbage dumps, our schools, shops, houses, offices, and everything inside them, and we would do so in both the wealthy North and the impoverished South. The economic activity such renovating could generate would be enormous. Better yet, it would be labor intensive, and so address the problem of poverty that is the irreducible other half of the environmental challenge.

Restoring our embattled environment could become the biggest stimulus program for jobs and business in history. It is odd, and unfortunate, that our environmental movements have not been more effective in publicizing this possibility; the elements of such a program are scattered throughout their literature. Still, there is no time like the present. A political program that made the economic case for environmental restoration in clear and inspiring language could capture the popular imagination and rally it to victory.

Some environmentalists have suggested that the race to the moon in the 1960s serve as the model for the race now needed to save the earth. It's a good idea, and not simply because that earlier race sent back pictures of this blue planet that revolutionized humanity's understanding of itself and its place in the cosmos. The race to the moon showed how a clear mission and deadline can focus resources and fire public enthusiasm. It also demonstrated something rarely acknowledged these days: that certain overarching public challenges cannot be left up to the workings of the marketplace; government must play a central, leading role.

Besides the moon race, it seems to me that another model worth emulating is the New Deal that President Franklin Roosevelt launched in the 1930s to propel the U.S. economy out of depression. After all, the environmental crisis is as much an economic challenge as anything, and the New Deal helped overcome the gravest economic challenge in modern American history. What's more, the problems afflicting today's global economy are strikingly similar to those the New Deal was created to solve in the 1930s. As analysts from Karl Marx to John Maynard Keynes have noted, capitalist economies have an inherent tendency toward stagnation and inequality. The routine workings of the system generate more wealth than can find profitable investment outlets and too little money at the bottom to generate sufficient overall demand to keep the system churning forward. As journalist William Greider reports in his book, *One World, Ready or Not: The Manic Logic of Global Capitalism,* such pillars of the world economy as the steel, auto, aircraft, and consumer electronic industries are all struggling today in the face of insufficient demand for their products. Perversely, the most common corporate reaction is to cut costs by shedding relatively high-wage workers in the United States, Europe, and Japan in favor of their cheaper counterparts in the Third World. Of course, this only worsens the underlying problem of insufficient demand.

The basic function of the New Deal was to restore sufficient demand to the economy by raising what can be called the social wage. New Deal policies raised the economy's collective purchasing power by guaranteeing workers a minimum wage and the right to strike for more; by putting unemployed people to work in government-funded public works projects; by providing direct cash payments to tide over the unemployed until they found work; and by establishing the universal pension plan for the elderly known as Social Security. In short, the New Deal redistributed society's surplus wealth, shifting a portion away from the rich, where it languished unproductively, toward the poor and working classes. Their spending of that surplus duly boosted overall demand and, along with the explosion of military spending during World War II, pulled the economy out of depression and prepared it for the unprecedented prosperity of the 1950s and 1960s.

Why not revive these New Deal policies but apply them in a green and global fashion? The program could even be called the Global Green Deal. It would rely on market mechanisms to the maximum extent possible, while realizing that government must also establish "rules of the road" that compel markets to respect rather than harm the environment. In particular, governments must reform skewed tax, subsidy, and economic accounting systems so that the market internalizes environmental values. Governments should also increase public investment to help nascent industries like solar power achieve commercial takeoff. Priming the pump with steady purchases by the Pentagon in the 1960s was what got the computer industry up and running, and the Clinton administration did much the same in the 1990s by having the federal bureaucracy shift its purchases from virgin to recycled paper. By requiring that the fifty thousand vehicles the U.S. government buys every year be fuel cell or hybrid powered rather than traditional gasoline powered, for example, Washington could help create market demand for green cars, demand that private capital could then step up and accommodate.

There would be a certain poetic justice in this, for the government's lavish subsidization of conventional automobiles throughout the twentieth century is no small cause of our current problems. Imagine what a similar level of government commitment could accomplish for the Global Green Deal! Mass transit would no longer be the unwanted stepchild of government policy but its proud focus, yielding transportation systems of such excellence that people would *want* to patronize them. Heat waves would no longer kill 800 people a summer in Chicago's inner city because they were too poor to afford air-conditioning; their apartment buildings would be renovated with the advanced energy efficiency methods that make air-conditioning unnecessary, with the renovations providing jobs for the neighborhood residents who now look in vain for work. Water would no longer be wasted in such criminally extravagant volumes; farmers would be encouraged to install superefficient drip irrigation systems. And so on, through all of the many technologies that need environmental reforming.

This shift to environmentally friendly technologies would set a compelling example for China, India, Brazil, and the other Southern nations whose participation in the global environmental cleanup is essential. No longer would solar energy and other environmentally progressive technologies be seen as backwards and economically second rate. This breakthrough, in turn, could set the stage for a massive transfer of these technologies from North to South. Installing efficient equipment and processes throughout China's energy system, for example, could reduce its energy consumption by 50 percent. It so happens that the United States is the world's leader in energy-efficient technology. If the Clinton administration and Congress wanted to, they could help China buy lots of that technology, and do the atmosphere and American companies and workers a favor at the same time. President Clinton is plainly aware of this possibility, judging from his comments to [Zhou Dadi and] other Chinese environmentalists in Guilin in 1998. But the agreement the administration signed with China pledged a mere $50 million credit line for energy efficiency in China—pretty small beer. It's worth remembering that Northern governments subsidize these kinds of business deals all the time, notably in the fields of weapons and military equipment. But if Northern taxpayers are going to be subsidizing corporations, shouldn't the corporations be doing something socially useful, like averting climate change, rather than arming the world to the teeth?

A Global Green Deal can succeed only if rich and poor countries alike participate, but there is no getting around the fact that the rich will have to take the lead, both logistically and financially. Not only will rich nations have to adopt environmental reforms at home before expecting poor nations to follow suit, they will also have to help pay for most of the environmental reforms of the poor. If they don't, many of those reforms simply will not happen, not only because some Southern leaders are malign despots interested solely in lining their personal pockets, but because poverty makes a more urgent claim on limited public funds in the South than long-term environmental considerations do. The most straightforward way to increase the funds available to Southern governments would be through rapid and sizable debt relief. "In Africa," Oxfam International has reported, "where one out of every two children doesn't go to school, governments transfer four times more [money] to northern creditors in debt payments than they spend on the health and education of their citizens." If Africa's population growth is to be brought under control, health and education spending is essential. So, therefore, is debt relief.

All this would cost money, of course. But contrary to conventional wisdom, there is lots of money available—we're just spending it foolishly at the moment. In the United States, military spending remains bloated at 85 percent of Cold War levels nearly ten years after the Berlin Wall fell. Amid such excess, even minor redeployment of resources can yield large gains. "Take the price-tag for safeguarding two-thirds of the Amazon rainforest: $3 billion, according to a 1989 estimate," wrote Jeremy Leggett in 1990. "Cancel just six U.S. 'Stealth' bombers and you have the cash to do it." If even half of the estimated $500–900 billion in environmentally destructive subsidies now being doled out by the world's governments were pointed in the opposite direction, the Global Green Deal would be off to a roaring start.

None of this will happen without a fight, however. Amory Lovins likes to say that the role of government is to "steer, not row." But it must steer in a fundamentally different direction than at present, and that will upset those who profit from the status quo. Taking from the rich and giving to the poor may have made Robin Hood a hero to the poor in ancient England, but it made Franklin Roosevelt a hated man among the American upper class of the 1930s. With honorable exceptions (such as the Scandinavian countries), most rich nations have been equally reluctant to offer more than crumbs to today's poor. But this is shortsighted in the extreme, just as the opposition to Roosevelt was. The truth is, Roosevelt was doing the elite a service. Without his reforms, the economy would have remained mired in depression and the mass anger finding an outlet in street protests might have evolved into outright revolution.

Today, the danger threatening the rich is not so much mass revolt as inescapable environmental meltdown. Like it or not, the rich need the poor if they are to save their own skins. Without the poor's cooperation, there is no hope of dealing with global warming, forest loss, and many other environmental hazards that will punish rich and poor alike. In some ways, then, the most basic question to ask about the human environmental future is, will we learn to share? We teach sharing to our children, and it is a central message of most of our religions. Indeed, as the Dinka remind us, sharing has been a basic survival tactic for most humans for most of our history. Can we rediscover it now?

The most difficult sharing will not be of money—the rich have plenty of that—but of environmental space. As Maurice Strong explained at the Earth Summit, the wealthy nations must substantially reduce their production of greenhouse gases and other such harmful activities in order to make space available so that Southern nations fighting poverty can increase *their* greenhouse emissions. Many Americans will resist such sharing, in part because they simply do not realize how lavish their lifestyles are compared with the rest of the world's. In taking for granted such luxuries as unlimited hot water at the turn of a tap, to say nothing of cars bigger than many people's houses, Americans inadvertently exhibit the sort of arrogance and self-centeredness that has made people hate the rich since time immemorial. The irony is that Americans are in fact a generous people, or at least like to think of themselves that way; they are simply oblivious to how wasteful and selfish their lives appear to outsiders. Countless millions of people in the South still live like the malnourished Dinka of Sudan, the unemployed workers of Shenyang, the landless squatters of rural Brazil. In the face of such deep, pervasive poverty, how can Americans and the rest of the world's comfortable class insist that they cannot cut back on their own consumption?

After all, cutting back is not the same thing as sacrificing. Most Americans have far more material possessions than they, or anyone else, could possibly need. Indeed, one of the best things extensive Third World travel can do for an American is to help him or her appreciate just how wealthy Americans are, and how life can still be deeply satisfying under modest material circumstances. Long before there was electricity, motor cars, and the other "necessities" of our modern lives, humans were living lives of consequence—raising children, building cities, creating art, pursuing knowledge. On the other hand, Third World travel also cautions one against romanticizing the nontechnological life. It's all very well to say modern man's separation from nature is an act of foolish arrogance, but a certain amount of separation is valuable, especially when it's forty below outside.

The point is to strike a proper balance, and that requires change from individuals as well as institutions. It may be emotionally gratifying to blame corporations and governments for our environmental problems, and god knows they deserve it, but individuals cannot escape responsibility for their actions. "The problems are our lives," argues the writer Wendell Berry. "In the 'developed' countries, at least, the large [environmental] problems occur because all of us are living either partly wrong or almost entirely wrong. It was not just the greed of corporate shareholders and the hubris of corporate executives that put the fate of Prince William Sound into one ship [the Exxon *Valdez*]; it was also our demand that energy be cheap and plentiful. . . . To fail to see this is to go on dividing the world falsely between guilty producers and innocent consumers."

A Global Green Deal that put people to work restoring our environment would, I believe, yield enormous economic and social benefits to the vast majority of earth's current inhabitants, to say nothing of their descendants. At a time when many sectors and regions of the world economy are in danger of sinking into depression, a Global Green Deal could stimulate enough economic activity to keep the system from crashing down around us. It could also reverse current trends toward widening in-equality and environmental overload. Such a fundamental shift in direction will not happen by itself, however. Politics must be committed.

Which returns me to the story of the Big Trees. If John Muir and a great number of other nature-loving citizens had not thrown themselves into defending the giant sequoias after the Discovery Tree was cut down, today's remaining seventy-five sequoia groves would have fallen to the logger's axe long ago. Muir's tireless advocacy helped establish Yosemite, Kings Canyon, General Grant, and numerous other national parks in California. (Even so, 35 percent of the original acreage of giant sequoias was logged from the late 1800s through the 1950s, along with untold amounts of sugar and ponderosa pines.) From the time the North Grove went to public auction in 1877, it took fifty years of activism before the grove was decisively removed from harm's way. On July 5, 1931, Calaveras Big Trees State Park was dedicated and declared public land forever.

The loss of the Discovery Tree can never be made right. In its place, the Big Stump now stands as a grotesque monument to the irreversible, shortsighted idiocy of some humans, just as nuclear waste dumps and other environmental scars will in the future. The rest of the sequoia grove, however, remains intact and spectacularly beautiful—living proof that humans can, with intelligence, cooperation, hard work, and perseverance, learn to live amid the natural bounties of this planet.

The moral of this story? The outlook is uncertain, the hour late, the earth a place of both beauty and despair. The fight for what's right is never ending, but the rewards are immense. Humans may or may not still be able to halt the drift toward ecological disaster, but we will find out only if we rouse ourselves and take common and determined action.

I wish us godspeed.

M.H.
Love Creek, California
August 21, 1998

Acknowledgments

This book was researched and written over the course of seven years, during which time I was helped by literally hundreds of people all over the world. Some offered a bed for a night or three; others supplied the personal contacts and travel advice crucial to a successful life on the road; still others provided the moral and professional support needed to get my writing published. I can never adequately thank everyone involved, but without them this book would not have been completed.

My deepest gratitude is to the late William Shawn, to whom this book is dedicated. It was Shawn, back when he was still the editor of *The New Yorker,* who first gave me faith in my future as a writer. After leaving *The New Yorker,* Shawn edited my book, *On Bended Knee: The Press and the Reagan Presidency* (Farrar, Straus & Giroux, 1988). In 1991, just before I left the United States on my journey around the world, Mr. Shawn and I had what turned out to be our last lunch together at the Algonquin Hotel in New York. I described my idea of making a global tour to investigate the environmental future. At the time, I had only the vaguest notion of how to go about such an inquiry, much less what kind of book might emerge from it. Shawn listened intently to my rambling discourse and then, in that timorous but unmistakably resolute voice of his, pronounced the project "the most important book anyone could be writing." Hyperbole, to be sure, but enormously inspiring for a writer to hear at the start of such an ambitious and uncertain project. Shawn's vote of confidence was all the more remarkable considering that my topic was not exactly a sure-fire commercial hit. But then intellectual substance and social value were what mattered most to Shawn, not the glitz and commercialism that now pervade so much of American magazine and book publishing. I miss Shawn, and the example he set, very much, and I hope this book, despite its shortcomings, approaches his original vision of it.

After Shawn passed away in 1992, I despaired of finding another editor who would embrace this project. But with the appearance of John Sterling, my despair vanished. John instantly saw what made this book different from other environmental books and how it should be published. John's faith and enthusiasm were a bedrock of support as I struggled to complete the manuscript, and his many keen editorial suggestions vastly improved the finished product. John's colleagues at Broadway Books have also been unflagging in their cheerful, professional nurturing of the book through the many stages of production and publication.

Thoughout this saga, my agent Ellen Levine has been a fierce and gifted advocate, magnificent in every respect. Diana Finch has handled overseas sales of the book with aplomb and high skill, connecting me with one excellent publisher after another.

I owe special thanks to the magazine editors who published my work during my travels; without their support, I could not have afforded to stay on the road long enough to research this book with the thoroughness required. My warm thanks go to Elise O'Shaughnessy at *Vanity Fair*, Chip McGrath at *The New Yorker*, Doug Foster and David Weir at *Mother Jones*, Richard Williams at *The Sunday Review* of *The London Independent*, Colin Harrison at *Harper's*, Jeff Morley at the *Washington Post*, Susan Brenneman at the *Los Angeles Times*, and Jack Beatty and William Whitworth at *The Atlantic*. I extend very special thanks to my friend Elise, who not only arranged for my visit to China but also gave a careful, critical read to the entire manuscript, a generosity that went far beyond the call of duty. I also thank the Transnational Institute of Amsterdam and the Plowshares Fund of San Francisco for the travel grants they provided.

To the interpreters whose skills in language and diplomacy gave me access to so many different cultures and individuals around the world, my gratitude and admiration are boundless. Dabbling in new languages was, for me, one of the great pleasures of this project, but there is a big difference between getting by in a foreign language and engaging in nuanced communication. For a journalist, the latter is essential, and I was lucky to work with interpreters who carried it off with intelligence and style. I am also grateful to everyone who took time to be interviewed for this book or who provided source material for it. Any errors of fact or interpretation are mine alone.

There remain so many other people who deserve thanks that, were I to list them all, these Acknowledgments would be longer than some Academy Awards speeches. A very incomplete roster includes Jasper Becker, Ray Bonner and Jane Perlez, Jana and Vasek Chaloupecky, Mark Cohen, Tom Devine, Mark Dowie and Wendy Schwartz, David Fenton and Beth Bogart, Bill Finnegan, Philip Frazer, Susan and Henri George, Natalia Gevorkyan, Todd Gitlin, Jean-Francis and Annette Held, Steve Hubbel, Christina Koenig, Howard Kohn, Denny May and Betsy Taylor, Shoon Murray, Mikko and Pia Pyhalla, Sidney Rittenberg, Fiametta Rocco, Roji, Sergio Sauer, Mark Schapiro, Orville Schell, Bettina Schwere and Paul Slovak, Renata Simone, Larry and Carla Smith, Walter Spofford, William Styron, Jaroslav Veis, Ed Vulliamy and Harvey Wasserman. Finally, to my family and friends, whose unstinting love and support made my lonely hours on the road both more bearable and more painful, a big hug of thanks and celebration.

Notes

Prologue

2   The history and environmental effects of the Chongqing Paper Factory were described in an author's interview with an official of the local Environmental Protection Bureau who, unfortunately, must remain nameless; the official was also the source of the quote "to cover the entire river. . . ."

5   China's official 1996 population was confirmed in an interview with the Chinese embassy in Washington, D.C.

5   The Chinese economy was estimated as the world's seventh largest by the World Bank in its *World Development Indicators* CD-ROM, February 1997. The projection of reaching number one by 2010 was made by *Newsweek* in its March 3, 1997, issue. That Chinese incomes more than doubled between 1979 and 1997 was reported in *China 2020* (Washington, D.C.: World Bank, 1997), pp. 3–4.

5   That five of the world's ten most polluted cities are in China was reported in the *New York Times*, September 19, 1997, citing World Bank data. (Such rankings are inevitably somewhat imprecise, given the paucity and inconsistency of data. For a comprehensive listing of such data for leading cities around the world, see *World Resources 1998–99*, World Resources Institute [New York: Oxford University Press, 1998], pp. 264–265.) The frequency of acid rain in Guangdong was confirmed by Ma Xiaoling, a senior official with the South China Institute of Environmental Sciences, in an author's interview. That 80 percent of coal went unwashed was noted by Yu Yuefeng, staff director of the National People's Congress environment committee, in an author's interview. The World Bank's estimate of two million deaths per year attributable to air and water pollution is found in *Clear Water, Blue Skies* (Washington, D.C.: World Bank, 1997), Table 2.3 on p. 21. That lung disease was responsible for one-fourth of China's deaths was reported in *World Resources 1994–95*, World Resources Institute (New York: Oxford University Press, 1994), p. 74. The loss of farmland was reported in "Environmental Change as a Source of Conflict and Economic Losses in China," Václav Smil, *Occasional Paper Number 2*, American Academy of Arts and Sciences and the University of Toronto, December 1992, pp. 27–28.

6   China's ranking in CFC production is reported in *Clear Water, Blue Skies*, op. cit., p. 11; its carbon dioxide ranking is listed in *World Resources 1994–95*, op. cit., Table 23.1, pp. 362–363.

7   The quote "suffer from a strange complacency . . ." is from *The Naked Ape*, Desmond Morris (New York: McGraw Hill, 1967), p. 240.

7   That 99 percent of species have gone extinct is reported in, among many other sources, *The Beak of the Finch: A Story of Evolution in Our Time*, Jonathan Weiner (New York: Alfred A. Knopt, 1994), p. 134.

8 Christopher Flavin's "inconceivable" comment is from an author's interview.

9 The discovery of the ozone hole is based on numerous sources, especially "The Annals of Chemistry," Paul Brodeur, the *New Yorker,* June 9, 1986; *Healing the Planet: Strategies for Resolving the Environmental Crisis,* Paul R. Ehrlich and Anne H. Ehrlich (Reading, Mass.: Addison-Wesley, 1991), pp. 113–129; *Ozone Diplomacy,* Richard Benedick (Cambridge, Mass.: Harvard University Press, 1991); and *Europe's Environment: The Dobris Assessment,* David Stanners and Philippe Bourdeau, eds. (Copenhagen: European Environmental Agency, 1995), pp. 523–529.

10 The discussion of climate change is drawn from a great many sources but above all *Climate Change 1995,* Working Groups I, II, and III, Intergovernmental Panel on Climate Change (Cambridge: Cambridge University Press, 1996). The 50–70 percent figure is found on p. 9 of the report's so-called Second Assessment Report. The quote from James Hansen was reported in the *New York Times,* June 24, 1988.

10 The rate and dangers of species loss are sketched in *In Search of Nature,* E. O. Wilson (Washington, D.C.: Island Press, 1996), pp. 171–174.

11 The 2050 date for full restoration of the ozone layer is reported in *Scientific Assessment of Ozone Depletion: 1994,* World Meteorological Association (Geneva: WMO, 1995). The overly optimistic assumptions underlying this date are dissected in "A Sky Full of Holes," Mark Dowie, the *Nation,* July 8, 1996. The EPA's estimated twelve million extra cancers was reported in the *New York Times,* April 5, 1991.

12 The 1 percent annual increase in carbon emissions and the doubling of atmospheric concentrations by 2100 is drawn from climate change 1995, op. cit.; see especially the Second Assessment Report, p. 9.

12 Mostafa Tolba's quote is from his introduction to *UNEP Profile* (Nairobi: UNEP, 1990), p. 2.

13 The report by the U.S. National Academy of Sciences and the Royal Society of London was called *Population Growth, Resource Consumption, and a Sustainable World* (London: 1992), p. 1.

13 The propaganda campaign of the fossil fuel lobby was exposed in *The Heat Is On: The High Stakes Battle Over Earth's Threatened Climate* (Reading, Mass.: Addison-Wesley, 1997).

14 The issue of worsening of the ozone hole in the 1990s, and government response to it, is ably summarized in "Learning from the Ozone Experience," Hilary F. French, in *State of the World 1997,* Lester R. Brown et al. (New York: W. W. Norton, 1997), especially pp. 152–156. The size of the ozone hole in 1995–1997 was reported in the *New York Times,* October 12, 1997.

Chapter One

21 That one in eleven African children die before their first birthday is documented in *World Development Indicators 1998* (Washington, D.C.: World Bank, 1998), p. 4, which reports an infant mortality rate of ninety-one out of one thousand.

23 The IPCC's projections of climate change's effects on Africa are found in *Climate Change 1995,* Contribution of Working Group II (Cambridge: Cambridge University Press, 1996), p. 9.

24 Paul Epstein's quote is from an author's interview.

25 The FAO's figures on increased malnourishment in Africa between 1970 and the early 1990s are found in a press release titled "Agriculture and Food Security" issued during the World Food Summit in Rome, November 13–17, 1996.

   The World Food Program's estimates for hunger in 1992 came from an author's interview with WFP officials, including Steven Green.

   For an excellent overview of the causes of and potential solutions to hunger in Africa, see *The Greening of Africa: Breaking Through in the Battle for Land and Food,* Paul Harrison (Nairobi: Academy Science Publishers, 1996).

26   The $12 billion that African nations paid annually in debt service, and the alternative potential, is based on figures provided by UNICEF and reported in Oxfam International's policy report "Multilateral Debt: The Human Costs," February 1996, p. 9.

34   The Nuer's mythic story about Stomach is recounted in *The Nuer,* E. E. Evans-Pritchard (Oxford: Oxford University Press, 1940), p. 83.

35   Alexander de Waal's excellent study of the Darfur famine is called *Famine That Kills* (Oxford: Oxford University Press, 1989). See p. 112 for the "tie your stomach" quote. See also his fascinating criticism on pp. 20–23 of so-called disaster tourists: journalists, relief workers, and others whose too-brief visits to impoverished areas serve mainly to reinforce simplistic, preconceived notions about what famine entails—a warning I found useful indeed during my travels among the Dinka.

45   The estimates of 1.5 million deaths and 85 percent population displacement during the Sudanese civil war come from "The Longest War in the World," Bill Berkeley, *New York Times Magazine,* March 3, 1996, and from the *Times* report of July 24, 1998. The developments of 1997 and 1998 were reported in the *Times* of March 3, 1998; March 18, 1998; May 5, 1998; and May 7, 1998.

46   That eleven thousand children die daily from starvation is based on statistics compiled by the UN's World Food Program that were reported in the *New York Times,* November 13, 1996. The hunger of Indian children was reported in the *New York Times,* August 14, 1997. That 40 percent of rural people in the Third World live in absolute poverty was documented in a 1992 report by the UN's International Fund for Agricultural Development, "The State of the World Rural Poverty," reported in the *San Francisco Chronicle,* November 25, 1992. That six hundred million city residents live in dangerous or nonexistent housing is reported in *Human Development Report 1997* (New York: Oxford University Press, 1997), p. 29. That 20 percent of humanity exists on $1 a day is documented in *Poverty Reduction and the World Bank* (Washington, D.C.: World Bank, 1996), chapter 1. The FAO's 841 million estimate was reported in the *New York Times,* July 22, 1996.

47   Mabub ul-Haq's remark is from a statement he made at an Earth Summit press conference that was attended by the author.

47   The quote from Maria da Graca de Amorim is from an author's interview.

Chapter Two

51   The rhinoceros scene comes from *My African Journey: Sabbatical of a Lifetime,* Winston Churchill (London: Mandarin, 1990), pp. 14–16.

52   "It is no good trying . . ." is from ibid., pp. 49–51.

54   "The best of all methods . . ." is from ibid., p. 84.

55   "Even in the early morning . . ." is from ibid., p. 92.

56   "The river was a broad sheet . . ." is from ibid., p. 103.

58   "Contract suddenly until . . ." is from ibid., p. 97.

58   "I cannot believe that modern science . . ." is from ibid., p. 101.

59–60 That the earliest *Homo sapiens* existed some two hundred thousand years ago in eastern Africa is not a proven fact, but it is currently the hypothesis with the strongest supporting evidence. See in particular the studies of the ancestral "Adam" in *Nature,* November 23, 1995, and of "Eve" in *Nature,* January 1987, Vol. 325, pp. 31–36. For a fine, broad review of the evidence of the deep human past in Africa, see also *Africa: A Biography of the Continent,* John Reader (New York: Alfred A. Knopf, 1998).

      The discoveries of the Leakeys, and the quote from Richard Leakey, are noted in *The Origin of Humankind,* Richard Leakey (New York: Basic Books, 1994), especially p. 77.

60   The ecological problems besetting Lake Nakuru were reported on the "ITN World News" broadcast of April 11, 1996.

PAGE

64   "During the drought, when big . . ." is from *The Beak of the Finch: A Story of Evolution in Our Time,* Jonathan Weiner (New York: Alfred A. Knopf, 1994), p. 78.

65   The discussion of how apes evolved into modern humans is drawn from many sources, including Weiner's *The Beak of the Finch,* ibid.; Leakey's *The Origin of Humankind,* op. cit.; *Guns, Germs and Steel: The Fates of Human Societies,* Jared Diamond (New York: W.W. Norton, 1997), *The Ascent of Man,* J. Bronowski (London: BBC, 1973); *The Fossil Trail: How We Know What We Think We Know About Human Evolution,* Ian Tattersall (New York: Oxford University Press, 1995); *African Exodus: The Origins of Modern Humanity,* Christopher Stringer and Robin McKie (New York: John Macrae/Holt, 1997); as well as the excellent reporting of John Noble Wilford in the *New York Times,* especially November 19, 1996, April 28, 1998, and May 12, 1998. For the minority view that modern humans evolved not solely in Africa, see *Race and Human Evolution,* Milford Wolpoff and Rachel Caspari (New York: Simon & Schuster, 1997).

66   The discussion of the Huaorani is based on *Savages,* Joe Kane (New York: Knopf, 1995).

67   The discussion of the Agricultural Revolution and the technological innovations and ecological collapses that followed it is based on Diamond's *Guns, Germs and Steel,* op. cit.; *A Green History of the World,* Clive Ponting (New York: St. Martin's, 1991); *The Sleepwalkers,* Arthur Koestler (New York: Grosset & Dunlap, 1963); and *Deutsches Musuem: Guide Through the Collections,* Deutsches Museum (Munich: C.H. Beck, 1988).

69   "The decline and eventual collapse . . ." is from Ponting, op. cit., p. 401.

70   "Since the rise of settle societies . . ." is from ibid., p. 315.

70   The history of the Scientific and Capitalist revolutions is drawn from many sources, especially Diamond's *Guns, Germs and Steel,* op. cit.; Ponting's *A Green History of the World,* op. cit.; and Koestler's *The Sleepwalkers,* op. cit.

72   The development of medical technology in the nineteenth century is described in *The Greatest Benefit to Mankind: A Medical History of Humanity,* Roy Porter (New York: W. W. Norton, 1998).

73   "High volume, high quality production . . ." is from *The Lunatic Express: An Entertainment in Imperialism,* Charles Miller (London:Macmillan, 1971), pp. 276, 448.

73   "Four-fifths of the best land . . ." is from *The Tree Where Man Was Born,* Peter Matthiessen (London: Picador, 1984), p. 39.

74   "Travel even for a little while . . ." is from *My African Journey,* Churchill, op. cit., p. 27.

76   That one in seven African children die before their fifth birthday is documented in *World Development Indicators 1998* (Washington, D.C.: World Bank, 1998), p. 4, which reports an under-five mortality rate of 147 out of 1,000.

77   "After the beautiful cascades" and "let the Nile begin its long . . ." are from *My African Journey,* Churchill, op. cit., pp. 74–75.

Chapter Three

81   Hobsbawm's quote is found on p. 12 of his book *The Age of Extremes: A History of the World, 1914–1991* (New York: Vintage, 1996).

85   Benzene's dangers were described by the World Health Organization as follows: "No safe level for airborne benzene can be recommended, as benzene is carcinogenic to humans and there is no known safe threshold level." See *Air Quality Guidelines for Europe,* WHO Regional Publications European Series Number 23 (Copenhagen: WHO, 1987).

86   Bangkok's average traffic speed is documented in a useful special report published in the *International Herald Tribune,* September 21, 1994.

87   The information about Thailand's environmental problems comes from my interview with Normita Thongtham, supplemented by an article in *Far Eastern Economic Review,* September 19, 1991, p. 44, and by the official *Thailand Country Report to the United Nations Conference on Environment and Development,* June 1992. The information from the Asian Development Bank comes from its 1996 annual report, p. 43.

PAGE

87   The data on Thailand's per capita income come from note 24 on p. 494 of *One World, Ready or Not: The Manic Logic of Global Capitalism*, William Greider (New York: Simon & Schuster, 1997).

87   The productivity losses caused by traffic in Thailand are documented in Peter Midgley's *Urban Transport in Asia: An Operational Agenda for the 1990s*, World Bank Technical Paper Number 224 (Washington, D.C.: 1994).

87   The information on refrigerator and air conditioner factories is found in *Far Eastern Economic Review*, September 19, 1991, p. 44.

88   The information about Thailand's development model and its consequences comes from Greider, op. cit., pp. 276–277, 337–355.

88   The Khor Chor Kor forest program is described in Greider, op. cit., p. 352.

89   The King's "talk, talk, talk" quote is found in the *New York Times*, June 10, 1996.

89   The CNN report was broadcast on August 31, 1996.

90   Thomas Friedman's article was published in the *New York Times*, March 20, 1996.
        For further information about Bangkok's traffic problems, see the report in the Internet magazine *Salon*, April 15, 1997.

91   The one billion cars in 2020 projection is included and explicated in Odil Tunali's "A Billion Cars: The Road Ahead," *World Watch*, January/February 1996.

91   The DOE data come from Energy Information Administration, "Annual Energy Outlook 1997," DOE/EIA 0383 (97), p. 5. Houghton's quote is found in the *Guardian*, March 25, 1995.

92   The information on climate change comes from *Climate Change 1995: The Science of Climate Change, Contribution of Working Group I to the Second Assessment Report of the Intergovernmental Panel on Climate Change*, J. T. Houghton et al., ed. (New York: Cambridge University Press, 1996). See also *Climate of Hope: New Strategies for Stabilizing the World's Atmosphere*, Christopher Flavin and Odil Tunali (Washington, D.C.: Worldwatch Institute, 1996), especially pp. 10–20.

92   The information on the fossil fuel lobby's propaganda campaign is found in *The Heat Is On: The High Stakes Battle Over Earth's Threatened Climate*, Ross Gelbspan (Reading, Mass.: Addison-Wesley, 1997). An earlier version of the story appeared in the December 1995 issue of *Harper's*, which is why I cite 1995 as the year Gelbspan exposed the campaign.

93   The information on the IPCC's findings and recommendations comes from the *Climate Change 1995* report, op. cit., as well as useful summaries by William K. Stevens in the *New York Times*, September 10 and September 18, 1995. See also my own articles in the *New York Times*, April 8, 1995, and the *Washington Post*, January 21, 1996.

94   The information on the car's role in greenhouse gas production comes from the following sources: *Driving Forces: Motor Vehicle Trends and Their Implications for Global Warming, Energy Strategies, and Transportation Planning*, James J. MacKenzie and Michael P. Walsh (Washington, D.C.: World Resources Institute, 1990); "One Billion Cars," *World Watch*, January/February 1996; *Climate of Hope*, op. cit.; "Population Issues: A Briefing Kit," UN Population Fund, July 1990, p. 9.

94   That cars are the single largest source of air pollution in the world is documented in *World Resources 1992–93*, World Resources Institute and the UN Development Program (New York: Oxford University Press, 1992); see especially p. 203.

95   On health costs, see *State of the World 1993* (Washington, D.C.: Worldwatch Institute, 1993), p. 122. The Mobil advertisement appeared in the *New York Times*, April 27, 1995.

95   The study "Clean Air and Public Health: Health Effects and Economic Costs of Air Pollution in the Northeast," was released at a conference at Harvard on July 11, 1995, and was cited in *Asphalt Nation: How the Automobile Took Over America and How We Can Take It Back*, Jane Holtz Kay (New York: Crown, 1997), p. 366.

95   The resources requirements of building cars are found in *Stuff: The Secret Lives of Everyday Things* (Seattle: Northwest Environment Watch, 1997), pp. 34, 37.

95   The Heidelberg study is summarized *Asphalt Nation,* op. cit., p. 93.

95   The story of the Huaorani is told in *Savages,* Joe Kane (New York: Knopf, 1995). The destruction of ecosystems in Ecuador by Texaco's oil operations documented by Kane continues; see Diana Jean Schemo's story in the *New York Times,* February 1, 1998.

96   The ravages of the world's waterways by oil drilling and shipping are described in *The Environmental Impact of the Car* (Seattle: Greenpeace, 1992), especially p. 36.

96   For the auto deaths in India, Brazil, and South Africa, see the *New York Times,* August 14, 1997; December 28, 1997; and January 23, 1998.

96   The data on car deaths are found in *World Resources 1996–97,* World Resources Institute et al. (New York: Oxford University Press, 1996), p. 87, and drawn from an author's interview with the public affairs department of the Federal Highway Administration.

97   Car figures in Bangkok appear in the *International Herald Tribune,* September 21, 1994.

97   Data on air pollution in Asian cities are drawn especially from *World Resources 1996–97,* op. cit., pp. 81–104 and especially the data on pp. 154–156, except for the dates on Guangzhou, which came from author's interviews with environmental officials there.

97   Future car sales in Asia are reported in the *New York Times,* June 6, 1996.

97   Mumford's warning appeared in *The Highway and the City* (New York: Mentor Books, 1963).

97   The $100 billion annual congestion loss is reported in *Smart Highways: An Assessment of Their Potential to Improve Travel* (Washington, D.C.: General Accounting Office, 1991).

98   Eighteenth Report:Transport and Environment, Royal Commission on Environmental Pollution (London: Her Majesty's Stationery Office, 1994), passim, especially chapters 1 and 14.

98   London's exceeding of WHO guidelines and the urgings of the Major government were reported by Phillip Davis on National Public Radio. The 70 percent traffic increase in London is documented in *Europe's Environment,* op. cit., p. 267; the 200 percent increase in Paris is found on p. 275.

98   Information on Paris traffic and air pollution is found in the *International Herald Tribune,* April 28, 1994, and the *New York Times,* July 14, 1995, and October 2, 1997.

100   The 1991 smog alert was reported in the *International Herald Tribune,* October 2, 1991. The deaths in 1987 were reported in the *Baltimore Sun,* July 6, 1995.

100   The smog emergencies in Athens were reported in the *International Herald Tribune,* November 1, 1992, June 16, 1994, July 4 and 9, 1996, September 11, 1997, and the *Baltimore Sun,* July 6, 1995.

101   The data on Italy's cars and traffic come from *Panorama* magazine's special report, November 9, 1995. The description of traffic problems in Rome comes from personal observation and from the following sources: *La Repubblica,* November 12, 1995; *Il Messaggero,* October 28, 1994, and November 22, 1994.

103   Czech auto ownership is reported in *World Watch,* January/February 1996, p. 27.

103   Information on car distribution internationally is found in *The State of the World 1993,* op. cit., p. 125, and the *UNESCO Courier,* October 1990, pp. 22–26.

104   The opinion poll about modern inventions was reported by the Associated Press on December 25, 1995. The data breaking down the purposes of U.S. car travel are reported in *Asphalt Nation,* op. cit., pp. 19–20. The car ownership of Americans is documented in *World Resources 1996–97,* op. cit., p. 82, and by the Department of Transportation report cited in the *New York Times,* September 21, 1997.

105   The Toles cartoon was published in the *Los Angeles Times,* July 26, 1989.

106   The story of the destruction of the rail-based mass transit system in the United States and the role of National City Lines therein is told in "America on Wheels," a PBS documentary produced by Jim Klein and Martha Wilson, aired June 1996, and most fully in "The Great Transportation Conspiracy," Jonathan Kwitney, *Harper's,* February 1981.

107   The Robert Lund quote and the role of Frances DuPont in the passage of the Interstate Highway Act come from "America On Wheels," op. cit.

107   The figures on U.S. commuters' use of mass transit come from the *Wall Street Journal*, June 29, 1993. The information on relative efficiency and subsidies of trains and autos is found in *State of the World 1993*, op. cit., pp. 121, 126, and 135.

108   The relative share of world autos in the United States is found in *Driving Forces*, op. cit., p. 11.

108   The Mexico City story was reported in the *New York Times*, February 4, 1996. The data on mental impairment among children was cited in *The Environment Support of the Car*, op. cit.

109   The role of the auto within the U.S. and world economies is based on "Hypercars: Materials and Policy Implications," Amory Lovins et al. (Snowmass, Colo.: Rocky Mountain Institute, 1995), and on an interview with Lovins by the author.

110   The history of the struggle over seat belts and other auto safety features is told in Ralph Nader's classic *Unsafe at Any Speed: The Designed-in Dangers of the American Automobile* (New York: Grossman Publishers, 1965), especially pp. 112–123 and 147–164.

111   The superefficient cars produced by auto makers are described in *Power to Change: Case Studies in Energy Efficiency and Renewable Energy* (Amsterdam: Greenpeace International, 1993) and in "Hypercars: The Next Industrial Revolution," Amory Lovins et al. (Snowmass, Colo.: Rocky Mountain Institute, February 1995) and in the annotated version of the article Lovins wrote for the *New York Times* on December 3, 1990; see note 8, which cites sources in the trade press.

111   The Ford Pinto story, including Iacocca's role, was exposed by investigative reporter Mark Dowie in *Mother Jones*, November 1977. Iacocca, who had earlier uttered the immortal phrase "Safety doesn't sell" to explain why Ford did not pay more attention to safety issues, saw the Pinto as a revolutionary advance in auto marketing. Iacocca's vision was to produce a car that would weigh under two thousand pounds and cost under $2,000; thus, he opposed the inclusion of safety features that might jeopardize these goals. For example, the tendency of Pintos to explode when struck from behind could have been treated with the addition of a part that would have cost $11 per car. But Ford decided against it after calculating that it would cost less in the long run to defend lawsuits brought by crash victims than to install the needed part. Whether Mr. Iacocca was personally aware of the crash tests done on the Pinto is unknown, though he denied as much at the time.

112   The data on fuel efficiency are found in the *New York Times*, November 28, 1995, and in *The Environmental Impact of the Car*, op. cit.

113   Lovins described his explorations into hybrids in an interview with me in May 1995. Additional details can be found in his book (with L. Hunter Lovins and Ernst Von Weizsacker) *Factor Four: Doubling Wealth, Halving Resource Use* (London: Earthscan, 1997), pp. 4–10.

113   The discussion of fuel cell electric cars and the explosion of activity following the Tokyo Motor Show is based on articles in the *Economist*, October 25, 1997, and the *New York Times*, April 22, 1997; October 8, 1997; October 21, 1997; and January 5, 1998.

115   The information about mid-1980s prototypes comes from *Factor Four*, op. cit.

116   Germany's spending plans are reported in *The State of the World 1993*, op. cit., p. 130. For reports on antiroad activism in the United Kingdom, see *The Independent*, June 9, 1998, and the Internet magazine *Salon*, June 5, 1997. The successes of Portland are described in *This Place on Earth: Home and the Practice of Permanence*, Alan Thein Durning (Seattle: Sasquatch Books, 1996) and in *World Resources 1996–97*, op. cit., p. 92; those of Curitiba, in ibid., pp. 120–121 and, more comprehensively, in *Hope, Human and Wild*, Bill McKibben (New York: Little, Brown, 1995), pp. 60–120. For the status of California's cars and air, see the *New York Times*, April 21, 1998. Carter Brandon's comments came in an in-

terview with the author. Europe's new air pollution guidelines were reported in the *New York Times*, July 3, 1998.

118   Friedrich Durrenmatt's quote is reported in the *UNESCO Courier*, October 1990, p. 23.

118   The *Scientific American* article, written by Colin Campbell and Jean Laherrere and titled "The End of Cheap Oil," appeared in the March 1998 issue.

Chapter Four

120   The conditions of Leningrad during the siege of World War II are described in the vivid, moving documentary film *Russia's War: Blood on the Snow*, produced by Victory Series and IBP Films Distribution, Ltd., broadcast on PBS, July 14, 1997.

122   The description of Leningrad's water problems is based on author's interviews with Constantin Yerukhin, member of the City Council; Anatoly Konstantinov, general director of the Association for Ecology and Culture; Aleksey Lushnikov, manager of the industry committee of the Leningrad "Polustrovo" mineral water works; Ivan Blokov, cochairman of the Leningrad Green Party; Serge Tsvetkov, a geophysicist and founder of the Delta Environmental Organization in Leningrad; and U. K. Sevenard, chairman of the interministerial party committee overseeing the construction of the dam across the Gulf of Finland.

123   That forty million people lived in areas of ecological crisis was estimated in an author's interview with Alexei Yablokov, deputy chairman of the Committee on Ecology in the Supreme Soviet of the U.S.S.R. and the widely acknowledged "ecological conscience" of the nation. The rest of the description of the environmental crisis is based on *Ecology & Perestroika: Environmental Protection in the Soviet Union*, Eric Green (New York: American Committee on U.S.–Soviet Relations, 1991), p. ix.

125   The translation of "Mayak," the quote about "most polluted spot on earth," and other information related to the Mayak complex is drawn from *Making the Russian Bomb: From Stalin to Yeltsin*, Thomas B. Cochran, Robert S. Norris, and Oleg A. Bukharin (Boulder, Colo.: Westview Press, 1995); see especially pp. 71–137. Dr. Cochran was a member of the specially invited team of foreign scientists who, in 1989, were the first outside visitors to the Mayak complex.

125–   The information on the Mayak disasters is drawn from ibid., as well as *Nuclear Waste-*
126   *lands: A Global Guide to Nuclear Weapons Production and Its Health and Environmental Effects*, Arjun Makhijani, Howard Hu, and Katherine Yih, eds. (Cambridge, Mass.: MIT Press, 1995), chapter 7. Both books drew heavily on the three-volume study undertaken on the orders of Mikhail Gorbachev, "Commission for Investigation of the Ecological Situation in the Chelyabinsk Region," 1991 (referred to in my text as "the Gorbachev report"), and on the paper "Estimate of the risk of leukemia to residents exposed to radiation as a result of a nuclear accident in the Southern Urals," by M. M. Kossenko, M. O. Degteva, and N. A. Petrushova, *PSR Quarterly*, December 1992, pp. 187–197. An advance copy of the Gorbachev report was shared with me during my visit to the Soviet Union, as were the results of the paper by Kossenko et al.

127   The sixty-six thousand figure was provided in my interview with Drs. Kossenko and Degteva.

128   The "ABC disease" code name was explained in the author's interview with Drs. Kossenko and Degteva.

129   The description of the Chernobyl accident is drawn from a number of sources including *The Truth About Chernobyl*, Grigory Medvedev (New York: Basis Books, 1991) and *Ablaze: The Story of the Heroes and Victims of Chernobyl*, Piers Paul Read (New York: Random House, 1993).

130   The desire to stonewall by some Soviet officials is described in *Ablaze*, op. cit., pp. 174–175. It should be noted that Mikhail Gorbachev later wrote in his memoirs about the Politburo meeting of April 28, "I absolutely reject the accusation that the Soviet

PAGE

leadership intentionally held back the truth about Chernobyl. We simply did not know the whole truth yet." See his *Memoirs* (New York: Doubleday, 1995), p. 189.

130   The radiation released at Chernobyl is documented in the WHO study, Health Consequences of the Chernobyl Accident (Geneva: WHO, 1995).

130   See *The Radiological Consequences in the USSR of the Chernobyl Accident: Assessment of Health and Environmental Effects and Evaluation of Protective Measures* (Vienna: International Atomic Energy Agency, 1991). Among the many critiques of the IAEA report, one of the sharpest appeared in the *Ecologist*, Vol. 21, No. 6, November/December 1991, p. 267, by Frank Barnaby.

130   The article by Michael Specter appeared in the *New York Times*, March 31, 1996. The two previous stories published by the *Times* about the WHO study appeared on November 21, 1995, and November 29, 1995. John Gofman's estimates of Chernobyl-related cancers was reported in the *New York Times Magazine*, April 14, 1991.

131   Alexander Penyagin made his comments in an interview with the author.

131–   The information on the Chelyabinsk environmental situation is drawn from a pre-
132   publication version of the Gorbachev Commission study, as translated for me in Chelyabinsk.

135   Cochran's "lethal dose" quote comes from *Soviet Nuclear Warhead Production*, Thomas B. Cochran and Robert S. Norris (Washington, D.C.: Natural Resources Defense Council, 1990), p. 20.

135   Ibid.

137   The study by Kossenko, Degteva, and Petrushka is cited, for example, in *Nuclear Wastelands*, op. cit., pp. 327–328.

138   The information on the Hanford site are drawn from "Black Water, Poisoned Lives," Mark Hertsgaard, [London] *Independent on Sunday*, December 15, 1991, and from *Atomic Audit: The Costs and Consequences of U.S. Nuclear Weapons Since 1940*, Stephen I. Schwartz (Washington, D.C.: Brookings Institution Press, 1998), p. 361.

138   Hanford's dismal environmental record is recounted in *Nuclear Wastelands*, op. cit., especially pp. 220–224.

138   The U.S. government's nuclear-related deceptions and conduct of experiments is described in *Nuclear Wastelands*, op. cit.; see especially pp. xv–xvii, 178–183, and 280–284.

139   The recommendation of "reeducation" programs to "correct" people's thinking was made in a military planning memorandum, as documented in *Fallout: An American Tragedy*, Philip Fradkin (Tucson: University of Arizona Press, 1989), p. 97.

140   The information on Oppenheimer, the other scientists, and the policies and ideology of the early nuclear era is drawn from *The Nuclear Barons*, Peter Pringle and James Spigelman (New York: Avon, 1983), passim.

141   The quotes from *Nuclear Wastelands* (op. cit.) are from pp. xiii and xxi.

141   The thyroid cancers were reported in the *New York Times*, August 2, 1997. The background on the choice of the Nevada test site, the number of worldwide nuclear tests, and the 2.4 million expected cancer deaths from nuclear testing are documented in *Radioactive Heaven and Earth*, International Physicians for the Prevention of Nuclear War and the Institute for Energy and Environmental Research (New York: Apex Press, 1991). See chapter 3, especially p. 40, and chapter 4.

142   The radioactive contamination of the Kola Peninsula is described in an on-the-spot report by Fred Barbash, published in the *Washington Post*, October 11, 1996, and a story in the *New York Times*, October 8, 1996.

142   Bill Arkin's comments were made in an interview with the author.

143   The internal Department of Energy report was disclosed in the *Washington Post*, April 17, 1993.

143   For the government's admission of explosion risks, see the report by Matthew Wald in the *New York Times*, December 25, 1992. Makhijani's comment came in an interview with the author.

143   The Department of Energy's $200 billion estimate was reported in the *New York Times,*
October 24, 1997. The information on the nuclear cleanup in the former Soviet Union
was provided by Makhijani in an interview with the author.

144   The figures on nuclear-weapons-related waste are found in *Nuclear Wastelands,* op. cit.,
pp. 582–584.

144   The information on Yucca Mountain comes from *Nuclear Wastelands,* op. cit., pp.
259–260, and from the *New York Times,* June 20, 1997.

145   The quotes from nuclear industry executives are found in *Nuclear Inc.: The Men and Money
Behind Nuclear Energy,* Mark Hertsgaard (New York: Pantheon, 1983), pp. 175–176.

146   General Horner's quote is drawn from *Nuclear Wastelands,* op. cit., p. xx. The twelve occasions when the United States threatened to use nuclear weapons are listed and documented in Daniel Ellsberg's introduction to *Protest and Survive,* E. P. Thompson and Dan
Smith, eds. (New York and London: Monthly Review Press, 1981).

146   The near nuclear war between India and Pakistan was reported in the *New Yorker,*
March 29, 1993.

146–   The paragraphs describing the dynamics of the arms race in the 1980s and the initia-
148   tives of Mikhail Gorbachev are drawn from *On Bended Knee: The Press and the Reagan Presidency,* Mark Hertsgaard (New York: Farrar, Straus & Giroux, 1988), chapter 12.

149   General Butler's remark was part of a statement sixty top military officials from
around the world released on December 5, 1996, urging elimination of nuclear
weapons. The statement was reported in *World Watch,* March/April 1997, p. 6.

149   The near nuclear war of 1995 was reported in the *New York Times,* July 6, 1997.

149–   The Blair and Schell quotes come from the excerpt from *The Gift of Time* that appeared
150   in the *Nation,* February 2, 1998.

150   The agreement between Clinton and Yeltsin about START III was reported in the *New
York Times,* March 24, 1997.

150   Background information on the loose nukes problem in Russia comes from articles in
the *New York Times,* July 1, 1997; April 20, 1996; March 13, 1996; and August 18, 1994. See
also the Op-Ed pieces by Jessica Mathews in the *Washington Post,* October 31, 1995, and
A. M. Rosenthal in the *New York Times,* November 22, 1996. Theodore Taylor's remarks
were reported in the *New York Times,* December 3, 1995.

151   The article by vanden Huevel and Cohen was published in the *Nation,* August 11, 1997.
The "joke" told by Russians was reported in the *New York Review of Books,* March 27, 1997.

152   Vladimir Nechai's suicide was reported in the *New York Times,* November 1, 1996. See
also the Op-Ed piece on November 15, 1996.

152   The number of U.S. and Russian nuclear weapons remaining and their costs are reported in *Atomic Audit,* op. cit., pp. 1–2. Regarding the revamping of nuclear doctrine,
see *The Gift of Time,* op. cit.; Brian Hall's article "Overkill Is Not Dead," the *New York Times
Magazine,* March 15, 1998; and the proposals that U.S. Senator Sam Nunn and scholar
Bruce Blair made in the *Washington Post,* June 22, 1997. For background on the
U.S.–Russian program to enhance nuclear security, see the *New York Times,* April 20,
1996. India's accusations about America's nuclear hypocrisy were reported in the *New
York Times,* June 18, 1998.

154   Makhijani's four-step program was outlined in an interview with the author.

154   The information on the fifteen unsafe nuclear reactors in the former Soviet bloc was provided by William Chandler of the Department of Energy in an interview with the author.

Chapter Five

157   The information on China's demographic history is drawn from an author's interview
with Judith Banister and from Banister's book, *China's Changing Population* (Stanford, CA:
Stanford University Press, 1988).

158   The data on past and present life expectancies in China were provided in an author's interview with Gu Baochang, associate director of the official China Population Information and Research Center. Data on life expectancies in Europe during the Industrial Revolution were provided by the Office of Population Research at Princeton University. In France, for example, life expectancy in the 1750s averaged 27.9 years; in the 1780s, 29.8 years; and in the 1820s, 38.8 years. The number of people living in poverty in China was, by 1996, at least sixty million to one hundred million, according to the World Bank, though that figure refers to people living in extreme poverty, on the edge of starvation; the bank estimated that some 350 million Chinese were subsisting on less than $1 a day. See the *New York Times*, October 26, 1996. See also the UNDP's *Human Development Report 1997* (New York: Oxford University Press, 1997), pp. 49–50.

For background on spitting in China, see the *Irish Times*, March 3, 1997, which notes the long history of spitting in China, the many public campaigns launched against it, and the role of new pollutants like ozone in worsening the problem.

161   The 70 percent figure for biomass reliance was cited in the author's interviews with numerous Chinese experts, including Wang Qingyi of the Ministry of Coal. See also World Resources Institute, *World Resources 1994–95* (New York: Oxford University Press, 1994), p. 66, and Tong Shusheng, "Present Status and Future Prospect on the Development of Biomass Energy in China," in *The Development of New and Renewable Sources of Energy in China*, ed. Chinese Solar Energy Society (Beijing: China Science and Technology Press, 1991), cited in *China's Energy and Environment in the Roaring Nineties: A Policy Primer*, Jessica Hamburger (Washington, D.C.: Pacific Northwest Laboratory, 1995), p. 71.

The information on months of energy shortages in rural China comes from Václav Smil's *China's Environmental Crisis* (Armonk, N.Y.: M. E. Sharpe, 1993), p. 103.

162   The health effects of coal smoke are described in the World Bank's report *Clear Water, Blue Skies: China's Environment in the New Century* (Washington: World Bank, 1997), especially pp. 2, 17–18 and table 203 on p. 21. The causal ranking of coal smoke is based on an author's interview with Xu Zhaoyi, chief epidemiologist of the Liaoning Environmental Protection Bureau and on his article in *Lung Cancer*, 14 Supplement 1 (1996) S149–160.

162   The 75 percent figure from He Kebin also was provided in an author's interview.

166–   The figures on China's energy supplies and the role of coal come from *China Energy An-*
167   *nual Review 1996*, published by the Department of Resource Conservation & Comprehensive Utilization, State Economic and Trade Commission, Beijing, 1997, passim.

167   The information on electricity shortages and the quote from Lang Siwei came from an author's interview.

167   Zhou Dadi's remark came during an author's interview.

167   The data on home appliance ownership in China come from *China Energy Annual Review 1996*, op. cit., p. 106.

168   The 9 percent growth of electricity demand from 1984 to 1994 is documented in *China Energy Annual Review 1996*, p. 91. The 7 percent projection was supplied in an author's interview with Pan Baozheng, a senior engineer with the State Science and Technology Commission and a power sector expert with China's administrative center for Agenda 21. The quote from Lang Siwei came in an author's interview.

168   The power plant construction plans described by Pan Baozheng came in an author's interview. The comparison to Louisiana is based on figures supplied by Jim Owen of the Edison Electric Institute in Washington, D.C.

168   Regarding China's coal production, Wang Qingyi, a senior official in China's Coal Ministry, said in an interview with the author that production would increase nearly 50 percent by 2010, to 1.8 billion tons. Noting that the World Bank had projected total production of three to four billion tons by 2050, Wang said he and his ministry col-

leagues were confident that three billion tons would be the maximum. The comment by Chandler also comes from an author's interview. The World Bank's prediction referred to by Wang was made in *China: Issues and Options in Greenhouse Gas Emissions Control* (Washington, D.C.: World Bank, 1994), Summary Report, p. 22.

169   The sulfur dioxide figures for Taiyuan come from *China 2020: Development Challenges in the New Century* (Washington, D.C.: World Bank, 1997), p. 72. For Chongqing, the average of 320 was provided in an author's interview with officials of the local Environmental Protection Bureau; the six hundred figure comes from *Clear Water, Blue Skies*, op. cit., p. 22.

169   The data on acid rain losses in China come from *Clear Water, Blue Skies*, op. cit., pp. 6, 22, and 104.

169   The $2.8 billion figure was reported in *Journal of Commerce*, January 5, 1994. The quote from Wang Wenxing came in an author's interview. The effects on Japan and South Korea are reported in *Powering China: The Environmental Implications of China's Economic Development*, Daniel C. Esty and Robert Mendelsohn, working paper in progress, Yale University, 1998.

169   The relative greenhouse potency of different fossil fuels is documented in *Climate Change 1995*, op. cit., Working Group II, p. 589. Specifically, coal releases 25 kilograms of carbon dioxide per gigajoule (GJ) of energy use, petroleum 20 kilograms, and natural gas 14 kilograms. A gigajoule is about how much energy the average car consumes to travel 100 kilometers. The 17 percent figure for petroleum is found in *China Energy Annual Review 1996*, op. cit., p. 102. The 165 billion tons of coal reserves is cited in William Chandler's testimony before the Committee on Energy and Natural Resources of the U.S. Senate, March 11, 1992. The expansion of China's carbon dioxide emissions between 1990 and 2025 and its effect on the global emissions picture are calculated in "U.S.–China Cooperation for Global Environmental Protection: A Primer on Economics and Energy Efficiency in China," William U. Chandler, Zhou Dadi, and Jessica Hamburger (Washington: Battelle, Pacific Northwest Laboratories, 1993). The 50–70 percent reductions recommended by the IPCC scientists are found in the IPCC's Second Assessment Report, Policymaker Summary, 1995, p. 9.

170   The impact of climate change on China is summarized in *China: Issues and Options in Greenhouse Gas Emissions Control*, Summary Report, op. cit., pp. 1, 10–12.

170   The disappearance of Benxi from satellite photos, as well as its subsequent reappearance, the attempted cleanup, and the high TSP levels that remained were all described by Chen Qi, director of the Liaoning Provincial Environmental Protection Bureau in an author's interview. Chen said that the government had invested five hundred million yuan (approximately $62.5 million) in the cleanup of Benxi since 1989, some of which was used to close polluting factories and transfer their workers to new jobs. One result was that emissions of TSP in Benxi fell from fifty tons per square kilometer every month in 1989 to forty tons a month in 1996. By contrast, said Chen, Norway emitted a mere eight tons per square kilometer over the entire twelve months of 1996, "so we still have a long way to go."

171   The extraordinarily high TSP levels in Taiyuan in winter are noted by William Chandler of the Battelle, Pacific Northwest Laboratories, in his testimony before the Committee on Energy and Natural Resources of the U.S. Senate, March 11, 1992.

172   All the information about Chongqing's air pollution, especially its sulfur dioxide levels, provided by Peng Zhong Gui came in an author's interview.

174   The 4.5 percent figure comes from *Clear Water, Blue Skies*, op. cit., p. 94.

180   The "Is your stomach too full?" quote came during an author's interview with a journalist who must remain nameless.

180   Orville Schell's quote came during an author's interview.

PAGE

180   Liang Conjie's remarks came during an author's interview.

181   The increase in environmental coverage by China's official news media was documented by Friends of Nature in the report made available to the author—"Survey on
Environmental Reporting in Chinese Newspapers (1995)"—and was confirmed sepa-
rately in interviews with Chinese journalists and officials. The quote from Yu Yuefeng
and the description of the environmental tours conducted by NEPA come from an author's interview with Yu.

182   The information on coal washing in China and on China's limited water supply and
the remarks by Yu Yuefeng came from an author's interview with Yu. That China's per
capita availability of water is one-third of the world average was derived from data in
World Resources: A Guide to the Global Environment, World Resources Institute (New York: Oxford University Press, 1996), pp. 306–307.

182   The quote from Wang Wenxing comes from an author's interview.

182   The 1 percent extra cost of electrostatic precipitators was described in an author's interview with William Chandler. The 1985 requirement and 90 percent TSP reduction
came from an author's interview with Wei Fengshun, deputy director of China's National Environmental Monitoring Center.

182   The dangers of small TSP and the idea of locating power plants at coal mine sites were
described in an author's interview with Wei Fengshun.

183   The history and data concerning China's increase in energy efficiency, especially during the 1980s, were provided in author's interviews with Wang Qingyi, Zhou Dadi,
William Chandler, Mark Levine, and Jonathan Sinton.

184   The additional improvements in energy efficiency available to China were described in
author's interviews with Zhou Dadi, Mark Levine, Jonathan Sinton, and Robert Taylor. The World Bank dispersed $63 million to China in March 1998 to help set up the
energy service companies described by Mr. Taylor.

184   China's relative investment in energy efficiency on the one hand and energy supply
expansion on the other were described in author's interviews with Zhou Dadi and
Mark Levine.

185   The belief that market discipline makes continued government investment in energy
efficiency unnecessary was mentioned in author's interviews with a number of
sources, including Mark Levine and a Western consultant (not Levine, but rather the
source of the "holier than the pope" quote) who must remain nameless. The thinking behind Chinese power plant purchases was described in an author's interview with
Pan Baozheng.

185   The Sinton quote and the figure of 430,000 industrial boilers in China came in an author's interview.

186   The paragraph on alternative fuel sources is based on a number of sources, including
author's interviews with William Chandler, Zhou Dadi, Pan Baozheng, Robert Taylor,
and Mark Levine. The data cited are found in *China's Energy Annual Review 1996,* op. cit.,
and *China: Issues and Options in Greenhouse Gas Emissions Control,* op. cit. The former report
documents on p. 102 that coal accounted for 75 percent of China's energy consumption in 1994; crude oil, for 17.4 percent; natural gas, 1.9 percent; and hydropower, 5.7
percent. The latter report calculates that nonfossil fuels (solar, wind, nuclear, and hydropower) could under the most optimistic scenario "provide nearly 40 percent of
China's electricity by 2020, equivalent to about 16 percent of total energy" (p. 28).

186   President Clinton described his meeting with Jiang Zemin to Thomas Friedman, who
reported it in the *New York Times,* April 17, 1996.

186   The "Global warming is not on our agenda" quote came in an author's interview. Besides the unnamed official in Chongqing, other officials and scientists in China who
considered sulfur dioxide a more urgent problem than carbon dioxide included Wang

Wenxing of the Chinese Research Academy of Environmental Sciences and an un-named senior official of the Environmental Protection Bureau in Guangzhou.

187 The 10 percent figure for China's energy consumption compared with that of the United States comes from "U.S.–China Cooperation for Global Environmental Protection," op. cit., p. 8. The "hold the world for ransom" quote comes from an author's interview with a Western consultant who must remain nameless.

187 Zhou Dadi's "the straw that breaks the camel's back" quote comes from an author's interview.

Chapter Six

195 The World Bank's health package was described by bank president Lewis Preston in the *International Herald Tribune*, September 21, 1994.

195 The comparison of Brazilian and industrial infant mortality rates comes from *Brazil: The Once and Future Country*, Marshall C. Eakin (New York: St. Martin's, 1997), p. 108.

The forty thousand daily children's deaths is reported in the World Health Organization's 1992 report *Our Planet, Our Health* (Geneva: World Health Organization).

The UNDP's income ratios are found in its *World Development Report 1997* (New York: Oxford University Press), p. 9.

196– The quote from Durning's *How Much Is Enough?: The Consumer Society and the Future of the*
197 *Earth* (New York: W.W. Norton, 1992) is from p. 51. The developed nations' consumption of energy is reported in *Climate Change 1995*, op. cit., Working Group II, p. 83. The data on U.S. fossil fuel consumption and carbon emissions is from *Stuff: The Secret Lives of Everyday Things*, John C. Ryan and Alan Thein Durning (Seattle: Northwest Environment Watch, 1997), p. 68. The use of fresh water by Americans and Senagalese is reported in *Sustaining Water: Population and the Future of Renewable Water Supplies*, Robert Engelman and Pamela LeRoy (Washington, D.C.: Population Action International, 1993), p. 15. The U.S.–China comparison comes from *World Population Data Sheet* (Washington, D.C.: Population Reference Bureau, 1996). The ratios of the environmental impacts linked to American, Brazilian, and Indian babies are found in *The Population Explosion*, Paul and Anne Ehrlich (New York: Simon & Schuster, 1990), p. 134. Paul Ehrlich's remark about the United States being the most overpopulated country in the world came in an interview with the author. The U.S. need to increase energy efficiency by 50 percent was noted in the opinion article by Bill McKibben in the *New York Times*, March 9, 1998.

197 Eberstadt's quote is from page 21 of *The True State of the Planet*, Ronald Bailey, ed. (New York: Free Press, 1995).

198 The drop in infant and child mortality rates is noted in "Population by the Numbers: Trends in Population Growth and Structure," Carl Haub and Martha F. Riche, in *Beyond the Numbers*, Laurie Ann Mazur, ed. (Washington, D.C.: Island Press, 1994), pp. 95–108, cited in "New Perspective on Population: Lessons from Cairo," Lori S. Ashford, in *Population Bulletin*, March 1995, Vol. 50, No. 1, p. 12.

198– The evolution of developing country governments' views of population policies is
199 well described in George D. Moffett's *Critical Masses* (New York: Viking, 1994); see especially chapter 7, and, regarding the Guatemalan educational example, pp. 162–163; the Council on Foreign Relations's report, *Negotiating Survival: Four Priorities After Rio*, is quoted on p. 2. The unemployment figures are provided in *Population, Resources and the Environment: The Critical Challenge*, UNFPA (New York: United Nations, 1991) and in a report by the UN's International Labor Organization that was reported in the *New York Times*, November 26, 1996. That 70 percent of population growth in the 1990s would take place in cities was predicted in the UNFPA's *Population Issues Briefing Kit 1994* (New York: UNFPA, 1994). Iran's policy shift was reported in the *New York Times*, September 8, 1996.

PAGE

199   The data on the state of the world's population in 1992 are drawn from an author's interview with Carl Haub of the Population Reference Bureau in Washington, D.C.

200   The 150 million figure is from an author's interview with Carl Haub.

201   Adrian Cowell's remark is found in his book, *The Decade of Destruction: The Crusade to Save the Amazon Rainforest* (New York: Henry Holt, 1990), p. 207.

202   The data on the size and relative rate of destruction of the tropical forests in Brazil and around the world came from "Tropical Forests: The Main Deforestation Fronts," Norman Myers, in *Environmental Conservation*, Vol. 20, No. 1, Spring 1993; and from "Fires in the Amazon: An Analysis of NOAA-12 Satellite Data, 1996–1997," and *Global Deforestation, Timber and the Struggle for Sustainability*, both by Stephan Schwartzman of the Environmental Defense Fund, 1997. The data on nontropical forests are found in *Vital Signs* (New York: W. W. Norton, 1998), p. 124. Forest loss of more carbon than absorbed was reported in Janet N. Abramovitz's *Taking a Stand* (Washington, D.C.: Worldwatch Institute, 1998), p. 14.

202   The information about the number and location of species in tropical forests is drawn from *The Diversity of Life*, E. O. Wilson (Cambridge: Harvard University Press, 1992), especially pp. 132 and 197, and from the *Global Biodiversity Assessment*, R. T. Watson et al. (Cambridge: UNEP and Cambridge University Press, 1995) and *Climate Change 1995*, Working Group II, op. cit., p. 26, 111–112.

202   The figures and quote from Wilson come from *The Diversity of Life*, op. cit., p. 280. The 1,000 to 10,000 times faster than background rate estimate is from *Global Biodiversity Assessment*, op. cit., Summary for Policymakers, p. 2.

203   In his autobiography, Wilson recounts the story of being one of seven Harvard professors asked by the editors of *Harvard Magazine* in 1980 to identify the single most important problem facing the world in the 1980s; four cited poverty, one the global nuclear threat, and one excessive government control. Wilson noted that, outside of all-out nuclear war, the loss of biodiversity would pose far greater dangers than the others, for they could be repaired in a matter of generations, whereas the loss of biodiversity would take millions of years to correct. See p. 355 of *Naturalist*, Edward O. Wilson (Washington, D.C.: Island Press, 1994). See also *The Sixth Extinction: Patterns of Life and the Future of Humankind*, Richard Leakey and Roger Lewin (New York: Doubleday, 1995), and *Human Population, Biodiversity and Protected Areas: Science and Policy Issues*, Victoria Dompka, ed. (Washington, D.C.: American Association for the Advancement of Science, 1996).

203   The WCU study was reported in the *New York Times*, April 9, 1998.

203   The nine of ten pharmaceuticals line comes from "No Middle Way on the Environment," Paul R. Ehrlich, Gretchen C. Daily, Scott C. Daily, Norman Myers, and James Salzman, *Atlantic Monthly*, December 1997, p. 101. Professor Janzen's remark appeared in *Time*, January 2, 1989.

204   The quote from "The Little Things That Run the World" is found on p. 144 of *In Search of Nature*, Edward O. Wilson (Washington, D.C.: Island Press, 1996).

205–   The quotes from Myers and Schwartzman are from author's interviews. Slightly dif-
206   ferent figures for land distribution in Brazil are found in *Brazil: The Once and Future Country*, by Marshall C. Eakin (New York: St. Martins, 1997), p. 106; citing official Brazilian census data, Eakin reports that in the mid-1980s, less than 1 percent of all farms accounted for more than 40 percent of all occupied farmland, while 50 percent of farms constituted 3 percent of occupied farmland.

       Harrison's 27 percent figure comes from p. 95.

206   The 539 assassinations were reported in *Rolling Stone*, "Rain Forest Journal," Tom Hayden, January 1, 1991, based on data compiled by the Pastoral Land Commission. The 411 killings in Pará were reported in *The Decade of Destruction*, op. cit., p. 138. For additional background on Brazil, see *The World Is Burning: Murder in the Rain Forest*, Alex Shoumatoff (Boston: Little, Brown, 1990); *The Fate of the Forest: Developers, Destroyers and*

PAGE

Defenders of the Amazon, Susanna Hecht and Alexander Cockburn (London: Penguin, 1990); and *Brazil: The Once and Future Country*, op. cit.

For more on Chico Mendes, see *The Decade of Destruction*, op. cit., especially pp. 133–134, 173–174, and 184–185; *The World Is Burning*, op. cit., especially pp. 67 and 87–90; and *The Fate of the Forest*, op. cit., especially pp. 206–216.

208   Schemo's report was in the *New York Times*, September 19, 1995. See also the *Times* of April 21, 1996. For the connection between the 1996 and 1998 massacres in Pará, see *U.S. News & World Report*, April 13, 1998.

209   Among the many reports on the rainforest fires in Indonesia see especially articles in the *New York Times*, September 25, 1997, November 2, 1997, and February 23, 1998, the latter of which predicted that the 1998 fires would be the worst ever. The greenhouse comparison between Indonesia forests and Europe's industry came from *Taking a Stand*, op. cit., p. 14.

For a comprehsive analysis of forest loss in Asia and throughout the world, see *The Third Revolution* and *Global Deforestation, Timber, and the Struggle for Sustainability*, both op. cit., as well as *State of the World Forests 1997* (Rome: FAO, 1997); *The Last Frontier Forests: Ecosystems and Economies on the Edge*, Dirk Bryant, Daniel Nielsen, and Laura Tangley (Washington, D.C.: World Resources Institute, 1997); *Plundering Paradise: The Struggle for the Environment in the Philippines*, Robin Broad with John Cavanagh (Berkeley: University of California Press, 1993), especially chapter 3; and *Population Growth, Poverty, and Environmental Stress: Frontier Migration in the Philippines and Costa Rica* (Washington, D.C.: World Resources Institute, 1992).

210   The fires in the Amazon were described by Diana Jean Schemo of the *Times* on November 2, 1997, following reports on September 12, 1996, and October 12, 1995.

210   The watering down of Brazil's environmental enforcement law was described in the *New York Times*, January 29, 1998, and in an author's interview with Schwartzman.

211   The zero population growth statement comes from *Population Summit of the World's Scientific Academies* (Washington, D.C.: National Academy Press, 1994).

211   The data on India's per capita land area are found in *Conserving Land: Population and Sustainable Food Production*, Robert Engelman and Pamela LeRoy (Washington, D.C.: Population Action International, 1995).

211   The Engelman and LeRoy quote, as well as the information on areas of water shortage around the world, come from *Sustaining Water*, op. cit., as well as "A Second Update" to the report issued in December 1997, p. 3.

212   The Dennis Avery quote is found on p. 67 of *Critical Masses: The Global Population Challenge*, George D. Moffett (New York: Viking, 1994); see also chapter 2.

212   The percentage decline in malnourishment is noted in *Conserving Land*, op. cit., p. 12. Sen's statement is found in "Population: Delusion and Reality," *New York Review of Books*, January 22, 1994. Property ownership in Latin America, Kenya, et al is described in *The Third Revolution: Environment, Population and a Sustainable World*, Paul Harrison (London and New York: Penguin Books, 1992), especially chapter 9.

214   Pimentel's remarks came in an interview with the author. For sources on the data supplied in remainder of the paragraph, see the paper he coauthored, "Will Humans Force 'Nature' to Control Their Numbers Within the Limits of the Earth's Resources?" Pimentel et al., working paper in publication.

214   The figures on future food production needs, limitations imposed by irrigation, and erosion problems are found in the fact sheets accompanying the FAO's *Report of the World Food Summit*, November 13–17, 1996, except for the 0.17 hectares per capita and one-third of all crops grown on irrigated land data, which are found in *Conserving Land*, op. cit., pp. 4 and 20, respectively, and the 3.7 billion acres figure, which was published by the United Nations and cited in the *New York Times*, June 17, 1997. The World Bank's findings regarding the Punjab were reported in the *New York Times*, August 10, 1995.

PAGE

214– The data on grain and fish production came from *Vital Signs 1998*, op. cit., pp. 29
215 and 35.

215 The decline in per capita food production is noted in the *FAO Production Yearbook 1993*,
Vol. 47 (Rome: FAO, 1994). The data on fisheries also come from the FAO, but as ana-
lyzed by the Worldwatch Institute in its *Vital Signs 1997* report, p. 33, and its *State of the
World 1998* report, p. 60.

215 The 841 million figure and Kay Killingsworth's quote appeared in the *New York Times*,
November 13, 1996. For the full background, see *Report of the World Food Summit*, op. cit.,
including technical background documents 1–15 (Rome: FAO, 1996).

216 The data on Third World birth rates, contraceptive use, and pregnancy-related deaths
come from "The State of World Population," by UN Population Fund, 1997, especially
pp. 2, 33, and data tables. Regarding China and India, see also *Reproductive Rights and
Wrongs: The Global Politics of Population Control*, Betsy Hartmann (New York: Harper & Row,
1987).

218 For background on the Cairo conference, see *Program of Action* (New York: UNFPA,
1996); *Critical Masses*, op. cit.; "New Perspectives on Population," op. cit.; "The Cairo
Conference on Population and Development," C. Alison McIntosh and Jason L. Fin-
kle, *Population and Development Review*, Vol. 21, No. 2, June 1995; and *The State of World Population 1995* (New York: UNFPA, 1995).

For a comprehensive, insightful description of Kerala's successes in family planning and poverty alleviation, see *Hope, Human and Wild*, Bill McKibben (New York: Little,
Brown, 1995).

The comments of Nafis Sadik and Jyoti Shankar Singh came during author's interviews.

219 The article, "A Manageable Crowd," appeared in *The New Yorker*, September 12, 1994.

219 For contraceptive use, see *The State of World Population 1997* (New York: UNFPA, 1997).
The 850 million figure is found in *Critical Masses*, op. cit., p. 15. The 40 percent figure
comes from "Human Population Prospects: Implications for Environmental Security," Robert Engelman, Woodrow Wilson Center Report, Spring 1997, p. 52.

219– For data on developed and developing country funding of family planning pro-
220 grams, see "Family Planning Expenditures in 79 Countries: A Current Assessment,"
Shanti R. Conly, Nada Chaya, and Karen Helsing (Washington, D.C.: Population Action International, 1995), as well as the follow-up report issued in September 1996. The
vote in Congress was reported in the *New York Times*, September 12, 1996.

Chapter Seven

222 Mosher's story is found on pp. 260–261 of *The Broken Earth: The Rural Chinese* (New York:
Free Press, 1983).

224 The Central Committee's announcement is reported in Václav Smil's *China's Environmental Crisis*, op. cit., p. 27.

225 The figures on China's population growth come from the *China Statistical Yearbook 1997*.

225 Lavely's statistics were confirmed in a private communication with the author. The
data on India were provided in a private communication with Judith Banister, author
of *China's Changing Population* (Palo Alto: Stanford University Press, 1983). Nicholas
Kristof and Sheryl WuDunn's story is found on p. 230 of their *China Wakes: The Struggle
for the Soul of a Rising Power* (New York: Times Books, 1994).

226 The 37 percent figure is found in the FAO's "World Food Summit Report," Technical
Background Document 4, p. 29. The Western demographers I consulted included
William Lavely of the University of Washington and Judith Banister of the U.S. Cen-
sus Bureau.

226– The exiling of Ma Yinchu is described in Richard Bernstein's *From the Center of the Earth:
227 The Search for Truth About China* (New York: Little, Brown, 1982), p. 51, and in Smil's *China's*

Environmental Crisis, op. cit., p. 19. A variation of the story, along with the "two hands attached" quote, was told in an author's interview with Chen Qi, director of the Environmental Protection Bureau in Liaoning.

The description of the later, longer, fewer campaign is also based on Smil's account, as well as interviews with Gu Baochang and Judith Banister, who described more re cent developments.

Jiang Zemin's one-child policy remark was reported in the *Washington Post,* March 21, 1995.

228 That local officials are happy to pocket fines for "extra" births was confirmed in author's interviews with numerous peasant families; see also the *New York Times,* August 17, 1997. The surveys cited by Banister were mentioned in a private communication with the author.

228 The per capita resources figures come from *Clear Water, Blue Skies,* op. cit., p. 6.

228 The figures on water shortages are drawn from ibid., pp. 6 and 88, as well as from the *New York Times,* November 7, 1993, from the July/August 1998 issue of *World Watch,* and from author's interviews with Vermeer and Yu.

229 Lavely's "all-purpose alibi" quote is from an interview with the author.

229– Chen's "historical error" quote came in an interview with the author.
230

231 China's one car per five hundred inhabitants ratio is drawn from the World Bank's *Clear Water, Blue Skies* report, op. cit., p. 74.

233 The data on electricity production came from an author's interview at the Guangzhou Environmental Protection Bureau. The data on Hong Kong's investment and industrialization came from an author's interview with Ma Xiaoling, a senior official with the South China Institute of Environmental Sciences, who directed a cross-border project on environmental protection.

235 The 300 TSP reading of Guangzhou came from the author's interview described in the text. The 82 figure for Hong Kong is from *World Resources 1996–97,* op. cit., p. 154.

235 The description of rubbish scavenging in Victoria Harbor is based on the 1996 annual report of the Hong Kong government's Environmental Protection Department. The data on sewage disposal in the harbor came from an author's interview with a senior scientist in the department.

236 The figures on sewage disposal in the Pearl River and the river's supply of drinking water for Hong Kong came from an author's interview with Dr. Ma.

237 The figures on Guangzhou's population came from the author's interview described in the text.

238 The range of estimates of China's floating population came from various sources, including Judith Banister's paper "China: End of Century Population Dynamics," prepared for the conference "Tomorrow: Mainland Development Under the Ninth Five-Year Plan, June 1996, Kaohsiung, Taiwan." The data on Third World cities came from *Critical Masses,* op. cit., pp. 28–31.

244 The urban population projections are cited in *Clear Water, Blue Skies,* op. cit., p. 76.

244 The 7.9 square meter figure came from an author's interview with Lang Siwei of the Air Conditioning Institute of the Chinese Academy of Building Research.

245 The data on vehicle ownership and trends are found in *Clear Water, Blue Skies,* op. cit., pp. 73–85. The cancellation of mass transit projects was reported in the *New York Times,* May 4, 1996. The highway construction program was described by Chinese officials in interviews with the author and in the *Journal of Commerce,* January 18, 1995.

245 The summary of farmland loss in China is based primarily on two articles by Václav Smil: "Environmental Problems in China: Estimates of Economic Costs" (Honolulu: East-West Center, 1996), p. 56, and "Environmental Change as a Source of Conflict and Economic Losses in China," op. cit.

245   Lester Brown's original article was published in *World Watch*, September/October 1994, under the title "Feeding China." An expanded version was published in book form as *Who Will Feed China?* (New York: W. W. Norton, 1995).

246   China's reaction to Brown's analysis was reported in his book, ibid., as well as in the *China Daily*, December 7, 1996.

247   Václav Smil's criticisms of Brown came in an article he wrote in the *New York Review Of Books*, February 1, 1996, and in an author's interview.

247   The NIC analysis was reported in the *South China Morning Post*, April 26, 1998. A copy of the report, "China Agriculture: Cultivated Land Area, Grain Projections, and Implications," was made available to the author.

248–   Joshua Muldavin's comments are drawn from an author's interview and from Mul-
249   davin's article "The Political Ecology of Agrarian Reform in China," in *Liberation Ecologies. Environment, Development, Social Movements*, R. Peet and M. Watts, eds. (New York: Routledge, 1996).

249   The figure of 80 percent of China's rivers comes from the UN's Food and Agriculture Organization, as cited in *World Watch*, July/August 1998.

250   China's participation in the 1992 Earth Summit was described in author's interviews with Huang Jing, a deputy director of the Administrative Center for China's Agenda 21, and with Barbara Finemore, an official of the UN Development Program who worked with Chinese officials on the preparation of China's Agenda 21 document for the summit. China's role was also noted in *World Resources 1994–95*, op. cit., p. 64. The sustainable development speeches by Li Peng and Jiang Zemin were reported by the Xinhua News Agency on July 15 and July 16, respectively, of 1996. The "We can choke on the air" quote from Li, however, comes from a different speech and was recited to the author by Zhang Kunmin, the deputy administrator of NEPA.

250   Liang Conjie's description of environmental law breaking in China came during an author's interview.

251   The human cost of China's floods has been widely reported, including in the *New York Times*, September 15, 1996 and August 26, 1998. See also the e-mail newsletter of Three Gorges Probe, an NGO opposed to the giant dam project, for August 26, 1998. Li Yining's remark about "inadequate ecological protection" came during an author's interview. The *China Daily*'s estimate of 7 percent GDP equivalence for environmental costs was noted in the *South China Morning Post*, September 22, 1995. The World Bank's estimate is found in *Clear Water, Blue Skies*, op. cit., p. 23. Václav Smil's estimate comes from his essential study, "Environmental Problems in China: Estimates of Economic Costs," op. cit.

252   The description of China's environmental laws and attendant budgets by Ye Ruqiu came during an author's interview.

252–   The information on unprofitable state enterprises and unemployment in Chongqing
253   was provided by Hu Jiquan in an author's interview.

253   For background on the rising numbers of demonstrations and strikes in China, see the April 1997 *Harper's*, p. 16; the July 18, 1997, *New York Times;* the March 27, 1997, *New York Review of Books;* and especially the September 11, 1997, *Washington Post*. The "We Don't Want Democracy" poster was described by Zhang Ji-qiang of the W. Alton Jones Foundation in an author's interview.

253–   Background on the Tiananmen Square uprising and its significance comes from a va-
254   riety of sources, including an author's interview with Jasper Becker of the *South China Morning Post*. Deng Xiaoping's warning to party leaders after the uprising was reported to me by a Chinese official who declined to be named.

254   Wei Jingsheng's experiences in prison are described in his stirring book of letters from jail, *The Courage to Stand Alone* (New York: Viking, 1997).

254   The quotes about the Huai River pollution and popular unrest from an unnamed se-

PAGE

255   nior government official ("There were social revolts . . ." and "For years, no boy . . .") came in an author's interview. The exchange among Song Jian, the brave peasant, and the local and regional leaders was described by a Chinese environmental expert who must remain nameless; the same expert also explained the situation of peasants working at TVEs and the thinking of the government in placating them; in addition, parts of the Song Jian and Huai story were reported in the *Irish Times*, October 26, 1996.

255   The sixty thousand factory closings ordered as part of the State Council's environmental campaign in 1996 were described by Ye Ruqiu and Zhang Kunmin, both deputy administrators of the NEPA, in author's interviews. Estimates about the TVE's share of China's total pollution were provided in author's interviews with Zhang Kunmin (30 percent) and Hu Jiquan (5 percent) and in *World Resources 1994–95*, op. cit. p. 77.

256   The pledge to end state ownership was reported in the September 12, 1997, *New York Times.* The spread of social unrest was reported most vividly in the September 11, 1997, *Washington Post,* which recounted numerous incidents where angry citizens and riot police clashed, including one episode in which one hundred people were injured in Mianyang, Sichuan.

256   Chen Qi's quote came in an author's interview.

257   Ye Ruqiu's quote "I think the understanding . . ." came in an author's interview.

257   Zhang Kunmin's "a believer in sustainable development" and other quotes came from an author's interview.

257   "This is the terrible dilemma . . ." quote came from an author's interview.

259   The World Bank's quote about capital stock comes from *China 2020*, op. cit., p. 74. Smil's quote is from an author's interview.

Chapter Eight

260   "I have become . . ." is from *Earth in the Balance*, Al Gore (New York: Houghton Mifflin, 1992), p. 15.

263–   The description of events at the Earth Summit is based on the author's on-site obser-
264   vations and reporting. See also *A New Name for Peace: International Environmentalism, Sustainable Development, and Democracy*, Philip Shabecoff (Hanover and London:University Press of New England, 1996), chapter 11.

264   Agenda 21 was published commercially as *Agenda 21: The Earth Summit Strategy to Save Our Planet*, Daniel Sitarz, ed. (Boulder, Colo.: Earthpress, 1994).

266   "The fate of the earth . . .": Strong interview in *New World Journal*, Spring 1993.

266   For background on the Earth Summit + 5 conference, see the *New York Times*, April 7, June 17, and June 28, 1997, as well as the *Washington Post,* June 22, 1997.

266–   The remarks by William Ruckelshaus are found in his article "Toward a Sustainable
267   World," in *Scientific American*, September 1989.

267–   The scientists' warning about East Coast beaches was reported in the *New York Times*,
268   September 18, 1995. The *Financial Times* ad appeared on October 25, 1995. Unless otherwise noted, the additional information and quotes reported here come from *Climate Change and the Financial Sector*, Jeremy Leggett, ed. (Munich: Gerling Akademie Verlag, 1996).

274   "A fundamental change of approach": *Greenpeace Business*, February/March, 1998 p. 3.

275   "We are prisoners . . .": The interview by Marlise Simmons of the *Times* was reprinted in the *San Francisco Chronicle*, February 13, 1994.

277–   Among the studies estimating humanity's use of the global output of photosynthesis
278   are "Human Appropriation of the Products of Photosynthesis," Peter Vitousek et al., *BioScience* 34, no. 6, pp. 368–373, 1986. See also *Food, Energy and Society*, D. Pimentel and M. Pimentel (Boulder, Colo.: Colorado University Press, 1996).

PAGE

279   The information on sulfur dioxide emissions trading comes from the *New York Times*, March 23, 1996, an opinion article by Jessica Mathews in the *Washington Post*, June 17, 1996, and the *Nation*, March 24, 1997.

279   "When fifty-one of the one hundred biggest economies...": *The Top 200: The Rise of Global Corporate Power*, Sarah Anderson and John Cavanagh (Washington, D.C.: Institute for Policy Studies, 1996).

283   U.S. military spending levels were reported in the *New York Times*, October 28, 1996.

284   The fossil fuel lobby's PR campaign and the other information on the lead-up to Kyoto is drawn from various newspaper accounts, as well as Ross Gelbspan's report on the *Mother Jones* Web site, op. cit.

284   Lee Raymond's remarks and the fossil fuel lobby's borrowing of its strategy from OPEC were reported by Ross Gelbspan on the Web site of *Mother Jones*, December 16, 1997, and in the *New York Times*, December 12, 1997.

284   Clinton's climate change policy was announced on October 6, 1997, and reported in the *New York Times* the following day; McKibben's article appeared on October 8, 1997.

285   Lester Brown's "pathetic" quote came from an interview broadcast on ABC News the day before the Kyoto conference began, and was later confirmed in an author's interview. The EU and coal industry views were reported in Curtis Moore's article in "Outlook" of the *Washington Post*, December 14, 1997.

285–   The analysis of the Kyoto Protocol's loopholes is based on author's interviews with
286   various NGO observers of the talks, including John Passacantando of Ozone Action, Philip Clapp of the National Environmental Trust, and Bonizella Biagini of Climate Action Network, Europe, as well as the report by Seth Dunn in *World Watch*, March/April 1998. Gene Sperling's quote was reported in the *Washington Post*, December 12, 1997.

286   Gelbspan's remarks were made in his article "A Puny Beginning" for the *Mother Jones* Web site, op. cit. The remark by the Chinese delegate was reported in the press and relayed to me by Kelly Simms of the American NGO Ozone Action.

287   The investment plans of Royal Dutch Shell and BP were reported by Gelbspan in *Mother Jones*, op. cit., and in the *New York Times*, December 12, 1997.

287   Gore's TV labeling program was reported in the *New York Times*, January 9, 1998. The quote from *Earth in the Balance* is from p. 274.

288   "The maximum that is...": *New York Times Magazine*, July 23, 1995. Gore's "mortal threat" quote comes from *Earth in the Balance*, op. cit., p. 325.

Chapter Nine

292   The formal report on the Gallup poll was called *Health of the Planet*, Riley E. Dunlap, George H. Gallup Jr., and Alex M. Gallup (Princeton: George H. Gallup International Institute, 1993).

293   *Newsweek*'s poll was brought to my attention by Ross Gelbspan, who wrote about it in the June 1998 issue of the *Atlantic Monthly*.

293   The full citation for Gregg Easterbrook's book is *A Moment on the Earth* (New York: Viking Penguin, 1995). The very favorable review of the book in the *New York Times Book Review* of April 23, 1995 made no mention of its many factual errors, many of which also passed undetected by the legendary fact-checkers of *The New Yorker* when that magazine published an excerpt from the book on April 10, 1995, to mark the twenty-fifth anniversary of Earth Day. The review in *Natural History* by Jack C. Schultz, a professor of entomology at Penn State University, appeared in the August 1995 issue. The review in the Scientific American of February 1996 was written by Thomas Lovejoy, the Assistant Secretary for External Affairs of the Smithsonian Institution. The corrections compiled by the Environmental Defense Fund were published under the title "A Mo-

PAGE

ment of Truth: Correcting the Scientific Errors in Gregg Easterbrook's *A Moment on the Earth*," volumes one and two, available from EDF's national headquarters, 257 Park Avenue South, New York, NY 10010.

294 The quote "as an ugly bride . . ." and the accompanying quotes and information on reporting Beijing's pollution are from the *South China Morning Post,* March 29, 1998.

296 Lester Brown's remark about Pearl Harbor came in an author's interview.

296 That nine of the eleven years previous to 1998 ranked as the warmest in recorded history was reported by the National Climatic Data Center of the National Oceanic and Atmospheric Administration on NOAA's Web site January 12, 1998. The dead zone in the Gulf of Mexico was described in the *New York Times,* January 20, 1998.

297 I questioned Paul McCartney at a press conference in London on April 8, 1993.

298 Regarding "pain" and "sacrifice," see especially the *New York Times,* December 11, 1997, and the *Washington Post,* November 13, 1997. The *Newsweek* quote is from July 14, 1997.

299 The energy efficiency examples, for both countries and companies, are drawn from *Factor Four: Doubling Wealth, Halving Resource Use,* Ernst von Weizsacker, Amory B. Lovins, and L. Hunter Lovins (London: Earthscan, 1997), passim. Copyright restrictions make the book unavailable in the United States, but many of the same examples and arguments are to be found in *Natural Capitalism: The Coming Efficiency Revolution,* Amory B. Lovins, L. Hunter Lovins, and Paul Hawken (New York: Little, Brown, 1999).

300 The labor intensity of economic efficiency investments is documented in numerous studies. See in particular the studies summarized in "Creating Sustainable Jobs in Industrial Countries," Michael Renner, in *State of the World* (Washington, D.C.: Worldwatch Institute, 1992), pp. 138–154. See also the *New York State Energy Plan, Economic Development Staff Report* (Albany: New York State Energy Office, 1989), cited in *Energy for Employment* (San Francisco: Greenpeace, 1992).

The figure of one billion adults not gainfully employed was documented by the International Labor Organization of the United Nations and reported in the *New York Times,* November 26, 1996.

300 "The idea that reducing global warming . . ." is from an author's interview with Lovins, as are subsequent quotes from Lovins in the text, unless otherwise noted.

303 The reports on subsidies were described in the *New York Times,* June 23, 1997, except for the critique of the World Bank by the Institute for Policy Studies, entitled "Changing the Earth's Climate for Business" (Washington, D.C.: Institute for Policy Studies, 1997).

304 The study of ecosystems' economic value, "The Value of the World's Ecosystem Services and Natural Capital," Robert Costanza et al., appeared in *Nature,* Vol. 387, May 15, 1997.

305 Daly's "Tax bads, not goods" quote came in an interview with the author, but he has argued the case for environmental tax reform in detail in numerous articles, including "Ecological Tax Reform," Herman E. Daly and John Duffy, *Perspectives on Business and Global Change,* 1996, Vol. 10, No. 2.

305– The discussion of the politics and progress of environmental tax reform is based on
306 *Factor Four,* op. cit., pp. 198–209, and on *Getting the Signals Right,* David Malin Roodman, Worldwatch Paper 134 (Washington, D.C.: Worldwatch Institute, 1997).

306 The solar and wind power advances were reported in the *New York Times,* May 19, 1997, and December 7, 1997, and in *Greenpeace Business,* February/March 1997 and February/March 1998.

306 The quote "[B]y now we know . . ." is found in *A New Name for Peace,* op. cit., p. 210.

307 The quotes from Wangari Maathai and Martin Khor and the general description of activists' work at the Earth Summit are based on the author's firsthand observation and interviews.

PAGE

308 The activist critique of globalization is summarized in *The Case Against the Global Economy*, Jerry Mander and Edward Goldsmith, eds. (San Francisco: Sierra Club Books, 1996).

308 "We broke the back . . ." and the general description of mainstream environmentalists' endorsement of NAFTA are described in *Losing Ground: American Environmentalism at the Close of the Twentieth Century*, Mark Dowie (Cambridge, Mass.: MIT Press, 1995), pp. 184–188.

309 "The international listing . . ." quote from Dowie is from an author's interview.

309– The descriptions of the protests over the *Brent Spar* oil rig and the French nuclear tests
310 are based on the author's own contemporaneous reporting; see my article in the *Washington Post*, September 3, 1995. In the aftermath of the *Brent Spar* affair, the impression grew that Greenpeace had erred in its protest and that burying the rig at the bottom of the sea, as Shell originally planned, was in fact the environmentally best solution. The source of this impression was a public statement by a Greenpeace executive that was later disavowed by the group, but not before it attracted international media coverage. In the end, Greenpeace continued to maintain that a sea burial was both a public safety and an environmental hazard. For the aftermath of the controversy and the Norway angle, see *The Consequences of the Brent Spar Victory*, available from Greenpeace.

310– The description of Greenpeace's refrigerator campaign is based on *The Corporate Planet:*
311 *Ecology and Politics in the Age of Globalization*, Joshua Karliner (San Francisco: Sierra Club Books, 1997), chapter 7.

312 "There is definitely a race . . ." quote from Passacantando is from an author's interview.

313 For Havel's description of his recollection with Kundera and his description of antipolitical politics, see *Disturbing the Peace* (New York: Alfred A. Knopf, 1990), especially pp. 172–177 and xvi.

314 The information on Czechoslovakia's environmental situation came from author's interviews with Josef Vavrousek, Czechoslovakian environment minister, as well as with Ivan Dejmal and Bedrich Moldan, the environment ministers for the Czech Republic, and with numerous environmental activists.

315 Havel's speech to the environment ministers was given June 21, 1991, at the conference held in Dobris, Czechoslovakia. Among the other accomplishments of the conference was the extremely useful and comprehensive report *Europe's Environment: The Dobris Assessment*, David Stanners and Philippe Bourdeau, eds. (Copenhagen: European Environment Agency, 1995). The report was dedicated to Josef Vavrousek, who had inspired the conference and who died with his daughter in an avalanche while hiking in the mountains of Slovakia in 1995.

For Havel's critique of technological civilization and the system of impersonal power, see his essays "Power and the Powerless" and "Politics and Conscience" in *Living in Truth* (London: Faber and Faber, 1989), especially pp. 113–117 and 136–153. See also *Disturbing the Peace*, especially pp. 9–17.

316 The reference to Beckett I owe to Martin Garbus's article in *The Nation*, January 29, 1990. Havel's remark about seeing "from above" comes from *Disturbing the Peace*, op. cit., p. 176. His comment about the results of antipolitical politics is from "Politics and Conscience," in *Living in Truth*, op. cit., p. 156. The prison letter is reprinted in *Letters to Olga* (New York: Henry Holt, 1989), pp. 235–237.

Epilogue

319 Unless otherwise noted, the descriptions of the giant sequoia and their history are based on the official history, published in cooperation with the California Department of Parks and Recreation, *The Enduring Giants: The Epic Story of Giant Sequoia and the Big Trees of Calaveras*, Joseph H. Engbeck Jr. (Berkeley: University of California, 1973), supple-

mented by author's interviews with Mr. Engbeck and study of additional primary documents he recommended.

320   The "ambassadors from another time" quote is from *Travels with Charley,* John Steinbeck (London: William Heinemann, 1962), p. 164.

324   Edwin Markham's poem is reprinted on the guidebooks to the North Grove distributed by the California Department of Parks and Recreation.

325   The EPA's judgment of the safety of American lakes and rivers is cited in "Dishonorable Discharge: Toxic Pollution of America's Waters," Jacquelin D. Savitz, Christopher Campbell, Richad Wiles, and Carolyn Hartman (Washington, D.C.: Environmental Working Group, 1996).

326   "The climate system is an angry beast . . ." is from the *New York Times,* January 27, 1998. For analysis of the possibility of global warming leading to a planetary deep freeze, see "The Great Climate Flip-Flop," William H. Calvin, *Atlantic Monthly,* January 1998.

327   "The inhabitants of planet earth . . ." is from *The Heat Is On,* op. cit., p. 28.

327   The arguments of Theo Colborn, Dianne Dumanoski, and John Peterson Myers are found in *Our Stolen Future* (New York: Dutton, 1996). The assertion of a 50 percent drop in human sperm counts during the last two generations was disputed by some scientists at the time; however, a more recent study has confirmed "a significant decline in sperm density in the United States and Europe" but noted differences among various geographic regions. See "Have Sperm Densities Declined? A Reanalysis of Global Trend Data," Shanna H. Swan, Eric P. Elkin, and Laura Fenster, *Environmental Health Perspectives,* Vol. 105, No. 11, November 1997.

328   The quotes "potential profit . . . is limitless" and "We are talking about restructuring . . ." are from *Green Gold: Japan, Germany, the United States, and the Race for Environmental Technology,* Curtis Moore and Alan Miller (Boston: Beacon Press, 1994), pp. 39, 72.

330   Greider's analysis of the overcapacity plaguing the world economy is found in *One World, Ready or Not: The Manic Logic of Global Capitalism* (New York: Simon & Schuster, 1997), especially pp. 111–119 and 192–201.

331   That efficiency improvements could reduce China's energy consumption by 50 percent has been demonstrated by studies done by Zhou Dadi and his colleagues at the Beijing Energy Efficiency Center. Additional improvements in energy efficiency available to China were described in author's interviews with Zhou Dadi, Mark Levine, Jonathan Sinton, and Robert Taylor.

331   That President Clinton was aware of the economic and environmental gains available through efficiency investments was clear from his comments in a meeting with Chinese environmental experts, including Zhou Dadi, in the city of Guilin on July 2, 1998. A copy of Clinton's remarks was made available by the White House press office.

332   The information on overall Third World debt levels come from *Human Development Report 1997,* United Nations Development Program (Oxford: Oxford University Press, 1997), p. 11. For Oxfam's quote, see "Making Debt Relief Work: A Test of Political Will," Oxfam International Position Paper, April 1998.

332   "Take the price-tag . . ." is from *Global Warming: The Greenpeace Report,* Jeremy Leggett, ed. (Oxford: Oxford University Press, 1990), p. 471.

334   "The problems are our lives . . ." is from *What Are People For,* Wendell Berry (San Francisco: North Point, 1990), p. 198.

Index

Mark Hertsgaard is the author of *Nuclear Inc.: The Men and the Money Behind Nuclear Energy; On Bended Knee: The Press and the Reagan Presidency;* and *A Day in the Life: The Music and Artistry of the Beatles.* He has contributed to *The New York Times, The New Yorker,* National Public Radio, *Harper's, The Atlantic Monthly, Outside, Vanity Fair, The Nation,* and numerous other publications at home and abroad. He teaches nonfiction writing at Johns Hopkins University and lives near Washington, D.C.

- unity
- peace
- Love
- culture
- RAINBOW
- Body
- Health
- hapiness
- state of mind